KB144027

제2판

외국인이 좋아하는 한국요리 100선

한식의 맛

비법전수 100선

전경철(全京喆, JEON KYUNG CHUL)
김은영(金銀永, KIM EUN YOUNG)
전동철(全东哲, JEON DONG CHOL)

Transmission of a Hundred Secrets of Korean Dishes Flavor

韩食美 "味" 秘方传授 100选

外国人喜爱的韩国料理100选

BAEKSAN Publishing Co

외국인이 좋아하는 한국요리 100선

한식의 맛

비법전수 100선

책을 펴내면서

세계인이 놀랄 정도로 빠르게 변화하는 현실에서 세계화시대에 음식문화의 발전은 한국의 아름다운 전통음식문화를 계승하고 불경기 속에서 외식문화를 활성화하기 위한 것으로 의미 있는 일이라 하겠습니다. 고객의 만족과 현장 음식의 맛 개선을 위한 비법전수를 위하여 그동안 저자들은 특급호텔 조리장의 경험과 한국요리전문점, 전통음식연구소 등에서 배운 노하우 및 대학과 조리전문학교, 현장 강의 경험을 바탕으로 한국음식의 세계화를 위하여 기초부터 고급요리까지 쉽게 이해하고 현장에 적용될 수 있도록 조리과정을 최대한 상세하게 기록하였습니다. 그리하여 다양한 한국음식 장르 중에서 외국인들에게 선호도가 높은 대표적인 100가지만을 선택하여 이 책을 만들게 되었습니다.

또한 이 책이 국내는 물론 한국음식을 사랑하는 외국인들에게 도움을 주기 위하여 조리국가대표로서 대한민국 · 러시아 · 중국 · 미국 등의 국제요리대회와 일본 · 중국 · 홍콩 · 인도네시아 · 태국 · 싱가포르 · 호주 · 유럽 · 미국 등의 연수, 국제대회심사, 참관 등을 통하여 외국인들이 좋아하면서 한국을 대표하는 음식으로 메뉴화하기 위해 각국의 음식에 대한 연구를 토대로 현지에서 전문가에 따른 관능평가를 실시함으로써 한국 전통음식의 옛 모습을 유지하면서 고객이 선호하는 맛을 찾기 위해 노력하였습니다.

불경기의 현장에서 생존하기 위해서는 매뉴얼을 토대로 차별화한 노하우를 가지고 있는 것이 무엇보다 중요하지만 더욱더 중요한 것은 사회의 흐름과 음식문화의 변화에 따른 조리방법과 기법 그리고 맛의 변화 또한 성공의 노하우로 생각되어 각 메뉴마다 트렌드에 맞는 맛의 비법을 담고 원가관리를 위한 식재료의 효율적 사용과 조리과정을 담았습니다.

이 책은 한국음식 전문점을 운영하는 분들과 외국인들을 위하여 혼자 학습할 수 있도록 최대한 쉽게 사진으로 풀이하면서 내용을 최소화하려 노력했지만 아직은 미진한 부분이 있으리라 사료됩니다. 이 책이 더욱 좋은 책으로 자리 잡을 수 있도록 아낌없는 지도편달을 부탁드립니다.

끝으로 항상 좋은 책 만들기를 고집하는 백산출판사 사장님과 임직원 여러분께 깊은 감사를 드리며, 도움을 주신 현장의 전문가 여러분께 고마움을 전합니다.

2013년 새해 아침

대한민국 서울에서 저자 일동

前言

饮食文化在世界化范围内的发展速度让世人为之震惊。在删除现实中，为了传承韩国悠久的传统饮食文化以及在不景气的经济情况下激活外餐经济，传授能让顾客满意的料理秘法。著者们基于特级酒店厨师长的经验和在在韩国料理专业店、传统饮食研究所学习的技巧；并以大学和烹饪专业学校、现场讲课经验为基础；为了韩国饮食的世界化，详细记录了从基础到高级料理删除的容易理解及容易实际操作的烹饪过程，并在各种各样的韩国饮食中，选择了外国人比较喜欢的一百种代表性的料理后，编写了这本适合实际制作的韩国之美味秘法一百选。

并且，为了这本书能有助于韩国人以及关心韩国饮食的外国人，在编写之前著者作为烹饪国家代表参加韩国、俄罗斯、中国、美国等地的国际料理大赛，通过在日本、中国、香港、印度尼西亚、泰国、新加坡、澳洲、欧洲、美国等地进修、参与国际大赛评审和参观等，为了研究外国人喜爱的韩国饮食，著者参考了相应国家的饮食传统，并尽量征求当地专家的主观评价，在维持韩国传统饮食固有特点的同时，对料理进行了全新的优化组合。

为了在不景气的市场中生存，删除重要的是基于固有的特点，研制出具有差别化的技巧，但更重要的是，随着社会趋势和饮食文化的变化而改变的烹饪方法以及秘诀和味道。于是著者每个菜单的内容都很重视社会潮流的美味秘法与成本管理的关系，并合理安排饮食材料的使用和烹饪过程。

这本书为了能让韩国餐厅经营人和外国人删除在繁忙的日常生活中容易学习，用照片做了解释，尽量减少了内容，但还是有很多不足之处。为了这本书能更加贴近韩餐爱好者们的生活，希望能够得到更多的意见和指教。

最后，感谢为了出版有助于大家的好书而努力的白山出版社社长和任职员，非常感谢大家的关心，也非常感谢帮助编写这本书的现场专家。

2013年元旦早晨
著者在大韩民国首尔

With the Publication of This Book...

In reality where the development of food culture in this era of globalization is changing its aspects rapidly enough to surprise the world, these authors, through the opportunity to inherit Korea's beautiful traditional food culture and transmission of the secrets for customer satisfaction and improvement in on-site food flavor for re-vitalizing food service business in a slump, and also on the basis of the experience of a Chef at a five-star hotel, and know-how acquired from a traditional food research institute and a specialty store of Korean cuisine, and lecturing experience in a college, and specialized school in cooking, and on-site lecturing, have tried to record the cooking process in depth as much as possible from its basic to fine cuisine so that they could be easily understood and applied to the on-site food service ultimately for globalization of Korean food; through this long process, we, authors selected only a hundred representative dishes highly preferred by foreigners out of diverse Korean food genres, and came to create "On-site Korea's Flavor Secrets Transmission 100" in a book.

In addition, in order for this book to be helpful to the foreigners who love Korean food, let alone domestic people, and to make our Korean foods into the menu representing Korean foods through the participation in the international cuisine competition held in Korea, Russia, China and America, and training sessions, judges & ob-servation in the international cuisine contest, these authors have made efforts to change our Korean foods into the flavor preferred by customers while maintaining the old appearance of Korea's traditional foods by conduct-ing the sensory evaluation at local sites as much as possible according to experts.

It's important to possess the know-how consequent on differentiation based on a food manual in order to survive in the actual arena more than anything else, but these authors regard the cooking method & techniques and changing flavor according to the changes in food culture as know-how of success as well, and thus contained the efficient use of food ingredients for cost control together with the secrets of flavor to meet the trend of individual menus, and cooking process as important contents in this book.

We, authors, made efforts to compile this book as easily as possible for those who are running Korean food specialty store and foreigners in the busiest daily routine to do self-study while interpreting the cuisine by illustrations and minimizing the contents, but these authors think there still remains something to be desired, and sincerely ask for your unsparing advice and encouragement for this book to take a position as a better book.

In closing, we'd like to extend our heartfelt thanks to the president & executives and staff members of the publisher-Baeksan which always persists in making a good book, and the experts at the food arena who gave us help.

One beautiful autumn day-2013

This author writes it in Seoul, Korea

차 례 한식의 맛 비법전수 100선

구이

조림, 볶음, 편육

생채, 숙채

김치

후식

부록

目 次

传授韩餐
韩食美"味"秘方传授100选

碳烤类

煎菜，炒菜

凉拌菜

泡菜

甜点

附录

Transmission of a Hundred Secrets of Korean Dishes Flavor

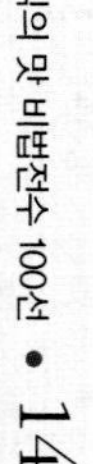

Gui

Jorim, Bokkeum, Pyeonyuk

Saengchae, Sukchae

Kimchi

Dessert

Appendix

한식 기본양념 만들기 韩食基本调料的做法 Basic sauce for Korean food

고추장구이양념장 辣酱烤用调味酱 Red pepper paste(gochujang) guyi sauce

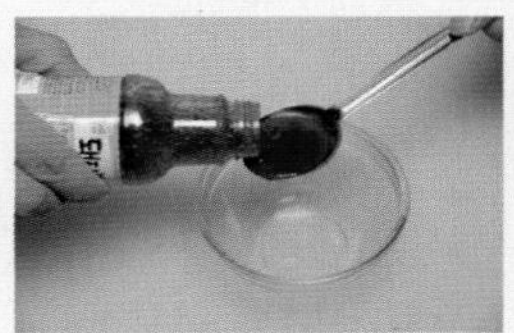

돼지불고기소스를 20g 넣는다.
放20g烤猪肉调味料
Put 20g of pork bulgogi sauce.

진간장을 5g 넣는다.
加5g浓酱油
Pour 5g of thick soy sauce.

황설탕을 15g 넣는다.
加15g黄糖
Put 15g of brown sugar.

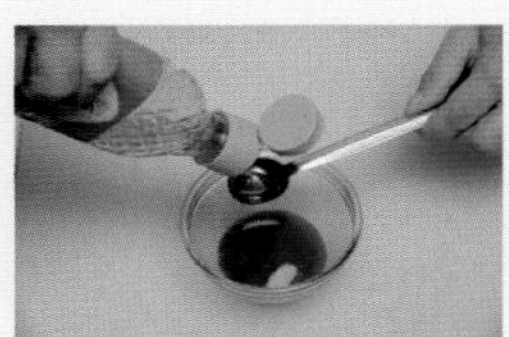

올리고당을 10g 넣는다.
加10g低聚糖
Pour 10g of oligosaccharide.

다진 마늘 5g, 배즙 20g, 양파즙 15g을 넣는다.
加5g蒜泥，20g梨汁，15g洋葱汁
Put 5g of chopped garlic, 20g of pear juice and 15g of onion juice.

후춧가루 3g, 수삼즙 5g을 넣는다.
加3g胡椒粉，5g水参汁
Put 3g of pepper and 5g of fresh ginseng juice.

고추장 15g을 넣는다.
加15g辣椒酱
Put 15g of pepper red paste (gochujang).

참기름 10g을 넣는다.
加10g香油
Add 10g of sesame oil.

김치양념장 泡菜调味酱 Kimchi sauce

고춧가루 130g에 까나리액젓 5g을 섞는다.
在130g辣椒粉中加5g玉筋鱼酱搅拌
Mix 130g of red pepper powder and 5g of sand eel sauce.

새우젓 3g을 넣는다.
加3g虾酱
Add 3g of salted shrimp.

밀가루풀 30g을 넣는다.
加30g面糊
Put 30g of flour paste.

다진 마늘 3g, 다진 생강 2g을 넣는다.
加3g蒜泥和2g姜沫
Put 3g of chopped garlic and 2g of chopped ginger.

통깨 2g을 넣는다.
加2g芝麻
Put 2g of sesame.

양파즙 15g을 넣는다.
加15g洋葱汁
Pour 15g of onion juice.

양념장을 잘 버무린다.
充分的搅拌调味酱
Mix the sauce well.

깻잎김치양념장 苏子叶泡菜 Sesame leaf Kimchi

깻잎을 깨끗이 씻는다.

把苏子叶洗干净

Wash sesame leaves thoroughly.

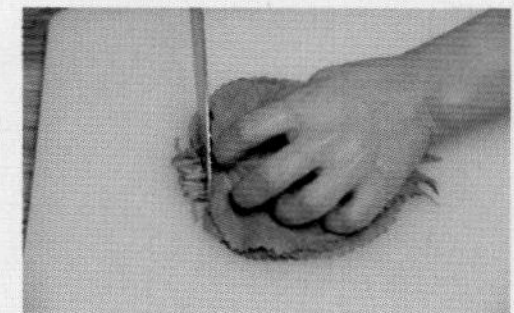

깻잎 줄기부위를 제거한다.

去掉苏子叶的茎部

Remove the stems of the sesame leaves.

고춧가루 15g, 까나리액젓 10g을 넣는다.

加15g辣椒粉和10g玉筋鱼酱

Put 15g of red pepper powder and 10g of sand eel sauce.

찬물 100g을 넣는다.

倒入100g凉水

Pour 100g of cold water.

진간장 15g을 넣는다.

加15g浓酱油

Add 15g of thick soy sauce.

설탕 15g을 넣는다.

加15g白糖

Put 15g of sugar.

다진 마늘 6g을 넣는다.

加6g蒜泥

Put 6g of chopped garlic.

송송 썬 실파와 대파를 넣는다.

加小葱葱沫和大葱

Put chopped spring onion and chopped leek.

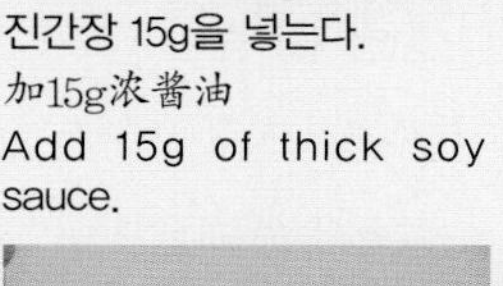

참기름 5g을 넣는다.

加5g香油

Add 5g of sesame oil.

통깨 5g을 넣는다.

加放5g芝麻

Put 5g of sesame.

양념장을 겹겹이 바른다.

苏子叶和调味酱呈现夹心状

Season each of the sesame leaves with the sauce.

숙성시킨 후 시식한다.

熟成之后可食用

Ferment the leaves and enjoy the dish.

된장구이양념장 酱烤用调味酱 Soybean paste (doenjang) guyi sauce

된장 20g에 설탕 15g을 넣는다.

在20g大酱中加15g白糖

Put 20g of soybean paste (doenjang) and 15g of sugar.

배즙 20g을 넣는다.

加20g梨汁

Put 20g of pear juice.

양파즙 10g, 청주 20g, 올리고당 30g을 넣는다.

加10g洋葱汁，加20g清酒，加30g低聚糖

Pour 10g of onion juice, 20g of rice-wine and 30g of oligosaccharide.

고춧가루 15g을 넣는다.

加15g辣椒粉

Put 15g of red pepper powder.

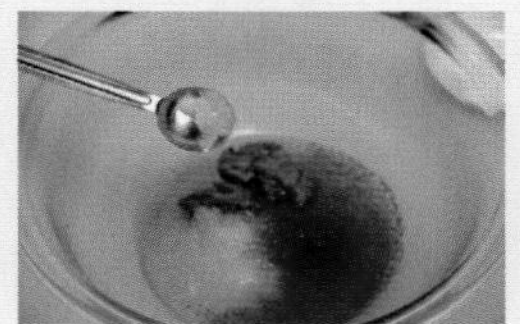

다진 마늘 5g을 넣는다.

加5g蒜泥

Put 5g of chopped garlic.

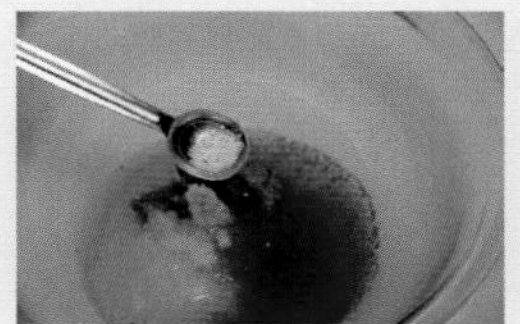

황기가루 2g, 수삼즙 3g을 넣는다.

加2g黄芪粉，加3g水参汁

Put 2g of milk vetch root powder and 3g of fresh ginseng juice.

골고루 저어준다.

均匀的搅拌

Mix the ingredients well.

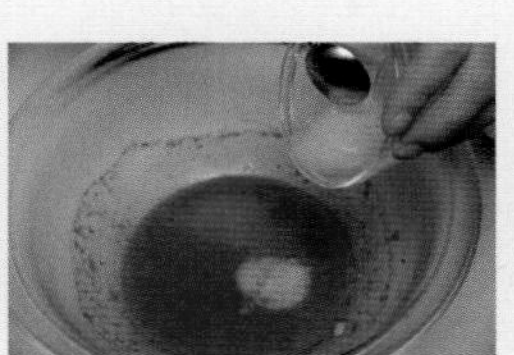

양념장에 달걀 노른자 20g을 넣는다.

在调料酱中加20g蛋黄

Put 20g of the yolk to the sauce.

무침양념장 凉拌菜调味料 Muchim sauce

고추장을 30g 넣는다.
放入30g辣椒酱
Put 30g of red pepper paste (gochujang).

홍초를 45g 넣는다.
加45g红醋
Pour 45g of red vinegar.

다진 마늘을 5g 넣는다.
加5g蒜泥
Put 5g of chopped garlic.

설탕을 8g 넣는다.
加8g白糖
Add 8g of sugar.

매실식초를 10g 넣는다.
加10g梅子醋
Pour 10g of plum fruit vinegar.

올리고당을 10g 넣는다.
加10g低聚糖
Pour 10g of oligosaccharide.

양념재료를 섞는다.
均匀的搅拌调味料
Mix the ingredients together.

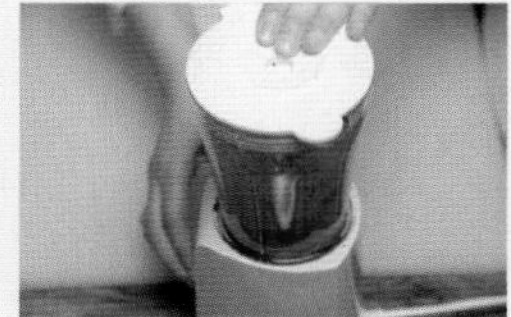

믹서기에 갈아 양념장을 만든다.
在搅拌机中搅拌后即可完成调料酱
Mix the ingredients again in a blender.

비빔면양념장 拌面调味酱 Bibimmyeon sauce

배즙 20g을 갈아준다.
准备20g梨汁
Make 20g of pear juice.

사과즙 10g을 갈아준다.
准备10g苹果汁
Make 10g of apple juice.

양파즙 5g을 갈아준다.
准备5g洋葱汁
Make 5g of onion juice.

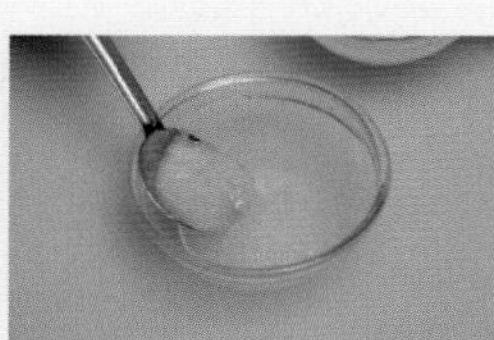

배즙, 양파즙, 사과즙을 섞는다.
搅拌梨，洋葱，苹果汁
Mix the juices the above together.

냉면육수 200g을 붓는다.
倒入200g冷面汤
Pour 200g of naengmyeon stock.

소금 2g을 넣는다.
加2g盐
Put 2g of salt.

고춧가루 30g을 넣는다.
加30g辣椒粉
Put 30g of red pepper powder.

다진 마늘 2g을 넣는다.
加2g蒜泥
Add 2g of chopped garlic.

설탕 2g을 넣는다.
加2g白糖
Sprinkle 2g of sugar.

물엿 2g을 넣는다.
加2g糖稀
Pour 2g of starch syrup.

청주 2g, 맛술 2g을 넣는다.
加2g清酒和2g料酒
Put 2g of rice-wine and 2g of Cooking wine.

불고기양념장 烤肉调味酱 Bulgogi sauce

간장 150g을 넣는다.

倒入150g酱油
Pour 150g of soy sauce.

청주 150g을 넣는다.

加150g清酒
Pour 150g of clean rice-wine.

마른 재료와 참기름 5g을 넣는다.
加5g香油和干材料
Put dried ingredients and 5g of sesame seed oil.

약불에서 끓인다.

在微火中煮
Boil the ingredients over a low heat.

설탕 15g을 넣는다.
加15g白糖
Put 15g of sugar.

올리고당 10g을 넣는다.
加10g低聚糖
Put 10g of oligosaccharide.

사과즙 30g을 넣는다.
加30g苹果汁
Pour 30g of apple juice.

불고기소스를 체에 내린다.
用过滤网过滤烤肉调味料
Sieve bulgogi sauce.

소갈비구이(불고기양념) 烤牛排骨（烤肉调味料） Beef rib guyi(bulgogi sauce)

소갈비소스 10g을 넣는다.

放入10g牛排骨调味料
Put 10g of beef rib barbecue sauce.

황기 3g, 구기자가루 3g을 넣는다.
加3g黄芪和3g枸杞
Put 3g of milk vetch root.

다진 마늘을 5g 넣는다.

加5g蒜泥
Put 5g of chopped garlic.

다진 생강을 3g 넣는다.

加3g生姜
Put 3g of chopped ginger.

다진 대파를 3g 넣는다.
加3g大葱葱沫
Put 3g of chopped leek.

올리고당을 5g 넣는다.
加5g低聚糖
Add 5g of oligosaccharide.

청주를 5g 넣는다.
加5g清酒
Pour 5g of rice-wine.

후춧가루를 0.5g 넣는다.
加0.5g胡椒粉
Sprinkle 0.5g of pepper.

참기름을 3g 넣는다.
加3g香油
Pour 3g of sesame oil.

맛술을 5g 넣는다.
加5g料酒
Put 5g of cooking wine.

통깨를 2g 넣는다.
加2g芝麻
Put 2g of sesame.

재운다.
熟成后完成
Season the meat with the sauce.

소갈비구이(소금구이양념장) 烤牛排骨(盐烤) Beef rib guyi(salt guyi)

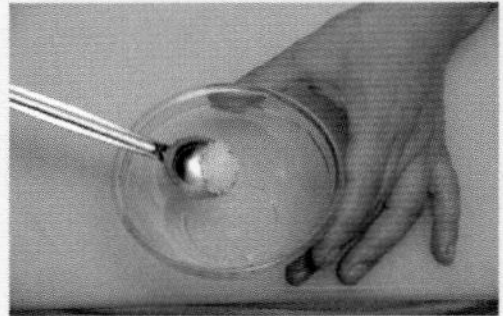

소금을 5g 넣는다.
加5g盐
Put 5g of salt.

배즙을 20g 넣는다.
加20g梨汁
Put 20g of pear juice.

양파즙을 15g 넣는다.
加15g洋葱汁
Put 15g of onion juice.

올리고당을 10g 넣는다.
加10g低聚糖
Add 10g of oligosaccharide.

다진 마늘을 5g 넣는다.
加5g蒜泥
Put 5g of chopped garlic.

다진 생강을 3g 넣는다.
加3g生姜
Put 3g of chopped ginger.

후춧가루를 0.5g 넣는다.
加0.5g胡椒粉
Put 0.5g of pepper.

청주를 5g 넣는다.
加5g清酒
Pour 5g of clean rice-wine.

참기름을 5g 넣는다.
加5g香油
Pour 5g of sesame seed oil.

황기 3g, 구기자 3g을 넣는다.
加3g黄芪和3g枸杞子
Put 3g of milk vetch root and 3g of chinese wolfberries (goji).

다진 대파 10g을 넣는다.
加10g大葱葱沫
Put 10g of chopped leek.

설탕을 5g 넣는다.
加5g白糖
Sprinkle 5g of sugar.

골고루 섞는다.
均匀的搅拌
Hand mix the ingredients well.

양념장을 바른다.
抹调料酱
Season the meat with the sauce.

완성한다.
完成
Grill it.

완성하여 담는다.
完成后盛上
Finish and serve it.

아귀찜양념장 炖鮟鱇鱼用调味酱 Monkfish sauce

고춧가루 30g, 국간장 10g을 넣는다.
加30g辣椒粉，加10g汤用酱油
Put 30g of chili powder and 10g of soy sauce.

후춧가루를 0.5g 넣는다.
加0.5g胡椒粉
Put 0.5g of pepper.

올리고당을 15g 넣는다.
加15g低聚糖
Pour 15g of oligosaccharide.

청주를 15g 넣는다.
加15g清酒
Pour 15g of clean rice-wine.

다진 마늘을 10g 넣는다.
加10g蒜泥
Put 10g of chopped garlic.

다진 생강을 5g 넣는다.
加5g姜沫
Put 5g of chopped ginger.

천연조미료를 5g 넣는다.
加5g天然调味料
Add 5g of natural seasonings.

홍합살을 다진다.
捣碎海虹肉
Chop mussel.

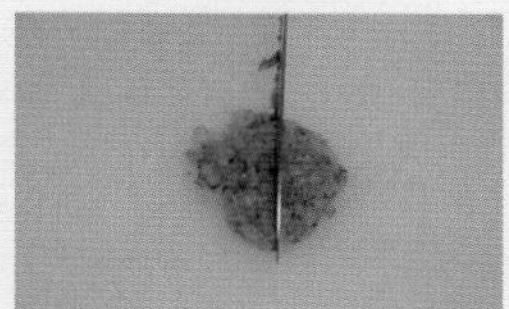

새우살을 다진다.
捣碎虾仁
Chop shrimp.

홍합살, 새우살을 넣는다.
放入海虹和虾仁
Put the mussel and the shrimp.

양념장을 잘 섞은 후 숙성시킨다.
搅拌均匀后熟成
Mix the ingredients with the sauce. Ferment them.

약고추장 调味辣酱 Sauteed hot pepper paste

다진 소고기 익은 것에 고추장을 20g 볶는다.
用20g辣椒酱炒牛肉沫（熟牛肉）
Roast chopped beef with 20g of pepper red paste (gochujang).

꿀을 15g 넣는다.
加15g蜂蜜
Put 15g of honey.

참기름을 8g 넣고 되직하게 완성한다.
加8g香油
Add 8g of sesame seed oil for a creamy consistency.

깨소금을 3g 넣는다.
加3g芝麻
Put 3g of roasted sesame seeds and salt.

육회양념장 生肉片 Yukhoe

고추장 15g, 국간장 3g을 넣는다.
加15g辣椒酱和3g汤用酱油
Put 15g of pepper red paste (gochujang) and 3g of soy sauce.

다진 마늘 7g을 넣는다.
加7g蒜泥
Put 7g of chopped garlic.

꿀 5g을 넣는다.
加5g蜂蜜
Add 5g of honey.

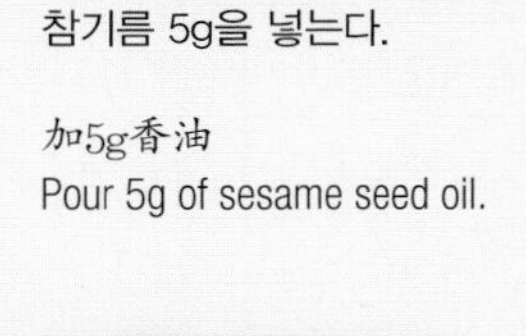

참기름 5g을 넣는다.
加5g香油
Pour 5g of sesame seed oil.

송송 썬 실파 3g을 넣는다.
加3g小葱葱沫
Put 3g of chopped green onion.

통깨 2g을 넣는다.
加2g芝麻
Put 2g of sesame seed.

소고기(우둔살) 채 썬 것에 양념장을 넣는다.
在牛肉片（牛臀肉）上倒入调味酱
Pour the sauce to julienne beef (rump).

배채와 같이 담는다.
与白菜盛在一起
Serve the dish with julienne pear.

튀김간장소스 油炸用酱油调味料 Twigim soy sauce

멸치육수를 끓인다.
煮鳀鱼汤
Boil anchovy stock.

멸치육수를 200g 붓는다.
倒入200g鳀鱼汤
Pour 200g of anchovy stock.

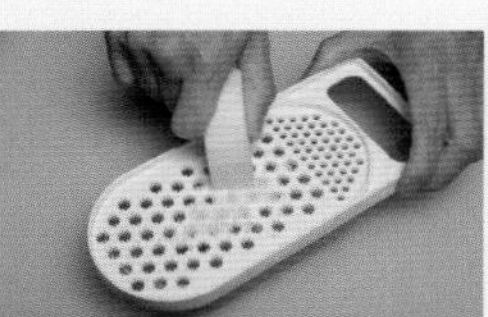

무를 갈아준다.
准备萝卜汁
Grate white radish.

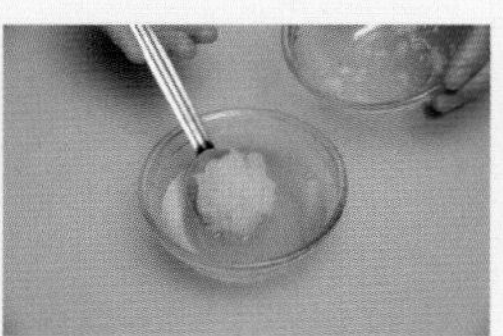

무즙을 30g 넣는다.
加30g萝卜汁
Pour 30g of white radish juice.

생강즙을 5g 넣는다.
加5g生姜汁
Add 5g of ginger juice.

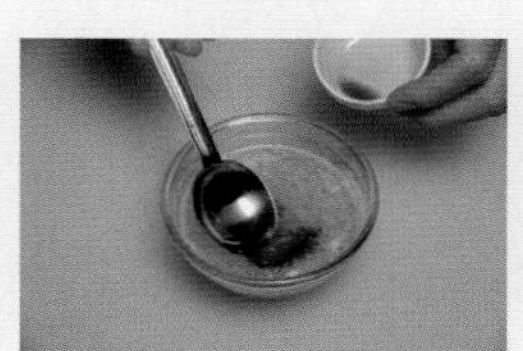

진간장을 8g 넣는다.
加8g浓酱油
Pour 8g of thick soy sauce.

고춧가루를 10g 넣는다.
加10g辣椒粉
Put 10g of chili powder.

송송 썬 실파와 레몬을 띄운다.
加葱沫和柠檬
Put chopped spring onion and a lemon.

한식 기본 썰기 制作韩食的基本切法 Basic cutting

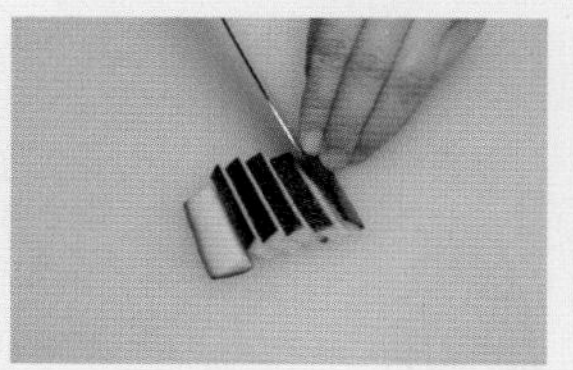

골패형썰기
切方片
Cut in a domino shape.

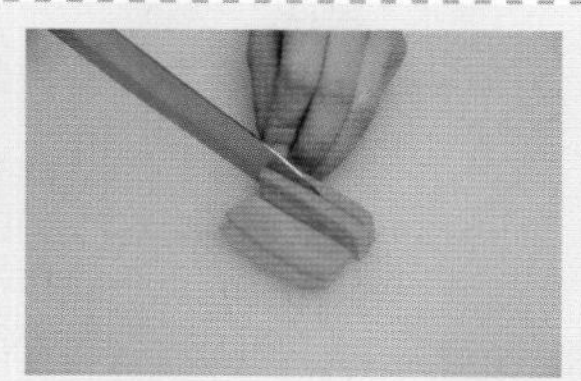

골패형썰기
切方片
Cut in a domino shape.

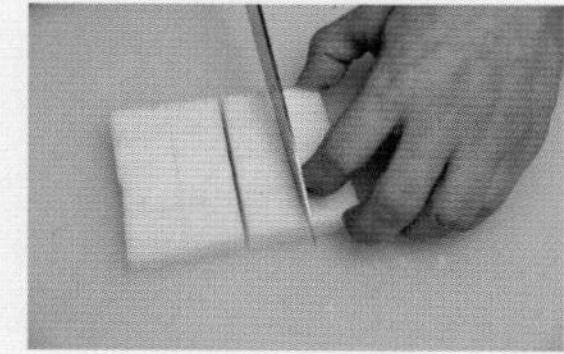

깍둑썰기
切块
Cut into cubes

깍둑썰기
切块
Cut into cubes

다지기
切沫
Chop

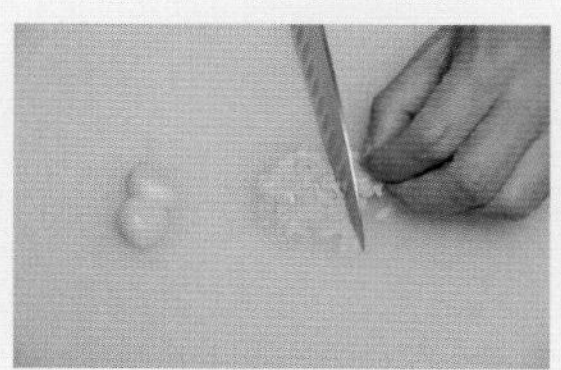

다지기
切沫
Chop

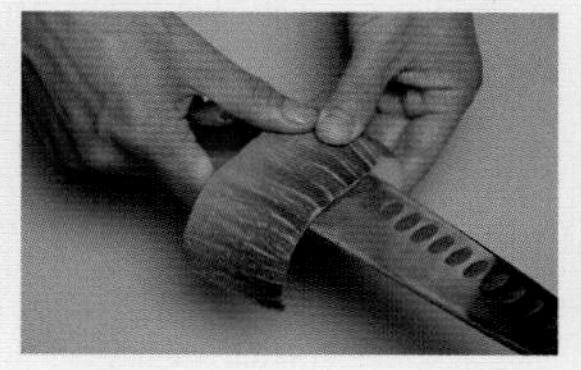

돌려깎기
削皮（取整块）
Cut into round shapes

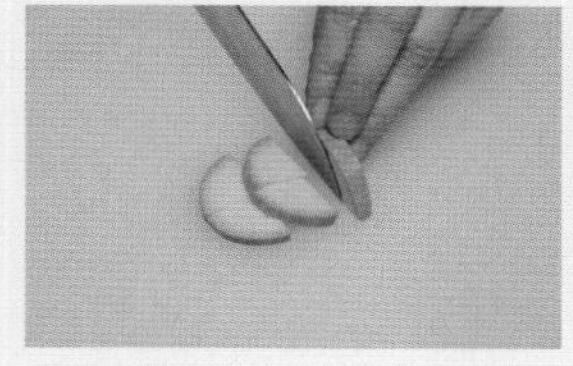

반달썰기
切扇面
Cut in a half-moon shape

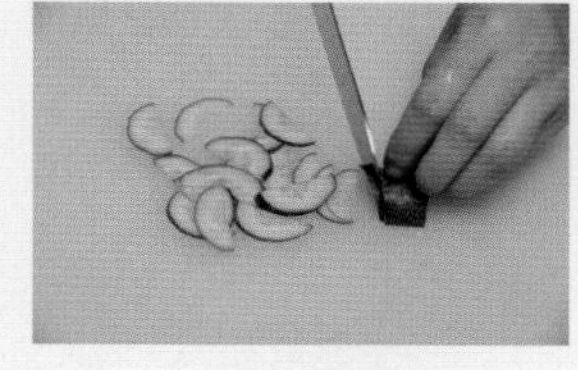

반달썰기
切扇面
Cut in a half-moon shape

어슷썰기
斜切
Cut into diagonal pieces

어슷썰기
斜切
Cut into diagonal pieces

채썰기
切丝
Julienne

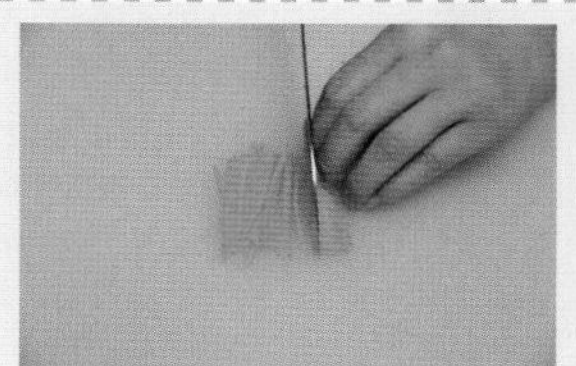

채썰기
切丝
Julienne

채썰기
切丝
Julienne

편썰기
切片
Slice

한식 양념류의 기본 韩食基本调料类 Korean sauces and condiments

3년 묵은 천일염
각종 미네랄이 풍부하고 요리의 풍미를 살려줌

新安岛产3年熟成盐
含有各种丰富矿物质，凸显料理的味道。

Sea Salt Aged for 3 years
Rich in minerals, Enhances the flavor of your dishes

순창 재래식 안심 생된장
나트륨 함량을 줄여 온 가족이 안심하고 먹을 수 있는 된장

淳昌 传统生大酱
降低含盐量，全家吃起来安心又健康

SUNCHANG Doenjang(Soybean Paste)
Fermented soybean paste with no added artificial preservatives

햇살담은 자연숙성 한식 국간장
잘 띄운 메주를 제대로 숙성시킨 한식간장

自然熟成韩式汤酱油
所用精选上等材料制成的纯韩式酱油

Naturally Brewed Soy Sauce for soup
Use in soup dishes or any other dish as a salt substitute

해물 감치미
시원한 국물요리를 원할 때

苷奇味 海鲜味复合调味粉
想做爽口的菜汤

Soup Stock Dry Powder – Seafood flavor
Creates a rich and flavorful soup stock

순창 양념듬뿍 쌈장
갖은 양념이 듬뿍 들어가 풍부한 맛

淳昌 包饭酱
多种配料融合，丰富的味道

SUNCHANG Ssamjang(Seasoned Soybean Paste)
Plenty of delicious ingredients for a traditional rich flavor

정통매실식초
100% 국산 매실, 국산 사과로만 만든 웰빙 식초

青梅食醋
100%韩国产青梅与韩国产苹果酿成的健康食醋

Plum Vinegar
Traditional vinegar made from plums and apples

맛선생 멸치 · 가쓰오
갈고 빻고 다진 맛있는 자연재료가 듬뿍

美味先生 银鱼味复合调味粉
各种自然材料的结合

Seasoning Mix – Anchovy and Bonito flavor
Plenty of delicious ingredients to create a flavorful dish

순창 떡볶이 고추장
다른 양념 필요없이 떡볶이 고추장 하나만으로 요리 가능

淳昌 炒年糕酱
不需其他调味酱，有炒年糕酱就能完成料理

SUNCHANG Topokki Gochujang(Spicy Rice Cake Stir-Fry Sauce)
Ready to use sauce without the need of adding other ingredients

사과식초
40년 전통의 식초 발효 기술로 만듦

苹果食醋
通过40年传统的技术 发酵酿成

Apple Vinegar
Fermented vinegar since 1971

순창 우리쌀로 만든 태양초 찰고추장 골드
쌀로 만들어 깔끔하게 매운맛

淳昌 辣椒酱(黄金版)
用大米制作，味道香辣醇正。

SUNCHANG Gochujang(Red Pepper Paste)
Made with sundried red peppers, rice and sea salt that gives a clean spicy taste

햇살담은 자연숙성 진간장
100% 자연숙성 진간장

自然熟成浓酱油
100%自然熟成浓酱油

Naturally Brewed Soy Sauce
100% naturally brewed

홍초 석류
맛있게 마시는 웰빙초

浓缩石榴红醋饮料
健康好喝的醋饮料

HongCho Drink Mix concentrate with Vinegar Pomegranate
A fruity, tangy, refreshing multipurpose drink mix concentrate that has all the health benefits of pomegranate and vinegar in one bottle

까나리액젓
12개월 자연 숙성한 까나리 사용

鳀鱼鱼露
使用上等12个月自然熟成鳀鱼制成

Sand Lance Sauce
Made with sand lance that was naturally aged for 12 months

소갈비 양념
과일을 갈아넣어 맛이 풍부하고,
연육작용 탁월

牛排烤肉酱
富含水果味道更加丰富，将肉更加鲜美嫩滑

Korean BBQ Galbi Sauce & Marinade for beef
With fruit purees for a zestier taste and tender meat

닭한마리 양념 백숙 – 삼계탕용
다른 추가양념 없이 간편하게 구수한 삼계탕 완성

参鸡汤酱
不需其他调味酱，简单完成浓香参鸡汤

Sauce for Korean Chicken Soup
Ready to use, rich and flavorful sauce

요리올리고당
설탕함량 0%, 설탕 대비 낮은 칼로리, 풍부한 식이섬유 함유

料理用低聚糖
0%的白糖含量，比白糖低卡路里且富含膳食纤维

Cooking Oligosaccharide
No added sugar and low calorie
compared to sugar

돼지불고기 양념
순창 쌀 고추장을 사용하여 한국적인 매운맛을 살림

猪肉烤肉酱
使用淳昌 辣椒酱突出了韩国风的辣味

Korean BBQ Bulgogi Sauce & Marinade for pork
With Sunchang Rice Gochujang that adds a touch of authentic Korean spiciness

닭한마리 양념 – 닭볶음탕용
다른 추가양념 없이 간편하게 매콤한 닭볶음탕 완성

微辣鸡肉烤肉酱
不需其他调味酱，简单完成辣炒鸡肉料理

Sauce for Korean Spicy Chicken Stew
Ready to use, spicy sauce

오푸드 유기 통참깨 참기름
100% 유기농 통참깨

有机香油
100%有机芝麻榨成

Organic Sesame Oil
Made with 100% organic sesame seeds

돼지갈비 양념
국산 명품 나주배 사용

猪排烤肉酱
添加了精选上等梨

Korean BBQ Galbi Sauce & Marinade for pork
With added high quality Korean pear

올리브유 재래김
향긋하고 바삭함

橄榄油传统海苔
又香又脆 韩国传统海苔

Seasoned Seaweed Sheets
Crispy and Tasty

소불고기 양념
자연숙성 양조간장을 사용, 맛과 향이 뛰어남

牛肉烤肉酱
使用自然熟成的酱油，味道香味更加突出

Korean BBQ Bulgogi Sauce & Marinade for beef
Made with naturally brewed soy sauce for exceptional taste

매운갈비 양념
순창 쌀 고추장과 국산 청양고추를 사용하여 화끈하게 매운맛

特辣排骨烤肉酱
使用淳昌 辣椒酱与韩国特产辣椒
更火辣的味道

Korean BBQ Spicy Galbi Sauce & Marinade
Made with Sunchang Rice Gochujang and Korean hot peppers

오푸드 유기 보리차
유기농 100% 보리차

有机大麦茶
100%有机大麦茶

Organic Barley Tea
100% organic

韩食美“味”秘方传授 100选 实践篇

Transmission of a Hundred Secrets of Korean Dishes Flavor 饭, 粥

한 식 의 맛 비 법 전 수 1 0 0 선

실기편

밥, 죽

(나물)비빔밥

(蔬菜)拌饭

(Herbal) Bibimbab

재료 및 분량(1인분)
멥쌀 230g, 소고기(등심) 150g, 도라지 80g, 고사리 80g, 애호박 100g, 당근 80g, 표고버섯 1장, 느타리버섯 80g, 청포묵 60g, 무 70g, 밤 20g, 달걀 1개, 통잣 3g, 소금 3g, 참기름 3g
***약고추장** : 소고기(등심) 15g, 고추장 20g, 꿀 15g, 참기름 8g
***불고기양념** : 진간장 10g, 설탕 6g, 마늘 3g, 참기름 3g, 깨소금 2g, 후춧가루 0.1g
***응용요리재료** : 약고추장은 다양한 종류의 비빔밥에 이용된다.

拌材料和份量(1人份)
粳米 230g, 牛肉(里脊) 150g, 桔梗 80g, 蕨菜 80g, 西葫芦 100g, 胡萝卜 80g, 香菇 1个, 平菇 80g, 青粉皮 60g, 萝卜 70g, 栗子 20g, 鸡蛋 1个, 松树籽 3g, 盐 3g, 香油 3g
***调味辣酱** : 牛肉(里脊) 15g, 辣椒酱 20g, 蜂蜜 15g, 香油 8g
***烤肉调味料** : 浓酱油 10g, 白糖 6g, 大蒜 3g, 香油 3g, 芝麻 2g, 胡椒 2g, 胡椒粉 0.1g
***应用料理材料** : 药辣椒酱可用于与各种米饭拌着吃。

Ingredients and portion(1 portion)
230g non-glutinous rice, 150g beef(sirloin), 80g bellflower roots, 80g dried bracken(gosari), 100g squash long, 80g carrot, 1 shiitake mushroom(mushroom pyogo), 80g oyster mushroom, 60g Mung bean jelly, 70g white radish, 20g chestnuts, 1 egg, 3g pine nut, 3g salt, 3g sesame oil
***Sauteed red pepper paste(gochujang)** : 15g beef(sirloin), 20g gochujang, 15g honey, 8g sesame oil
***Bulgogi sauce** : 10g soy sauce, 6g sugar, 3g garlic, 3g sesame oil, 2g crushed sesame, 0.1g ground black pepper
***Note** : You can add the sauteed hot pepper paste for other bibimbap dishes(a rice dish consisteing of beef, mushrooms and vegetables mixed with sauce).

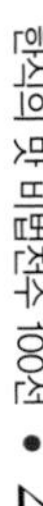

재료 손질 材料处理方法 Clean ingredients

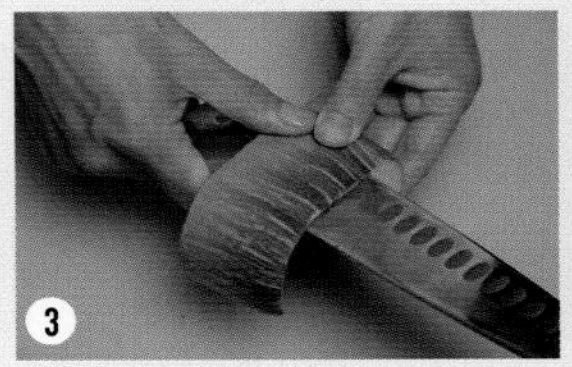

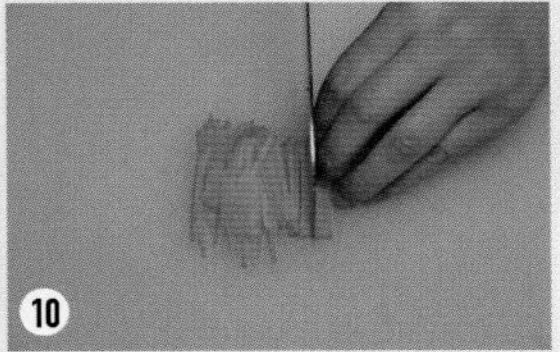

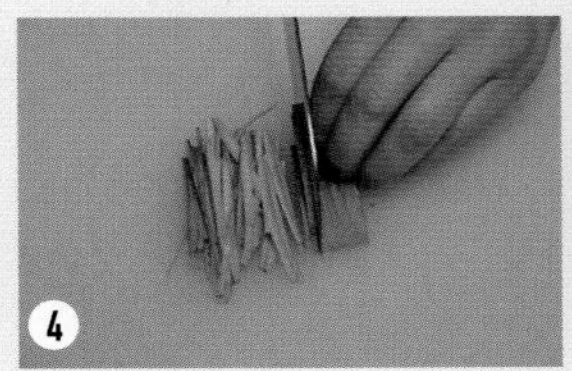

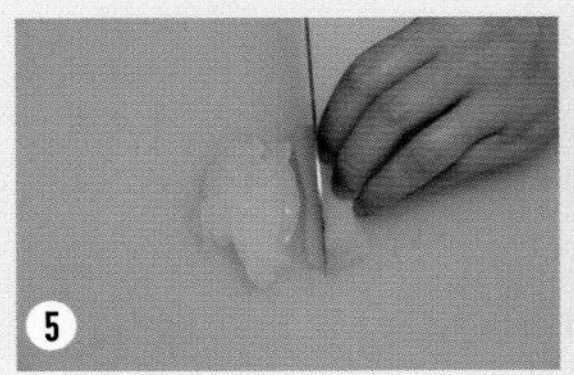

만드는 방법

1. 쌀을 불려 밥을 짓는다.
2. 고사리를 불린다.
3. 애호박은 돌려깎기한다.
4. 애호박 껍질은 0.3cm로 채 썬다.
5. 청포묵은 0.5cm로 채 썬다.
6. 끓는 물에 애호박, 청포묵을 데친다.
7. 애호박에 소금, 참기름을 간하여 볶는다.
8. 청포묵에 소금, 참기름으로 밑간한다.
9. 느타리버섯은 0.3cm로 찢는다.
10. 당근은 0.3cm로 채 썬다.
11. 끓는 물에 느타리버섯과 당근을 데친다.
12. 당근은 소금, 참기름으로 밑간하여 볶는다.

制作方法

1. 泡大米， 然后制作大米饭。
2. 泡蕨菜。
3. 将西葫芦的皮整块取下来。
4. 把切下来的西葫芦皮切成宽0.3cm的丝。
5. 将青粉皮切成 大小0.5cm。
6. 用烧开的水中焯一下西葫芦和青粉皮。
7. 炒西葫芦， 用盐和香油调味。
8. 用盐和香油给粉皮调味。
9. 把平菇撕成大小0.3cm。
10. 将胡萝卜切成大小0.3cm。
11. 用烧开的水焯一下平菇和胡萝卜。
12. 炒胡萝卜， 用盐和香油调味。

How to cook

1. Soak rice in water and boil rice.
2. Soak dried bracken(gosari) in water.
3. Cut the squash long into round shapes.
4. Julienne the peel of long squash into 0.3cm wide pieces.
5. Julienne the Mung bean jelly into 0.5cm wide pieces.
6. Blanch the squash long and the jelled Mung-Bean in boiling water.
7. Season the squash long and the jelled Mung-Bean with salt and sesame oil. Stir-fry the ingredients.
8. Season the jelled Mung-Bean with salt and sesame oil.
9. Tear oyster mushroom into 0.3cm wide pieces.
10. Julienne the carrot into 0.3cm wide pieces.
11. Blanch the oyster mushroom and the carrot in boiling water.
12. Season the carrot with salt and sesame oil. Stir-fry the ingredients.

약고추장 만들기
制作药辣椒酱。
How to make the light gochujang

만드는 방법

13. 느타리버섯은 소금, 참기름으로 밑간한다.
14. 무는 0.3cm로 채 썬다.
15. 밤은 0.3cm로 편 썬다.
16. 끓는 물에 밤, 무를 데친다.
17. 무는 소금, 참기름으로 밑간한다.
18. 불린 표고버섯은 0.3cm로 채 썬다.
19. 소고기는 0.3cm로 채 썬다.
20. 소고기, 표고버섯, 고사리를 불고기양념한다.
21. 볶아 익힌다.
22. 다진 소고기 익은 것에 고추장 20g을 넣어 볶는다.
23. 꿀 15g을 넣는다.
24. 참기름 8g을 넣고 되직하게 완성한다.

制作方法

13. 用盐和香油给平菇调味。
14. 将萝卜切成大小 0.3cm。
15. 将栗子切成大小为 0.3cm的片状。
16. 用烧开的水焯一下栗子和萝卜。
17. 用盐和香油给萝卜调味。
18. 将泡好的香菇撕成大小0.3cm。
19. 将牛肉切成大小0.3cm。
20. 为牛肉、香菇和蕨菜调味。
21. 炒熟把菜炒更绸后完成。
22. 将炒熟的牛肉放在20g辣椒酱中炒一下。
23. 加入蜂蜜15g。
24. 加入香油8g把菜炒更绸后完成。

How to cook

13. Season the oyster mushroom with salt and sesame oil.
14. Julienne the white radish into 0.3cm wide pieces.
15. Slice the chestnuts into 0.3cm wide pieces.
16. Blanch the chestnut and the white radish in boiling water.
17. Season the white radish with salt and sesame oil.
18. Julienne the soaked shiitake mushroom into 0.3cm wide pieces.
19. Julienne the beef into 0.3cm wide pieces.
20. Season the beef, the shiitake mushroom and the bracken(gosari) with bulgogi sauce.
21. Stir-fry them all.
22. Saute the finely julienne beef with 20g of gochujang.
23. Add 15g of honey to make the ingredients feel thick.
24. Add 8g of sesame oil.

완성하기
完成
How to finish

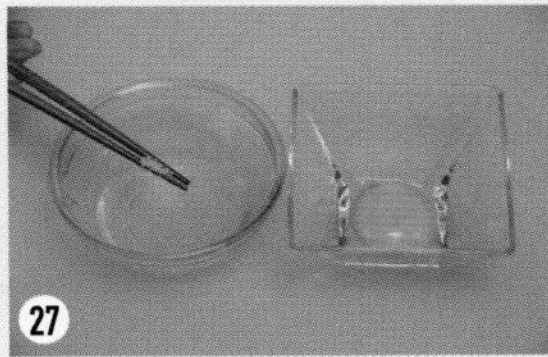

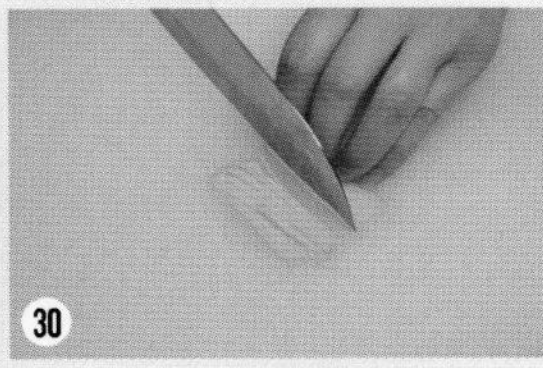

지단 만들기
鸡蛋的制作
For garnish

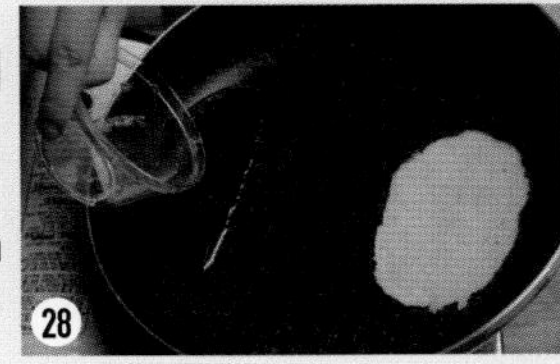

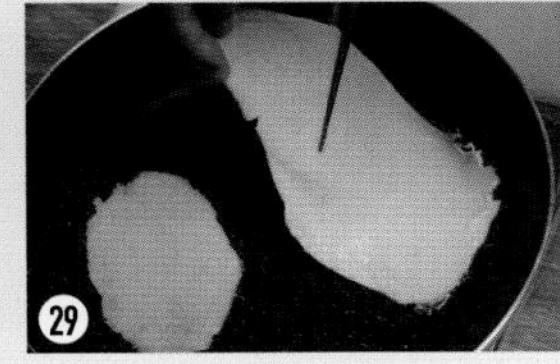

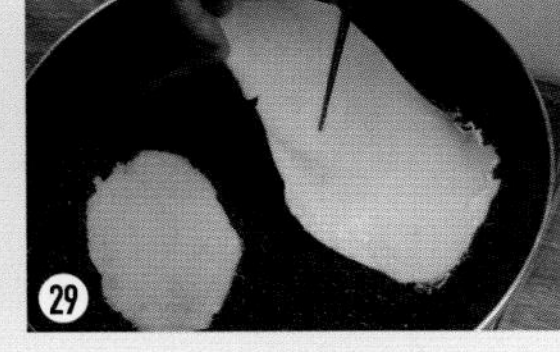

만드는 방법

25. 깨소금을 3g 넣는다.
26. 약불에 볶아서 완성한다.
27. 달걀은 황, 백으로 나눈다.
28. 얇게 지단을 부친다.
29. 약불에서 완성한다.
30. 지단을 0.2cm로 채 썬다.
31. 고명을 색스럽게 올린다.
32. 지단을 올려 완성한다.

制作方法

25. 加入芝麻盐3g。
26. 用小火炒熟。
27. 将鸡蛋的蛋清和蛋黄分开。
28. 把蛋清和蛋黄烙成薄薄的荷包蛋。
29. 用小火完成。
30. 将荷包蛋切成宽0.2cm的丝。
31. 用装饰菜点缀一下。
32. 将鸡蛋放上去即完成。

How to cook

25. Sprinkle 3g of sesame salt.
26. Stir-fry the ingredients all over a low heat.
27. Separate the egg yolk from the white.
28. Pan-fry the egg in very thin sheets.
29. Finish over a low heat.
30. Julienne the egg sheets into 0.2cm wide pieces.
31. Decorate the dish with garnish.
32. Make the dish look even more delicious with the last touch.

TIP
建议

밥 위의 고명은 계절에 따라 선택한다.
약고추장은 냉장 보관하여 편리하게 사용한다.

根据季节选择米饭上面的装饰菜。
请将药辣椒酱放在冰箱里保管，以便日后使用。

For garnish, you can choose different ingredients for each season.
Keep the sauteed red pepper paste(gochujang) in a refrigerator so that you can use it whenever you like.

단호박죽

南瓜粥

Sweet pumpkin porridge

재료 및 분량(2인분)
단호박 500g, 찹쌀가루 60g, 대추 2개, 해바라기씨 5g, 꿀 50g, 소금 5g
***새알심** : 멥쌀가루 150g, 소금 3g, 물 적량
***응용요리재료** : 계절에 따라 늙은 호박으로 활용

材料和份量(2人份)
南瓜 500g, 粘米面儿 60g, 大枣 2个, 葵花籽 5g, 蜂蜜 50g, 盐 5g
***丸子** : 粳米面儿 150g, 盐 3g, 水容量
***应用料理材料** : 根据季节使用熟透的南瓜。

Ingredients and portion(2 portions)
500g sweet pumpkin, 60g Glutinous rice flour, 2 jujubes, 5g sunflower seeds, 50g honey, 5g salt
***Non-glutinous rice balls** : 150g non-glutinous rice flour, 3g salt, water
***Note** : You can also use an old pumpkin.

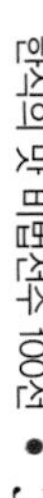

재료 손질 材料处理方法 Clean ingredients

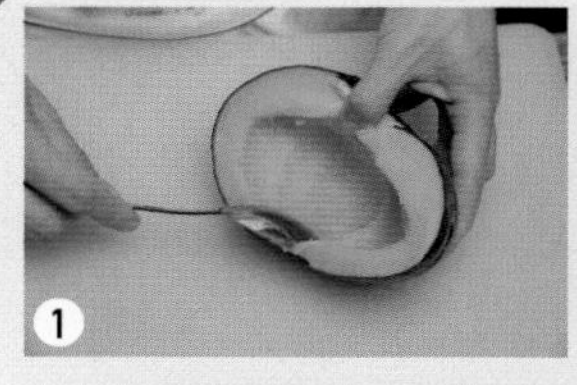

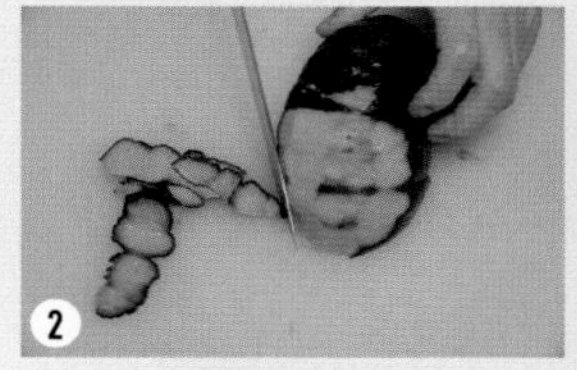

완성하기 完成 How to finish

만드는 방법

1. 단호박을 잘라 씨를 꺼낸다.
2. 단호박 껍질을 제거한다.
3. 단호박살을 15분 정도 찐다.
4. 익힌 호박을 체에 곱게 내린다.
6. 새알심은 끓는 물에 익힌다.
7. 익힌 새알심을 넣는다.
8. 꿀의 단맛과 약간의 소금으로 간을 한다.
9. 찹쌀가루로 죽의 농도를 맞춘다.

制作方法

1. 将南瓜切开并且取出南瓜籽。
2. 将南瓜去皮。
3. 蒸南瓜约15分钟。
4. 将蒸熟的南瓜取出来。
6. 将丸子放入烧开的水中煮熟。
7. 放入煮熟的丸子。
8. 用蜂蜜和盐调味。
9. 用粘米面儿调整粥的浓度。

How to cook

1. Cut the sweet pumpkin and take out the seeds.
2. Peel off the sweet pumpkin.
3. Steam the sweet pumpkin for about 15 minutes.
4. Sieve the steamed sweet pumpkin.
6. Cook the rice balls in boiling water.
7. Put the well-cooked rice balls.
8. Season the dish with honey and a pinch of salt.
9. Adjust the consistency of the dish with glutinous rice flour.

끓이기 煮粥 How to boil

5-1. 끓는 물 600g에 단호박을 350g 넣는다.

5-1. 在烧开的水600g中放入南瓜350g。

5-1. Put 350g of the sweet pumpkin into 600g of boiling water.

5-2. 단호박을 풀어준다.

5-2. 将南瓜碾碎。

5-2. Stir the sweet pumpkin in water.

5-3. 흰쌀가루에 소금 3g과 따뜻한 물로 반죽한다.

5-3. 在粳米面儿中加入盐3g和温水搅拌。

5-3. Mix non-glutinous rice flour with salt in warm water to make dough.

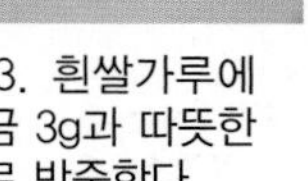

5-4. 반죽을 치댄 후 10분간 휴지시킨다.

5-4. 搅拌好以后，放置10分钟。

5-4. Work the dough. Set it aside for 10 minutes.

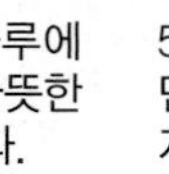

5-5. 새알심을 1cm로 빚는다.

5-5. 制作大小为1cm的丸子。

5-5. Make the dough into rice balls of 1cm in diameter.

5-6. 새알심 모양을 완성한다.

5-6. 丸子制作完成。

5-6. You now have the perfect little rice balls.

TIP 建议

5-3. 멥쌀가루는 질지 않고 되직하게 익반죽을 해야 완자모양이 완성된다.
7. 새알심이 끓는 물 위에 동동 떠오르면 완성

5-3. 搅拌粘米面儿时须保证粘米面儿既不要太稠也不要太稀，这样才能做出丸子模样。
7. 当丸子浮在烧开的水面上时即完成。

Step 5-3. Make sure that you have your non-glutinous rice dough thick. This way, you can make perfect little rice balls.
Step 7. Serve the dish after you see the rice balls floating on the surface.

전복영양죽

营养鲍鱼粥

Nutritious abalone porridge

재료 및 분량(2인분)
멥쌀 300g, 전복 2미(200g), 당근 30g, 표고버섯 20g, 실파 5g, 은행 5g, 참기름 30g, 소금 3g, 청주 5g, 김 2g
***육수** : 다시마 3g
***응용요리재료** : 소고기(등심), 인삼 등의 다양한 건강재료와 함께 사용 가능하다.

材料和份量(2人份)
百米 300g, 鲍鱼 2个(200g), 胡萝卜 30g, 香菇 20g, 小葱 5g, 银杏 5g, 香油 30g, 盐 3g, 白酒 5g, 紫菜 2g
***肉汤** : 海带 3g
***应用料理材料** : 牛肉(里脊)、人参等各种健康材料。

Ingredients and portion(2 portions)
300g non-glutinous rice, 2 abalones(200g), 30g carrot, 20g shiitake mushroom(mushroom pyogo), 5g green onion, 5g ginkgo nut, 30g sesame oil, 3g salt, 5g rice-wine, 2g sea laver
***Meat stock** : 3g dried sea tangle
***Note** : You can use other healthy ingredients as beef(tenderloin) and ginseng.

만드는 방법

1. 찬물에 다시마를 넣고 끓인다.
2. 물이 끓으면 걸러낸다.
3. 쌀을 씻어 2시간 이상 물에 불려 놓는다.
4. 당근을 0.2cm로 편 썬다.
5. 당근을 0.2cm로 다진다.
6. 표고버섯을 0.3cm로 편 썬다.
7. 표고버섯을 0.3cm로 다진다.
8. 실파를 송송 썬다.
9. 실파를 깔끔하게 준비한다.
10. 팬에 은행을 볶는다.
11. 은행을 0.2cm로 편 썬다.
12. 전복을 소금, 솔로 닦는다.

制作方法

1. 将海带放入凉水中煮。
2. 待水烧开后取出来。
3. 将大米洗好泡在水中2个小时以上。
4. 将胡萝卜切成大小为0.2cm的片状。
5. 将胡萝卜整理成大小为0.2cm的片状。
6. 将香菇撕成大小为0.3cm的片状。
7. 将香菇整理成0.3cm的片状。
8. 将小葱切成细丝。
9. 精心地准备好小葱。
10. 在平板锅上炒银杏。
11. 将银杏切成大小为0.2cm的片状。
12. 用盐和刷子清洗鲍鱼。

How to cook

1. Boil the dried sea tangle in cold water.
2. When the water boils, sieve the sea tangle.
3. Wash the rice. Soak it in water for more than two hours.
4. Slice the carrot into 0.2cm wide pieces.
5. Chop the carrot into 0.2cm wide pieces.
6. Slice the shiitake mushroom(mushroom pyogo) into 0.3cm wide pieces.
7. Chop the shiitake mushroom(mushroom pyogo) into 0.3cm wide pieces.
8. Chop the spring onion finely.
9. Clean the chopped green onion.
10. Pan-fry the ginkgo nut.
11. Slice the ginkgo nut into 0.2cm pieces.
12. Brush the abalones with salt.

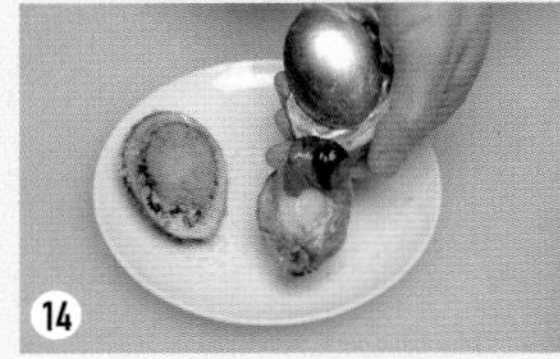

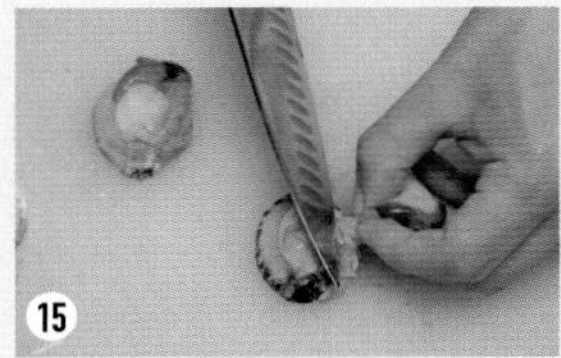
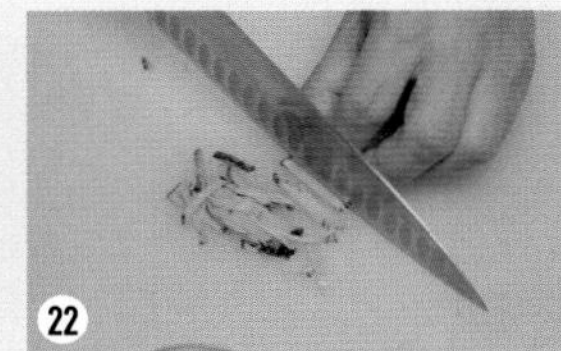
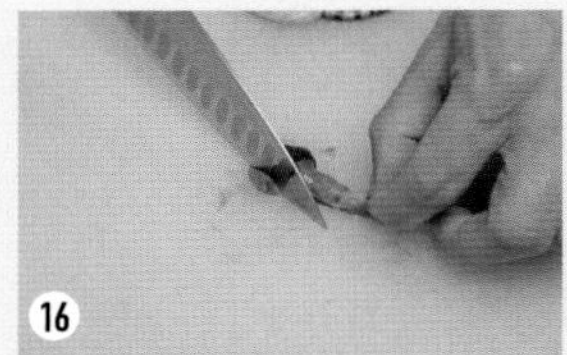
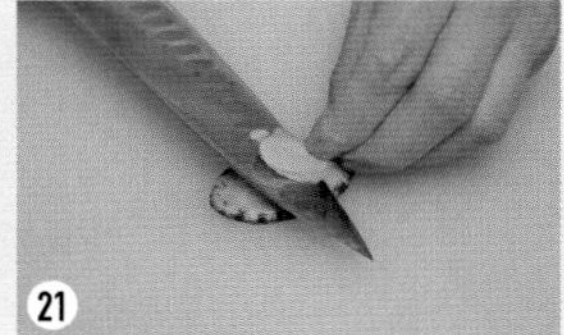

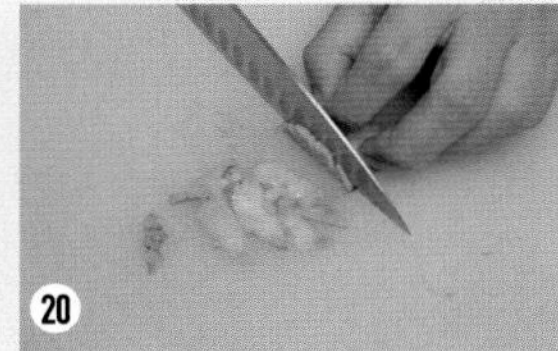

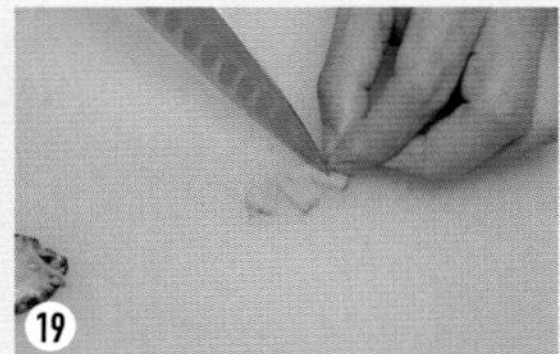

죽 끓이기
煮粥
How to finish

만드는 방법

13. 껍질 안쪽으로 살을 떼어낸다.
14. 내장이 터지지 않게 분리한다.
15. 내장을 분리한다.
16. 모래주머니를 제거한다.
17. 모래주머니를 깔끔하게 제거한다.
18. 입 부분을 제거한다.
19. 둥근 살은 모양내어 자른다.
20. 가로로 칼집을 낸 후 세로로 자른다.
21. 얇게 저며 썬다.
22. 곱게 채로 썬다.
23. 두꺼운 냄비에 참기름을 두른다.
24. 전복을 넣어 볶는다.

制作方法

13. 向着壳儿内侧取出鲍鱼肉。
14. 取出鲍鱼肉时小心不要让内脏破裂。
15. 将内脏分离。
16. 清除沙袋。
17. 干净地清除沙袋。
18. 清除嘴部分。
19. 将剩下的肉切成美观的形状。
20. 先横切，然后再纵切。
21. 切成薄薄的片状。
22. 切成美观的形状。
23. 在锅里加入香油。
24. 开始炒鲍鱼。

How to cook

13. Take the flesh off.
14. Carefully separate internal organs.
15. Pay extra attention with Step 14.
16. Remove the gizzard thoroughly.
17. Throw away the gizzard.
18. Remove the mouth.
19. Cut the circular flesh beautifully.
20. Make horizontal cuts and then, vertically cuts in the flesh.
21. Cut the flesh into thin slices.
22. Julienne the flesh finely.
23. Pour the sesame seed oil in a big pot.
24. Stir-fry the abalones in the pot.

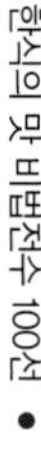

고명 준비
准备装饰菜
For garnish

만드는 방법

25. 전복 내장을 넣는다.
26. 불린 쌀을 넣는다.
27. 고소하게 볶는다.
28. 육수로 농도를 맞춘다.
29. 약한 불에서 저으면서 끓인다.
30. 채소를 넣고 소금, 청주로 간을 한다.
31. 김을 바삭하게 굽는다.
32. 곱게 채 썰어 준비한다.
33. 완성분을 볼에 담은 후 김가루를 뿌린다.

制作方法

25. 放入鲍鱼内脏。
26. 放入泡好的大米。
27. 继续炒，直至炒出香味来。
28. 用肉汤调整浓度。
29. 用微火煮。
30. 加入蔬菜，并用盐和白酒调味。
31. 将紫菜烤熟。
32. 将紫菜切成长条状备用。
33. 将粥盛在碗里后撒上准备好的紫菜。

How to cook

25. Put the internal organs.
26. Put the soaked rice.
27. Fry the ingredients all until they smell slightly sweet.
28. Adjust the consistency with the meat stock.
29. Stir the ingredients together. Boil them over a low heat.
30. Add vegetables. Season the ingredients with salt and rice-wine.
31. Toast sea laver until it gets crispy.
32. Julienne the sea laver finely.
33. Put the ingredients in a bowl. Sprinkle the sea laver powder on top.

TIP
建议

26. 약불에서 볶아야 바닥이 눋지 않는다.
27. 오래 볶아야 참기름의 고소한 맛이 밥알에 스며든다.
29. 바닥이 눋지 않도록 가끔씩 저어줘야 밥알이 살아 있다.

26. 用微火煮炒锅底才不会烧焦。
27. 只用炒的久一些，香油的香味才能进入饭粒中。
29. 为了不让锅底烧焦，偶尔搅拌一下，这样饭粒才不会黏成团。

Step 26. Stir-fry the ingredients over a low heat.
Step 27. Stir-fry the ingredients as long as possible with the sesame oil. This way, you can make your dish taste sweet.
Step 29. Stir the juk(porridge) from time to time so that the ingredients would not get burned in the pot.

채소버섯죽

蘑菇蔬菜粥

Vegetable mushroom porridge

재료 및 분량(2인분)
불린 쌀 200g, 쥬키니호박 60g, 당근 50g, 양파 50g, 브로콜리 50g, 표고버섯 1장, 실파 10g, 김 2g, 천연조미료 2g, 참기름 30g
***육수** : 다시마 5g, 밴댕이(디포리) 15g, 소금 3g, 천연조미료 3g
***응용요리재료** : 해물(새우, 홍합살, 굴, 낙지 등)을 함께 넣으면 해물죽으로 활용된다.

材料和份量(2人份)
泡好的大米 200g, 菜瓜 60g, 胡萝卜 50g, 洋葱 50g, 西兰花 50g, 香菇 1个, 小葱 10g, 紫菜 2g, 天然调味料 2g, 香油 30g
***肉汤** : 海带 5g, 干青鳞鱼 15g, 盐 3g, 天然调味料 3g
***应用料理材料** : 同时放入海鲜(虾, 海虹肉, 海蛎子, 八爪鱼等)可制成海鲜粥。

Ingredients and portion(2 portions)
200g soaked rice, 60g zucchini, 50g carrot, 50g onion, 50g broccoli, 1 shiitake mushroom(mushroom pyogo), 10g green onion, 2g sea laver, 2g natural seasonings, 30g sesame oil
***Meat stock** : 5g dried sea tangle, 15g Big Eyed Herring, 3g salt, 3g natural seasonings
***Note** : You can add other seafood ingredients(shrimp, mussel, oyster, small octopus, etc.) to make seafood juk(seafood porridge).

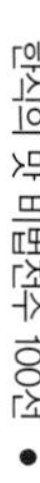

재료 손질 材料处理方法 Clean ingredients

고명 만들기 制作装饰菜 For garnish

만드는 방법

1. 불린 쌀을 준비한다.
2. 호박을 돌려깎은 후 다진다.
3. 브로콜리를 데친 후 다진다.
4. 모든 채소를 0.3cm로 다진다.
6. 해바라기씨는 볶는다.
7. 대추의 씨를 제거하고 말아 자른다.
8. 김을 바삭하게 굽는다.
9. 곱게 채 썰어 고명으로 올린다.

制作方法

1. 准备泡好的大米。
2. 将菜瓜去皮后剁碎。
3. 将西兰花用水焯一下后剁碎。
4. 将所有的蔬菜切成大小为0.3cm。
6. 将葵花籽炒一下。
7. 去掉大枣的核，切成小片。
8. 将紫菜烤熟。
9. 将紫菜切成细丝，用作装饰菜。

How to cook

1. Prepare soaked rice.
2. Cut the zucchini into round shapes. Julienne the zucchini.
3. Blanch the broccoli and chop it up.
4. Chop up all the other vegetables into 0.3cm wide pieces.
6. Roast the sunflower seeds.
7. Remove the seeds out of the jujube. Cut the jujube into round shapes.
8. Roast the sea laver until it gets crispy.
9. Julienne the sea laver finely. Use it as a garnish.

죽 끓이기
煮粥
Cook the porridge

5-1	5-2	5-3	5-4	5-5	5-6
5-1. 두꺼운 냄비에 참기름을 두른다.	5-2. 불린 쌀을 고소하게 볶는다.	5-3. 육수로 농도를 잡는다.	5-4. 준비한 채소를 넣는다.	5-5. 약불에서 나무주걱으로 저어준다.	5-6. 소금, 천연조미료로 양념한다.
5-1. 在锅里放入香油。	5-2. 将泡好的大米炒一下，直至炒出香味来。	5-3. 用肉汤调整浓度。	5-4. 放入准备好的蔬菜。	5-5. 用微火煮，同时用木制铲子搅拌。	5-6. 放入盐和天然调味料调味。
5-1. Pour the sesame oil in a big pot.	5-2. Roast the soaked rice until it smells sweet.	5-3. Adjust the consistency with the meat stock.	5-4. Put the vegetables in the pot.	5-5. Stir the ingredients over a low heat with a wooden spatula.	5-6. Season the ingredients with salt and natural seasonings.

TIP
建议

5-3. 불린 쌀의 3~4배 정도의 물(육수)을 부어 농도를 맞춘다.
5-4. 쌀이 완전히 익었을 때 넣어야 색깔을 살릴 수 있다.

5-3. 调整粥的浓度时，使肉汤的量相当于泡好的大米的3~4倍左右。
5-4. 只有在大米完全煮熟后加入才能保持新鲜的颜色。

Step 5-3. Adjust the meat stock consistency three to four times to the soaked rice.
Step 5-4. Put the vegetables after the rice is fully boiled.

콩나물밥

韩式肉丁豆芽饭

Bean sprouts-cooked rice

재료 및 분량(1인분)
불린 쌀 300g, 소고기(등심) 70g, 콩나물 30g, 무 30g, 진간장 8g, 마늘 5g, 후춧가루 0.1g, 참기름 5g, 다시마 3g
***비빔양념장** : 진간장 20g, 고춧가루 5g, 다진 마늘 3g, 다진 대파 3g, 참기름 3g, 통깨 3g
***응용요리재료** : 밥을 뜸 들일 때 손질된 굴, 홍합 등의 재료로도 활용

材料和份量(1人份)
泡好的大米 300g, 牛肉(里脊) 70g, 豆芽 30g, 萝卜 30g, 浓酱油 8g, 大蒜 5g, 胡椒粉 0.1g, 香油 5g, 海带 3g
***拌饭调味料** : 浓酱油 20g, 辣椒面儿 5g, 蒜泥 3g, 大葱末儿 3g, 香油 3g, 芝麻 3g
***应用料理材料** : 焖米饭时还可以灵活利用处理好的海蛎子和海虹等材料。

Ingredients and portion(1 portion)
300g soaked rice, 70g beef(sirloin), 30g bean sprouts, 30g white radish, 8g thick soy sauce, 5g garlic, 0.1g ground black pepper, 5g sesame oil, 3g dried sea tangle
***Bibim sauce** : 20g thick soy sauce, 5g chili powder, 3g chopped garlic, 3g chopped leek, 3g sesame oil, 3g sesame seed
***Note** : You can add clean oyster or mussel when you steam boiled rice.

재료 손질 材料处理方法 Clean ingredients

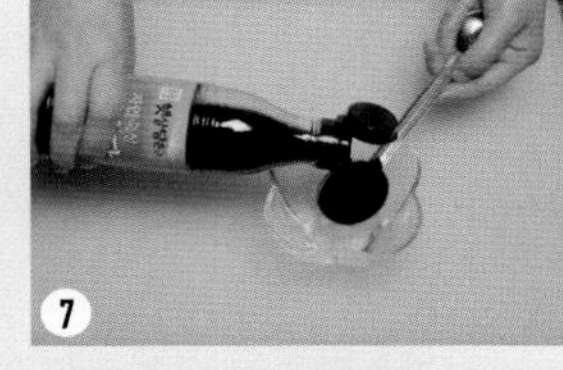

비빔양념장 만들기 制作拌饭调味料 Cook the bibim seasoning

만드는 방법

1. 소고기를 0.3cm로 채 썰어 진간장을 8g 넣는다.
2. 다진 마늘 5g을 넣는다.
3. 참기름 5g을 넣는다.
4. 후춧가루 0.1g을 넣는다.
6. 뜸 들인 밥을 위아래로 고루 섞는다.
7. 진간장 30g을 넣는다.
8. 고춧가루 5g, 통깨 3g, 다진 마늘 3g을 섞는다.
9. 다진 파 3g을 넣고 완성한다.

制作方法

1. 将牛肉切成大小0.3cm，加入酱油8g。
2. 加入蒜泥5g。
3. 加入香油5g。
4. 加入胡椒粉0.1g。
6. 将焖好的米饭上小搅拌。
7. 浓酱油30g。
8. 将辣椒面儿5g，芝麻3g，蒜泥3g混合。
9. 加入大葱末儿3g即完成。

How to cook

1. Julienne the beef into 0.3cm wide pieces. Pour thick soy sauce of 8g.
2. Put the chopped garlic of 5g.
3. Pour sesame oil of 5g.
4. Sprinkle ground black pepper of 0.1g.
6. Stir the steamed rice evenly.
7. Put thick soy sauce of 30g.
8. Put chili powder of 5g, sesame seed of 3g and chopped garlic of 3g.
9. Add chopped leek of 3g. Finish.

쌀 안치기 焖饭 Cook the rice

5-1. 불린 쌀에 양념한 소고기를 넣는다.

5-1. 在泡好的大米中放入调好味的牛肉。

5-1. Put the seasoned beef to the soaked rice.

5-2. 무를 0.3cm로 채 썰어 넣는다.

5-2. 将萝卜切成大小0.3cm。

5-2. Julienne the white radish into 0.3cm wide pieces.

5-3. 콩나물은 꼬리만 제거하여 사용한다.

5-3. 将豆芽尾去掉后备用。

5-3. Cut off the ends of the bean sprouts before you use them for cooking.

5-4. 다시마를 칼집 내어 넣는다.

5-4. 放入切好的海带。

5-4. Make cuts in the dried sea tangle.

5-5. 물을 쌀과 동량으로 넣는다.

5-5. 加入与大米的量相同的水。

5-5. Pour water as much as the rice.

5-6. 뚜껑을 닫아 콩나물을 완전히 익힌다.

5-6. 关上盖子，使豆芽完全熟透。

5-6. Boil the bean sprouts thoroughly with a lid on.

TIP 建议

1. 소고기 핏물을 제거하여 밥 색깔을 살린다. 5-4. 다시마는 밥을 부드럽고 윤기 있게 해준다.
5-5. 콩나물수분을 감안하여 물량을 조절한다. 5-6. 콩나물밥은 익을 때까지 뚜껑을 닫아 비린내를 없앤다.
비빔양념장은 하루 숙성하면 감칠맛을 낸다. (그 외 김가루, 다진 양파도 넣는다.)

1. 清除牛肉的血水，以使皮饭保持原色。 5-4. 海带起着使米饭柔软和滋润的作用。
5-5. 考虑到豆芽含有水分，适量地调节水量。 5-6. 直至豆芽饭焖好，盖子应一直保持关闭，这样可以去除异味。
拌饭调味料放置一天熟成后味道会更好。(此外，建议加入紫菜粉末和捣碎的洋葱。)

Step 1. Remove the blood out of the beef. This way, you can still keep the original color of the rice.
Step 5-4. Use the dried sea tangle to make the rice soft and oily.
Step 5-5. Adjust water in consideration of water content in the bean sprouts.
Step 5-6. Keep the lid on when you boil the bean sprouts to remove fishy smell of the bean sprouts.
For the bibim sauce, leave it to be fermented for a day. (you can add sea laver powder or chopped onion.)

韩食美"味"秘方传授 100选 实践篇

Transmission of a Hundred Secrets of Korean Dishes Flavor 面，年糕汤，面片汤

한 식 의 맛 비 법 전 수 100선

실기편

국수, 면, 떡국, 수제비

감자수제비

土豆面片汤

Potato sujebi

재료 및 분량(2인분)
감자 300g, 중력분 500g, 애호박 30g, 홍고추 10g, 풋고추 10g, 쑥갓 5g, 국간장 3g, 소금 5g
***육수** : 다시마 3g, 국멸치 4마리, 고추씨 5g, 홍고추 10g, 풋고추 10g
***응용요리재료** : 밀가루색을 뽕잎, 녹차, 도토리 등으로 맞춰 활용

材料和份量(2人份)
土豆 300g, 中力粉 500g, 西葫芦 30g, 红辣椒 10g, 青辣椒 10g, 茼蒿 5g, 汤用酱油 3g, 盐 5g
***肉汤** : 海带 3g, 汤用鳀鱼 4条, 辣椒籽 5g, 红辣椒 10g, 青辣椒 10g
***应用料理材料** : 依据面粉颜色 (桑叶, 绿茶, 橡栗)等灵活使用。

Ingredients and portion(2 portions)
300g potato, 500g medium flour, 30g squash long, 10g red chilli, 10g green chilli, 5g edible chrysanthemum(ssukgot), 3g soy sauce, 5g salt
***Meat stock** : 3g dried sea tangle, 4 anchovies, 5g pepper seed, 10g red chilli, 10g green chilli
***Note** : You can also use mulberry leaves, green tea or oak nut jelly(acorn) to add some color to your flour.

재료 손질 材料处理方法 Clean ingredients

1
2
3
4
5

끓이기
煮粥
How to boil

6
7
8
9
10
11
12

만드는 방법

1. 껍질 제거한 감자를 큼직하게 썬다.
2. 믹서기에 물을 감자의 30%만 넣고 갈아준다.
3. 중력분 300g에 소금 2g으로 간하고 간 감자로 반죽한다.
4. 애호박은 반달썰기 한다.
5. 홍고추, 풋고추는 0.2cm로 어슷썰기
6. 멸치육수를 붓는다.
7. 냄비에 반죽을 얇게 떼어낸다.
8. 국간장 3g으로 간한다.
9. 부재료에 소금 간을 한다.
10. 홍고추, 풋고추, 쑥갓을 넣는다.
11. 따뜻하게 담는다.
12. 완성한다.

制作方法

1. 将去皮的土豆切成大块。
2. 在搅拌机中加入相当于土豆的30%量的水后绞碎。
3. 在搅拌好的土豆泥当中放入300g中力粉和2g盐后开始和面。
4. 将西葫芦切成半月状。
5. 将红辣椒和青辣椒切成大小0.2cm。
6. 加入鳀鱼汤。
7. 将搅拌好的面团揪成薄薄的片状放入锅里。
8. 加入汤用酱油3g。
9. 用副材料和盐调味。
10. 加入红辣椒、青辣椒和茼蒿。
11. 盛放在容器里。
12. 完成。

How to cook

1. Peel off the potato. Cut the potato in chunks.
2. Pour water 30% as much as the potato into a blender. Grind them all in the blender.
3. Season medium flour with 2g of salt. Make dough with the ground potato.
4. Cut the squash long in a half-moon shape.
5. Diagonally cut the red chilli and the green chilli into 0.2cm wide pieces.
6. Pour anchovy stock.
7. Put thin slices of dough into a pot.
8. Pour 3g of soy sauce.
9. Season the ingredients the above with side ingredients and salt.
10. Put the red chilli, the green chilli and edible chrysanthemum(ssukgot).
11. Serve the dish warm.
12. Finish.

TIP
建议

3. 반죽을 되직하게 하여 20분 정도 휴지 → 탄력성 상승
7. 수제비 반죽은 → 손에 물을 발라가면서 떼어내면 표면이 매끄러워진다.

3. 搅拌面粉应保证面粉既不要太稠也不要太稀，并放置20分钟左右 → 弹力上升。
7. 将面团揪成薄片时 → 手上沾些水可以防止粘手。

Step 3. Make the dough thick. Set it aside for about 20 minutes → Flexibility of the dough improves.
Step 7. Make sujebi(hand torn noodle soup) dough with wet hands to make cooking easier.

떡국

韩式年糕汤

Tteokguk

재료 및 분량(2인분)
떡국떡 500g, 소고기(목심) 100g, 달걀 1개, 김 5g, 대파 5g, 국간장 3g, 소금 2g, 마늘 5g
***육수** : 다시마 5g, 국멸치 2마리, 감초 2g, 통마늘 4g, 양파 10g
***응용요리재료** : 떡은 조랭이떡도 가능하며, 소고기(양지) 육수 활용

材料和份量(2人份)
年糕 500g, 牛肉(颈部肉) 100g, 鸡蛋 1个, 紫菜 5g, 大葱 5g, 汤用酱油 3g, 盐 2g, 大蒜 5g
***肉汤** : 海带 5g, 汤用鳀鱼 2条, 泡菜 2g, 大蒜 4g, 洋葱 10g
***应用料理材料** : 年糕也可以使用料理用年糕, 肉汤使用牛肉(牛排)肉汤。

Ingredients and portion(2 portions)
500g long rice cake(garae tteok), 100g beef(chuck roll), 1 egg, 5g sea laver, 5g leek, 3g soy sauce, 2g salt, 5g garlic
***Meat stock** : 5g dried sea tangle, 2 anchovies, 2g licorice, 4g garlic, 10g onion
***Note** : You can use small rice cake(joraengyi tteok) instead of long rice cake(garae tteok). You can also make stock with beef(brisket).

재료 손질 材料处理方法 Clean ingredients

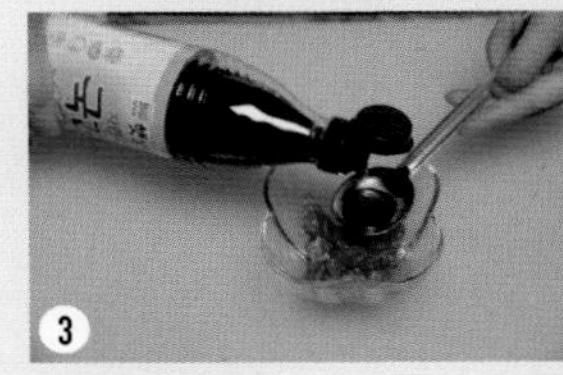

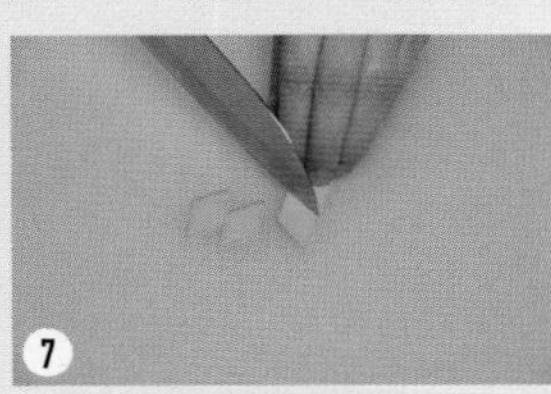

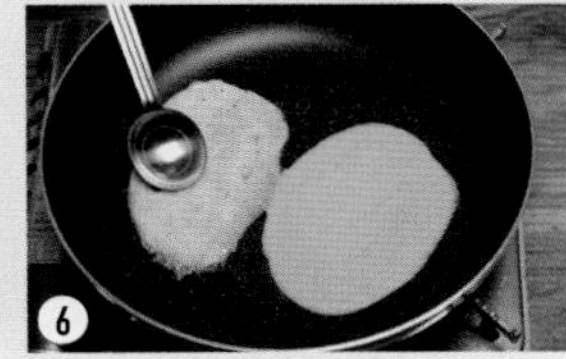

고명 만들기 制作装饰菜 For garnish

만드는 방법

1. 떡국떡은 찬물에 10분 정도 불린다.
2. 멸치육수를 끓인다.
3. 소고기 0.3cm로 편 썰어 국간장 3g에 양념
4. 다진 마늘 5g으로 양념한다.
6. 팬에 황, 백지단을 넣어 약불로 완성한다.
7. 지단은 2cm 마름모꼴로 자른다.
8. 마른 김을 구운 후 5cm로 채 썬다.
9. 따뜻하게 담는다.

制作方法

1. 将年糕放在凉水中泡10分钟左右。
2. 煮鳀鱼汤。
3. 将牛肉切成大小为0.3cm的片状，用汤酱油3g调味。
4. 用蒜泥5g调味。
6. 把蛋清和蛋黄用微火做成荷包蛋。
7. 将做好的鸡蛋薄片切成大小为2cm的细丝。
8. 将紫菜烤熟后切成大小5cm。
9. 盛放在容器里。

How to cook

1. Soak the long rice cake(garae tteok) in cold water for about 10 minutes.
2. Boil anchovy stock.
3. Slice the beef into 0.3cm wide pieces. Season the beef with 3g of soy sauce.
4. Season 5g of chopped garlic.
6. Pan-fry the egg yolk and the white into very thin sheet over a low heat.
7. Cut the sheets into diamond shapes of 2cm in diameter.
8. Roast sea laver. Julienne the sea laver into 5cm long pieces.
9. Serve the dish warm.

끓이기 煮粥 How to boil

5-1. 냄비에 참기름으로 소고기를 볶는다.

5-1. 在锅里放入香油后炒牛肉。

5-1. Stir-fry the beef with sesame seed oil in a pot.

5-2. 멸치육수에 양파 30g을 갈아서 육수로 붓는다.

5-2. 在鳀鱼汤里放入绞碎的洋葱30g后备好用作肉汤。

5-2. Grind 30g of onion and add it to the anchovy stock.

5-3. 끓는 육수에 떡국떡을 넣는다.

5-3. 在烧开的肉汤中加入年糕。

5-3. Put the long rice cake(garae tteok) to the boiling stock.

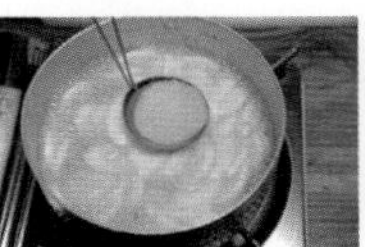

5-4. 거품을 제거한다.

5-4. 去除产生的泡沫。

5-4. Skim off the foam.

5-5. 소금 간을 한다.

5-5. 用盐调味。

5-5. Season the stock with salt.

5-6. 대파 0.3cm로 어슷썰기하여 넣는다.

5-6. 将大葱切成大小0.3cm后放入其中。

5-6. Diagonally cut the leek into 0.3cm wide pieces. Put the leek into the stock.

TIP 建议

3. 육수 색깔은 국간장으로 맞춘다.
5-1. 소고기를 50%만 익혀서 육수를 부어야 고깃결이 부드럽다.

3. 肉汤的颜色依靠汤用酱油来调节。
5-1. 待牛肉熟至50%时再加入肉汤，这样可以保持牛肉的柔嫩。

Step 3. Adjust the color of the stock with the soy sauce.
Step 5-1. Pan-fry the beef by 50% only and then, pour the stock. This way, you can make your beef soft.

평양물냉면

平壤冷面

Pyongyang Mulnaengmyeon

재료 및 분량(1인분)
냉면(칡면) 200g
***육수재료** : 소고기(양지) 200g, 대파 5g, 통마늘 3g, 진간장 3g, 소고기다시다 2g, 동치미국물 5g, 소금 3g
***고명재료** : 오이 10g, 배 3g, 무 10g, 고춧가루 2g, 소금 3g, 식초 3g, 설탕 2g, 삶은 달걀 1개, 소고기(양지) 100g, 향신채(대파 5g, 통마늘 2g)

材料和份量(1人份)
冷面(葛粉面条) 200g
***肉汤材料** : 牛肉(牛排) 200g, 大葱 5g, 大蒜 3g, 浓酱油 3g, 牛肉粉 2g, 腌萝卜泡菜汤 5g, 盐 3g
***装饰菜材料** : 黄瓜 10g, 梨 3g, 萝卜 10g, 辣椒面儿 2g, 盐 3g, 食醋 3g, 白糖 2g, 煮鸡蛋 1个, 牛肉(排骨) 100g, 香味菜(大葱 5g, 大蒜 2g)

Ingredients and amount(1 person)
200g Naengmyeon(Korean cold noodle)
***Stock** : 200g Beef(brisket), 5g leek, 3g garlics, 3g think soy sauce, 2g beef base, 5g Donchimi soup(radish water kimchi), 3g salt
***Topping** : 10g cucumber, 3g pear, 10g white radish, 2g red pepper powder, 3g salt, 2g sugar, 1 boiled egg, 100g beef(brisket), condiment herbs(5g leek, 2g garlic)

냉면육수 만들기 制作肉汤 Making meat stock for Korean cold noodles

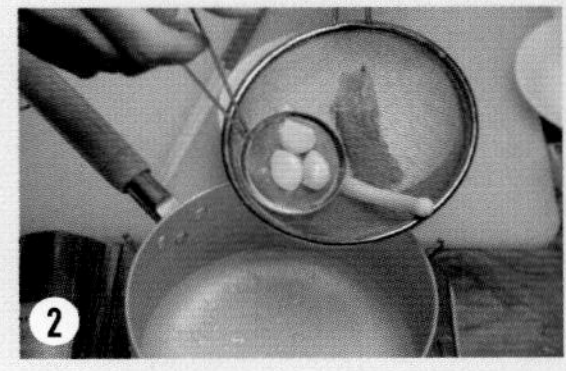

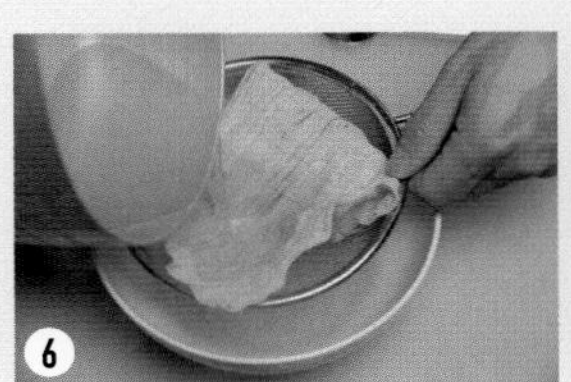

물냉면 고명준비 准备冷面装饰菜 Prepare topping for Mulneng-myeon

만드는 방법

1. 고기와 향채소를 넣고 30분 정도 끓인 후 건진다.
2. 진간장을 연하게 1인분(500g)에 2~3g을 넣는다.
3. 색감과 간을 맞춘다.
4. 소고기다시다를 1인분(500g)에 2g 넣어 맛을 낸다.
5. 육수는 투명하게 면보에 거른다.
6. 차갑게 식힌 후 육수 500g에 동치미국물 5g 정도 넣는다.
7. 1인분 500g에 3g 정도 소금으로 간을 한다.
8. 무는 0.2cm 두께로 편 썰어 3%의 소금에 30분간 절인다.
9. 오이는 반으로 어슷썰어 3%의 소금에 30분간 절인다.
10. 무의 물기를 제거한다.
11. 오이의 물기를 제거한다.
12. 무와 오이는 설탕 2g으로 단맛을 낸다.

制作方法

1. 放入肉和香味菜煮30分钟左右后取出来。
2. 一人分(500g)中放入2~3浓酱油。
3. 调节颜色和味道。
4. 一人分(500g)中放入2g韩国牛肉粉。
5. 过滤肉汤。
6. 肉汤冷却后每500g兑入5g左右的腌萝卜泡菜汤。
7. 一人分(500g)中放入3g左右的盐。
8. 将萝卜切成0.2cm厚的片状，在3%的盐中腌制30分钟。
9. 将黄瓜斜切成一半，在3%的盐中腌制30分钟左右。
10. 控去萝卜的水分。
11. 控去黄瓜的水分。
12. 萝卜和黄瓜中放入2g白糖调节甜味。

How to cook

1. Meats and vegetables and boil for 30 minutes and take them out.
2. Put thick soy sauce at 500g of stock.
3. Check the flavor and color.
4. Add the beef base.
5. Strain the stock through a cotton cloth.
6. Keep it cool and add 5g of dongchimi soup.
7. Season with 3g of salt.
8. Slice the radish into 0.2cm and salt them for 30minutes with 3% of salt.
9. Cut the cucumber into half and then diagonally, and do the same as the radish for 30 minutes.
10. Drain radish.
11. Drain cucumber.
12. With 2g of sugar, flavor the radish and cucumber.

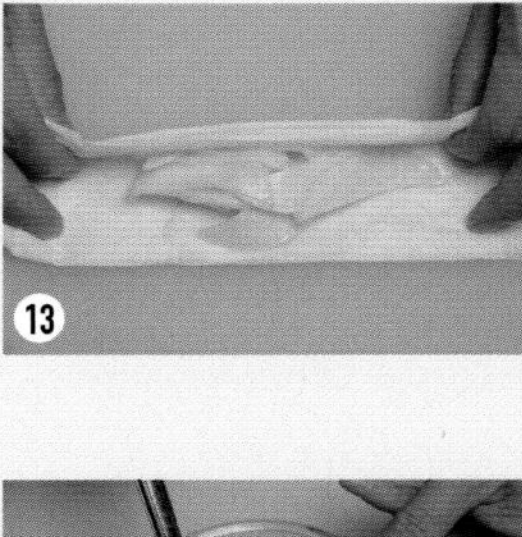

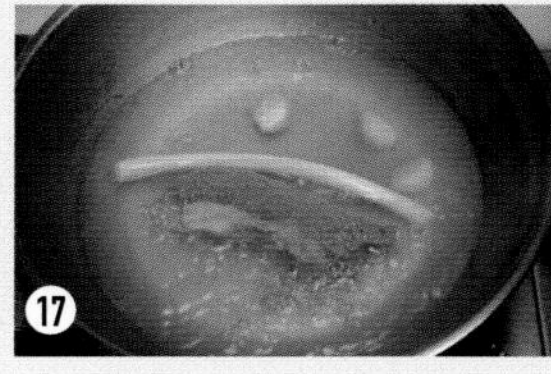

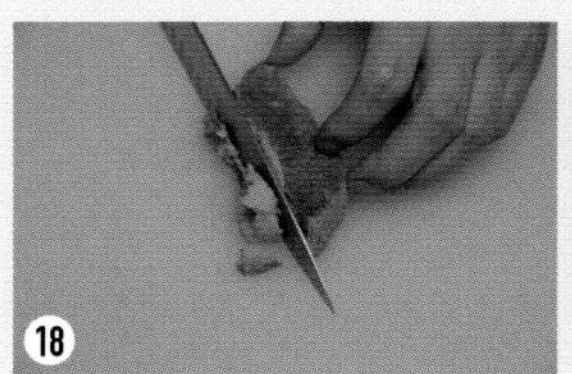

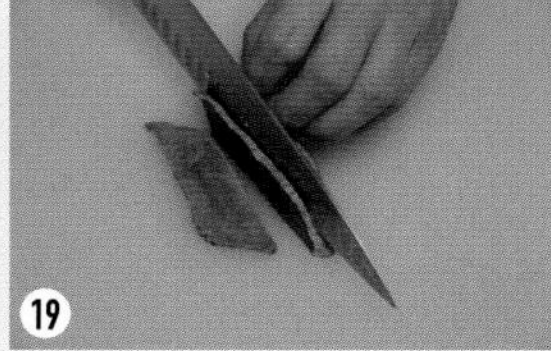

면 삶기
煮面
Boil noodles

만드는 방법

13. 고춧가루 2g으로 색감을 준다.
14. 삶은 달걀을 준비한다.
15. 달걀은 반으로 자른다.
16. 배는 두께 0.3cm로 반달썰기 한다.
17. 소고기는 향신채를 넣고 삶는다.
18. 삶은 소고기는 기름기를 제거한다.
19. 편육은 식은 후에 결 반대로 얇게 썬다.
20. 5가지의 고명을 준비한다.
21. 물을 끓인다.
22. 끓는 물에 면을 200g 넣는다.
23. 엉키지 않게 잘 저어준다.
24. 얼음물을 준비한다.

制作方法

13. 用2g辣椒面儿调节颜色。
14. 准备好煮好的鸡蛋。
15. 把鸡蛋切成一半。
16. 将梨切成厚0.3cm的半月状。
17. 放入香味菜煮牛肉。
18. 将煮好的牛肉的除去油分。
19. 肉块冷却后沿着肉的纹路将其切成薄片状。
20. 5种装饰菜准备完毕。
21. 烧水。
22. 在开水中放入200g面。
23. 为了不让面粘在一起，要不断的搅拌。
24. 准备好冰水。

How to cook

13. Add color with 2g of red pepper powder.
14. Get an egg boiled.
15. Cut it into half.
16. Slice pear by 0.3cm in half-moon shape.
17. Boil the brisket for topping with condiment herbs.
18. Remove the fat of cooked beef.
19. Cool the beef and slice in the opposite direction to the texture.
20. Done with 5 toppings.
21. Boil the water.
22. Put 200g of naengmyeon in to hot water.
23. Stir them not to get tangled.
24. Prepare ice water.

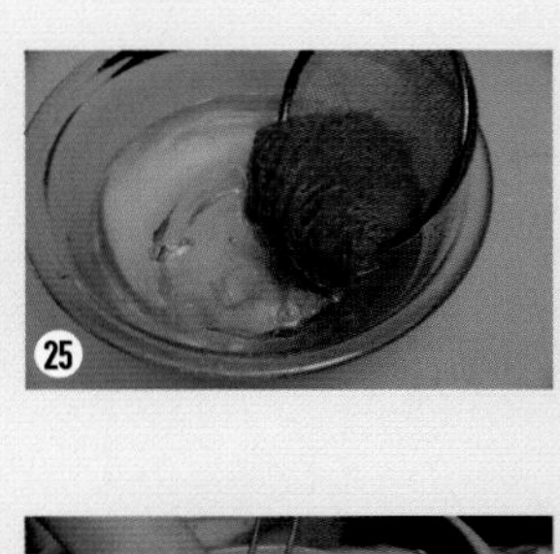

담기
盛装
Putting

만드는 방법

25. 2~3분 정도 삶아 익으면 얼음물에 넣어 씻어 준다.
26. 2차로 다시 씻어준다.
27. 얼음물에서 건져 면으로 모양을 만든다.
28. 물기를 제거한다.
29. 냉면을 용기에 담고 채소고명(무, 오이)을 올린다.
30. 소고기편육을 올린다.
31. 배편을 올린다.
32. 삶은 달걀을 올린다.
33. 차가운 육수를 7~8부 정도 붓는다.
34. 완성한다.

制作方法

25. 煮2~3分钟后，放入冰水中冲洗。
26. 冲洗第2次。
27. 从冰水中捞出面弄成美观的形状。
28. 控去水分。
29. 将冷面盛放在容器里，放上装饰菜(萝卜，黄瓜)。
30. 放上牛肉薄片。
31. 放上梨片。
32. 放上煮好的鸡蛋。
33. 加70~80%的冷却后的肉汤。
34. 完成。

How to cook

25. Boil for 2~3 minutes and when it cooked well, wash it in cold water.
26. Wash it again.
27. Take it out and make a shape.
28. Drain well.
29. Put it into a bowl and put radish and cucumber on it.
30. Put the beef slice on it.
31. Put pear slice on it.
32. Put egg on it.
33. Pour 8~9 glass of cold stock.
34. Finish.

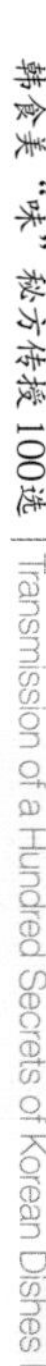

비빔냉면

拌冷面

Bibim-naengmyeon

재료 및 분량(1인분)
냉면(칡면) 160g, 냉면 고명재료(평양물냉면 참조)
***비빔양념장** : 배 20g, 사과 10g, 양파 5g, 소금 2g, 고춧가루 30g, 다진 마늘 2g, 설탕 2g, 물엿 2g, 청주 2g, 맛술 2g

材料和份量(1人份)
冷面(葛粉面条) 160g, 冷面装饰菜材料(参考平壤冷面)
***拌面调味酱** : 梨 20g, 苹果 10g, 洋葱 5g, 盐 2g, 辣椒面儿 30g, 蒜泥 2g, 白糖 2g, 糖稀 2g, 白酒 2g, 味精 2g

Ingredients and amount(a person)
160g Naengmyeon(Korean cold noodle), Topping ingredients(refer to Pyengyang Mulnaengmyeon)
***Sauce** : 20g pear, 10g apple, 5g onion, 2g salt, 30g red pepper powder, 2g chopped garlic, 2g sugar, 2g starch syrup, 2g rice wine, 2g cooking wine

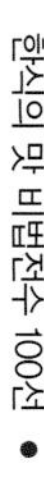

비빔양념장 拌面调味料 Bibim marinade

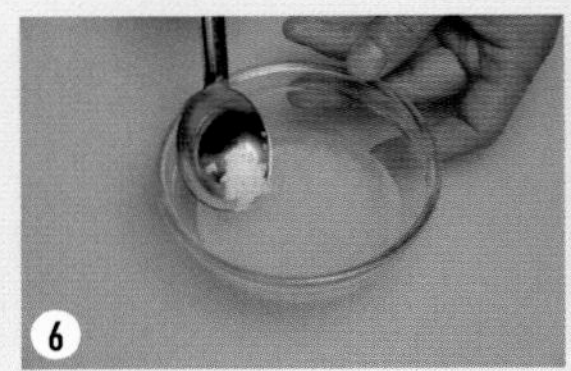

면 삶기
煮面
Boil noodles.

만드는 방법

1. 배즙 20g을 갈아준다.
2. 사과즙 10g을 갈아준다.
3. 양파즙 5g을 갈아준다.
4. 배즙, 양파즙, 사과즙을 섞는다.
5. 냉면육수 200g을 붓는다.
6. 소금 2g을 넣는다.
7. 고춧가루 30g을 넣는다.
8. 다진 마늘 2g을 넣는다.
9. 설탕 2g을 넣는다.
10. 물엿 2g을 넣는다.
11. 청주 2g, 맛술 2g을 넣는다.
12. 물을 끓인다.

制作方法

1. 制作梨汁20g.
2. 制作苹果汁10g。
3. 制作洋葱汁5g。
4. 将梨汁、苹果汁和洋葱汁混合。
5. 倒入冷面汤 200g。
6. 放2g盐。
7. 放5g辣椒面儿。
8. 放2g蒜泥。
9. 放2g白糖。
10. 加2g糖稀。
11. 加2g白酒, 2g味精。
12. 烧水。

How to cook

1. Grate pear.
2. Great apple.
3. Grind onion.
4. Mix them well.
5. Pour 200g of naengmyeon stock.
6. Add 2g of salt.
7. Put 30g of red pepper powder.
8. Add 2g of chopped garlic.
9. Add 2g of sugar.
10. Put 2g of starch syrup.
11. Put 2g of rice wine and 2g of cooking wine.
12. Boil the water.

면 담기
盛装
Put noodle

만드는 방법

13. 끓는 물에 면을 160g 넣는다.
14. 엉키지 않게 잘 저어준다.
15. 얼음물을 준비한다.
16. 2~3분 정도 삶아 익으면 얼음물에 넣어 씻어 준다.
17. 2차로 다시 씻어준다.
18. 얼음물에서 건져 면으로 모양을 만든다.
19. 물기를 제거한다.
20. 비빔양념장 30g을 올린다.
21. 면의 가운데 쪽에 올린다.
22. 참기름 2g을 올린다.
23. 통깨 1g을 뿌린다.
24. 채소고명(무, 오이)을 올린다.

制作方法

13. 在烧开的水中放入面160g。
14. 搅拌以防止粘结。
15. 准备好冰水。
16. 煮2~3分钟后，放入冰水中冲洗。
17. 冲洗第2次。
18. 从冰水中捞出面弄成美观的形状。
19. 控除水分。
20. 放上拌面调味酱30g。
21. 放在面的中央。
22. 加入香油2g。
23. 撒上芝麻1g。
24. 放上装饰菜(萝卜,黄瓜)。

How to cook

13. Cook the naengmyeon it hot water.
14. Stir well not to be tangled.
15. Get ice water.
16. Cook for 2~3 minutes and wash them in cold water.
17. Wash them again.
18. Take them out and make a shape.
19. Drain well and put it in a bowl.
20. Put the sauce on it.
21. On the middle of the naengmyeon.
22. Put 2g of sesame oil.
23. Drizzle 1g of sesame.
24. Put the vegetable toppings on it.

만드는 방법
25. 소고기편육, 배편, 삶은 달걀을 올린다.
26. 완성한다.

制作方法
25. 放上牛肉薄片、梨片和煮鸡蛋后完成。
26. 完成。

How to cook
25. Put the beef slice, pear slice, and boiled egg on it.
26. Finish.

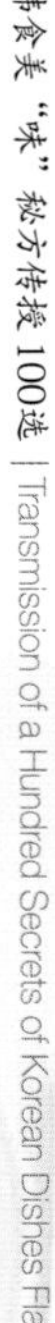

콩국수

豆浆冷面

Cold bean-soup noodles

재료 및 분량(1인분)
국수(소면) 90g, 흰콩(메주콩) 150g, 청오이 50g, 당근 20g, 배 20g, 방울토마토 2개, 통잣 5g, 콩가루 20g, 찹쌀가루5g, 소금 5g
***응용요리재료** : 검정콩, 검정깨도 갈아서 국물을 첨가할 수 있으며, 콩국물은 곤약면, 칼국수, 우무묵 등의 냉국수에 주로 활용

材料和份量(1人份)
面条(阳春面) 90g, 黄豆(酱曲豆) 150g, 青黄瓜 50g, 胡萝卜 20g, 梨 20g, 小柿子 2个, 松树籽 5g, 豆粉 20g, 粘米面儿 5g, 盐 5g
***应用料理材料** : 黑豆和黑芝麻碾碎后可放入冷面汤, 豆浆冷面汤主要用于由石菜花粉皮、魔芋粉丝和刀削面等凉面制作而成的冷面中。

Ingredients and portion(1 portion)
90g plain noodles, 150g white beans(soy beans), 50g cucumber, 20g carrot, 20g pear, 2 cherry tomatoes, 5g pine nut, 20g bean flour, 5g glutinous rice flour, 5g salt
***Note** : You can add ground black soy bean and black sesame seed to the soup. You can also use the bean soup for Konnyaku(jelly) noodles, kalguksu(handmade chopped noodles) and agar jelly.

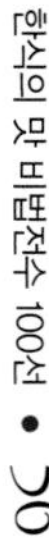

콩국물 만들기 制作豆浆 Clean ingredients

완성하기 完成 Work it through with the delicate touch

고명 준비 准备装饰菜 For other garnish

만드는 방법

1. 불린 콩에 찹쌀가루 15g을 넣고 끓인다.
2. 익은 콩에 콩가루 20g, 통잣 5g을 넣는다.
3. 소금 5g으로 간한다.
4. 믹서기에 곱게 갈아준다.
6. 1인분 사리를 담는다.
7. 당근채, 오이편, 배편, 방울토마토를 올린다.
8. 고명을 완성한다.
9. 얼음과 콩국물을 8부 붓는다.

制作方法

1. 在泡好的豆子中加入15g的糯米粉一起煮。
2. 在煮熟的豆子中加入20g豆粉和5g松仁。
3. 用5g盐调味。
4. 用搅拌机绞碎。
6. 盛1人份的面。
7. 放上胡萝卜丝，黄瓜片，梨片和小柿子。
8. 装饰菜制作完成。
9. 将冰块和冰镇豆浆倒入其中至80%左右。

How to cook

1. Soak the beans. Boil the bean with glutinous rice flour of 15g.
2. Put bean flour of 20g and pine nut of 5g to the cooked beans.
3. Season the ingredients the above with salt of 5g.
4. Mix the ingredients all in a blender.
6. Serve a single portion of the noodles in a plate.
7. Put the slices of carrot, cucumber and pear on top of the noodles. Add some cherry tomato.
8. Finish the garnish.
9. Pour the cold bean-soup of 80%. Add ice cubes.

국수 삶기 煮面条 Cook the noodles

사리모양 잡기 调整面条模样 For the noodle garnish

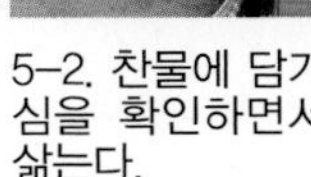

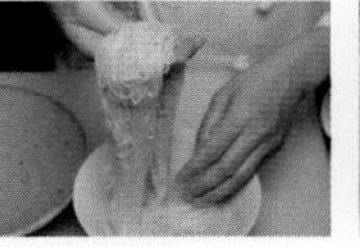

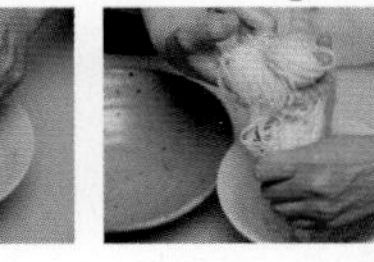

5-1. 끓는 물에 소면을 삶아준다.
5-1. 用烧开的水煮面条。
5-1. Boil plain noodles in boiling water.

5-2. 찬물에 담가 심을 확인하면서 삶는다.
5-2. 一边煮面条，一边放入凉水中品尝确认。
5-2. Put the noodles into cold water. Keep boiling as watching the noodles.

5-3. 국수를 양쪽으로 길게 나눈다.
5-3. 将面条向两侧分开。
5-3. Spread out the noodles.

5-4. 국수를 양쪽 방향으로 말아준다.
5-4. 将面条向两侧卷起来。
5-4. Roll the noodles from the both sides.

5-5. 마지막 국수가 풀리지 않게 말아준다.
5-5. 卷面条时注意使最后卷起来的面条不会松开。
5-5. Keep the roll tightly.

5-6. 원형으로 고정시킨다.
5-6. 将面条调整成圆形模样，然后固定。
5-6. Keep the roll in a circular shape.

TIP 建议

5-1. 국수를 삶을 때 중간에 찬물을 한번 부어서 탄력성을 주며, 3분 정도 삶으면 된다.
6. 국수가 붇지 않도록 과정을 신속하게 진행한다.

5-1. 在煮面条的过程中加入一次凉水，这样可以保持面条的筋道。
6. 为了使面条不会粘在一起，这个过程要快速地进行。

Step 5-1. When you boil the noodles, pour cold water in between to keep the noodles flexible. Boil the noodles for about 3 minutes.
Step 6. Cook fast so that the noodles will not get swollen up.

韩食美"味"秘方传授 100选 实践篇

Transmission of a Hundred Secrets of Korean Dishes Flavor 汤

한 식 의 맛 비 법 전 수 1 0 0 선

실기편

국, 탕

녹두삼계탕

绿豆参鸡汤

Mung bean samgyetang

재료 및 분량(1인분)
닭(삼계닭 1마리) 650g, 찹쌀 30g, 녹두 20g, 수삼 1뿌리, 황기 1뿌리, 대파 20g, 표고버섯 1개, 통마늘 4개, 은행 5g, 대추 2개, 감초 2g, 청주 15g, 다시마 5g
***채소육수** : 양파 50g, 당근 10g, 대파 10g, 대추 2개
***응용요리재료** : 전복 또는 낙지, 한방재료 등을 첨가하여 다양하게 삼계탕에 활용

材料和份量(1人份)
鸡(参鸡汤用鸡 1只) 650g, 粘米 30g, 绿豆 20g, 水参 1根, 黄芪 1根, 大葱 20g, 香菇 1个, 大蒜 4个, 银杏 5g, 大枣 2个, 甘草 2g, 白酒 15g, 海带 5g
***蔬菜汤** : 洋葱 50g, 胡萝卜 10g, 大葱 10g, 大枣 2个
***应用料理材料** : 添加鲍鱼或八爪鱼, 韩方材料等可制成各种参鸡汤。

Ingredients and portion(1 portion)
650g(samgye) chicken, 30g Glutinous rice, 20g mung bean, 1 fresh ginseng root, 1 milk vetch root, 20g leek, 1 dried shiitake mushroom(dried mushroom pyogo), 4 garlics, 5g ginkgo nut, 2 jujubes, 2g licorice, 15g rice-wine, 5g dried sea tangle
***Vegetable stock** : 50g onion, 10g carrot, 10g leek, 2 jujubes
***Note** : You can add abalones, small octopus and other Chinese herbal ingredients.

재료 손질 材料处理方法 Clean ingredients

끓여 완성하기 完成 How to boil the chicken

만드는 방법

1. 녹두와 찹쌀을 30분 정도 불린다.
2. 녹두는 껍질을 제거한다.
3. 두꺼운 뚝배기에 녹두를 넣는다.
4. 찹쌀 30g, 녹두 20g을 채운다.
5. 청주 15g을 넣는다.
6. 다시마를 넣는다.
7. 수삼은 머리를 제거한다.
8. 삼계닭 내장 쪽에 수삼, 황기를 채운다.
9. 다리모양을 고정한다.
10. 향신채를 넣는다.
11. 채소육수를 붓고 30분간 끓인다.
12. 완성하여 담는다.

制作方法

1. 将绿豆和粘米用水泡 30分钟左右。
2. 将绿豆去皮。
3. 将绿豆放入石锅中。
4. 加入粘米30g, 绿豆20g。
5. 加入白酒15g。
6. 加入海带。
7. 将水参去头。
8. 将水参和黄芪放入参鸡的内脏里。
9. 固定鸡腿的形状。
10. 加入调味菜。
11. 加入蔬菜汤后再煮30分钟。
12. 完成后盛菜。

How to cook

1. Soak mung bean and Glutinous rice for about 30 minutes.
2. Remove skin of mung bean.
3. Put the mung bean into a thick earthen pot.
4. Stuff the chicken with Glutinous of 30g rice and mung bean of 20g.
5. Add 15g of rice-wine.
6. Put the dried sea tangle.
7. Cut off the head of the fresh ginseng.
8. Stuff the chicken with the fresh ginseng and milk vetch root.
9. Fix the chicken's legs neatly.
10. Add condiment vegetable.
11. Pour vegetable stock. Boil the stock for 30 minutes.
12. Finish.

TIP
建议

삼계탕은 소금 간을 하지 않으며, 음식을 먹을 때 별도로 제공된다.
삼계닭 내장 쪽에 찹쌀, 한방재료 등의 모든 재료를 넣어 국물을 맑게 제공한다.

参鸡汤不用盐调味，食用时另行提供盐。
在参鸡汤内脏中放入粘米和韩方材料等是为了保持鸡汤为清汤。

Do not season the chicken with salt. Instead, serve the salt on a separate plate at a tale.
Make the soup cleanly by stuffing the chicken with Glutinous rice and Chinese herbal ingredients.

미역오이냉국

海带黄瓜凉汤

Cold seaweed-cucumber soup

재료 및 분량(2인분)
마른 미역 10g, 청오이 70g, 홍고추 15g, 풋고추 15g, 마늘 5g, 양파 30g, 실파 5g, 진간장 50g, 식초 50g, 올리고당 40g, 통깨 1g
***육수** : 다시마 5g, 국멸치 3마리
***응용요리재료** : 냉국 건더기 재료는 삶은 가지, 늙은 오이, 청포묵, 김치, 닭고기 등을 활용

材料和份量(2人份)
干海带 10g, 青黄瓜 70g, 红辣椒 15g, 青辣椒 15g, 大蒜 5g, 洋葱 30g, 小葱 5g, 浓酱油 50g, 食醋 50g, 低聚糖 40g, 芝麻 1g
***肉汤** : 海带 5g, 汤用鳀鱼 3条
***应用料理材料** : 制作凉汤时可利用蒸好的茄子, 老黄瓜, 粉皮, 泡菜, 鸡肉等。

Ingredients and portion(2 portions)
10g dried seaweed, 70g cucumber, 15g red chili, 15g green chili, 5g garlic, 30g onion, 5g green onion, 50g thick soy sauce, 50g vinegar, 40g oligosaccharide, 1g sesame seed
***Meat stock** : 5g dried sea tangle, 3 anchovies
***Note** : You can use boiled eggplant, old cucumber, green-lentil jelly, kimchi, and chicken as other cold soup.

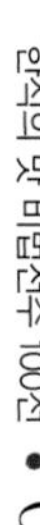

재료 손질 材料处理方法 Clean ingredients

완성하기
完成
How to finish

냉국 만들기
制作凉汤
How to cook cold soup

만드는 방법

1. 마른 미역을 불린다.
2. 끓는 물에 데친다.
3. 실파는 송송 썰어서 찬물에 헹군다.
4. 양파는 0.3cm로 채 썬다.
5. 멸치육수 400g에 진간장 50g을 풀어준다.
6. 식초 50g을 넣는다.
7. 올리고당 40g을 넣는다.
8. 청오이는 0.2cm로 채 썰어 넣는다.
9. 0.2cm로 어슷썬 홍고추와 다진 마늘을 넣는다.
10. 미역과 섞어준다.
11. 얼음으로 시원하게 완성한다.
12. 완성하여 담는다.

制作方法

1. 用水泡开干海带。
2. 放入烧开的水中焯一下。
3. 将小葱切成丝细后放入凉水中冲一下。
4. 将洋葱切成大小0.3cm。
5. 在400g鳀鱼汤中兑入50g酱油。
6. 加50g醋。
7. 加40g低聚糖。
8. 将青黄瓜切成大小 0.2cm后加入其中。
9. 加入0.2cm的红辣椒丝和蒜泥。
10. 与海带混合。
11. 最后加入冰块完成。
12. 完成后盛菜。

How to cook

1. Soak the dried seaweed.
2. Blanch the seaweed in boiling water.
3. Chop the green onion. Wash the green onion in cold water.
4. Slice the onion into 0.3cm wide pieces.
5. Pour thick soy sauce of 50g into anchovy stock of 400g.
6. Add vinegar of 50g.
7. Add oligosaccharide of 40g.
8. Julienne cucumber into 0.2cm wide pieces.
9. Diagonally cut the red chili into 0.2cm wide pieces. Put the red chili and the chopped garlic into the stock.
10. Put the seaweed.
11. Add ice cubes.
12. Finish.

TIP
建议

냉국 만들기 : 멸치육수 200g, 진간장 100g, 설탕 50g, 올리고당 50g, 식초 50g, 매실즙 10g, 소금 3g, 다진 마늘 5g의 원액으로 만들어 생수와 1:1로 희석하여 사용한다.

制作凉汤: 用鳀鱼汤 200g, 浓酱油 100g, 白糖 50g, 低聚糖 50g, 醋 50g, 梅子汁 10g, 盐 3g, 和蒜泥 5g的原液制成, 与水以1:1的比例稀释使用。

To cook cold soup : 200g anchovy stock, 100g thick soy sauce, 50g white sugar, 50g oligosaccharide, 50g vinegar, 10g plum juice, 3g salt, 5g chopped garlic. Make juice with all those ingredients together. Mix the juice with water at a ratio of 1:1.

버섯들깨탕

蘑菇苏子汤

Mushroom-perilla soup

재료 및 분량(2인분)
느타리버섯 30g, 팽이버섯 5g, 표고버섯 1장, 실파 5g, 깻잎 10g, 들깨가루 20g, 찹쌀가루 30g, 조랭이떡 50g, 까나리액젓 10g, 들기름 5g
***육수** : 다시마 5g, 국멸치 3마리, 무 5g, 통마늘 5g, 양파 5g, 대파 5g, 버섯기둥 3g
***응용요리재료** : 닭 가슴살, 감자, 말린 나물류 등으로 요리된다.

材料和份量(2人份)
平菇 30g, 金针蘑 5g, 香菇 1个, 小葱 5g, 苏子叶 10g, 野芝麻粉 20g, 粘米面儿 30g, 料理用年糕(小葫芦年糕) 50g, 玉筋鱼酱 10g, 芝麻油 5g
***肉汤** : 海带 5g, 汤用鳀鱼 3条, 萝卜 5g, 大蒜 5g, 洋葱 5g, 大葱 5g, 蘑菇杆 3g
***应用料理材料** : 可利用鸡胸脯肉、土豆、干菜等。

Ingredients and portion(2 portions)
30g oyster mushroom, 5g mushroom winter, 1 dried shiitake mushroom(dried mushroom pyogo), 5g green onion, 10g sesame leaf, 20g perilla seed flour, 30g glutinous rice flour, 50g small rice cake(joraengyi tteok), 10g sand eel sauce, 5g perilla seed oil
***Meat stock** : 5g dried sea tangle, 3 anchovies, 5g white radish, 5g garlic, 5g onion, 5g leek, 3g mushroom base
***Note** : You can also add chicken breast, potato and dried herbs.

육수 끓이기 制作肉汤 How to boil stock

재료 손질 材料出路方法 Clean ingredients

끓이기 煮 How to boil

완성하기 完成 How to finish

만드는 방법

1. 육수를 끓인다.
2. 조랭이떡을 데친다.
3. 들기름을 5g 넣는다.
4. 들깨가루 20g을 넣는다.
5. 골고루 볶는다.
6. 찹쌀가루를 볶는다.
7. 멸치육수 600g을 붓는다.
8. 조랭이떡을 넣고 끓인다.
9. 데친 모둠버섯을 넣는다.
10. 까나리액젓으로 간을 한다.
11. 깻잎을 넣는다.
12. 완성하여 담는다.

制作方法

1. 煮肉汤。
2. 将料理用年糕焯一下。
3. 加入5g芝麻油。
4. 加入20g野芝麻粉。
5. 均匀地炒熟。
6. 炒粘米面儿。
7. 加入600g鳀鱼汤。
8. 放入料理用年糕（小葫芦年糕）后开始煮。
9. 加入焯好的蘑菇。
10. 用玉筋鱼酱调味。
11. 加入苏子叶。
12. 完成后盛菜。

How to cook

1. Boil stock.
2. Blanch small rice cake(joraengyi tteok).
3. Put perilla seed oil of 5g.
4. Sprinkle perilla seed flour of 20g.
5. Roast the ingredients well.
6. Roast Glutinous rice flour.
7. Pour anchovy stock of 600g.
8. Put small rice cake(joraengyi tteok). Boil the ingredients.
9. Put the blanched mushrooms.
10. Season the ingredients with sand eel sauce.
11. Put sesame leaves.
12. Finish.

TIP 建议

들깨가루는 거피 제거한 것으로 준비하여 부드러운 맛을 낸다.
신선한 들깨가루를 사용해야 구수한 국물 맛을 낼 수 있다.

野芝麻粉使用去皮的野芝麻粉，可以使味道更柔嫩。
使用新鲜的野芝麻粉，可以使汤的味道更香酥。

Use the perilla seed flour made of perilla seed with the skin removed. This will make your dish taste soft.
Use fresh perilla seed flour to make the soup taste sweet.

북어국

干明太鱼汤

Dried pollack soup

재료 및 분량(2인분)
포북어 1마리, 무 200g, 콩나물 80g, 달걀 1개, 홍고추 10g, 풋고추 10g, 대파 10g, 실파 10g, 마늘 5g, 참기름 10g, 소금 1g, 후춧가루 0.5g
***육수** : 다시마 5g, 북어머리 1개, 무 200g, 국멸치 2마리, 통마늘 5g
***응용요리재료** : 북어굴국, 북어감자국 등의 다양한 재료로 요리된다.

材料和份量(2人份)
明太鱼 1条, 萝卜 200g, 豆芽 80g, 鸡蛋 1个, 红辣椒 10g, 青辣椒 10g, 大葱 10g, 小葱 10g, 大蒜 5g, 香油 10g, 盐 1g, 胡椒粉 0.5g
***肉汤** : 还嗲 5g, 明太鱼头 1个, 萝卜 200g, 汤用鳀鱼 2条, 大蒜 5g
***应用料理材料** : 利用各种材料可制成海蛎子明太鱼汤和土豆明太鱼汤等。

Ingredients and portion(2 portions)
1 dried pollack fillet, 200g white radish, 80g bean sprouts, 1 egg, 10g red chili, 10g green chilli, 10g leek, 10g green onion, 5g garlic, 10g sesame oil, 1g salt, 0.5g ground black pepper
***Meat stock** : 5g dried sea tangle, 1 dried pollack head, 200g white radish, 2 anchovies, 5g garlic
***Note** : You can also use the ingredients to cook a dish of polack fillet and oyster or polack fillet and potato soup.

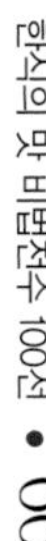

육수 끓이기 制作肉汤 How to boil stock

재료 손질
材料处理方法
Clean ingredients

끓이기
煮
How to boil

만드는 방법

1. 육수를 끓인다.
2. 콩나물 꼬리를 다듬는다.
3. 찢은 포북어를 달걀옷에 재운다.
4. 소금, 후춧가루로 버무린다.
5. 육수에서 건진 무를 0.5cm로 채 썬다.
6. 참기름 10g에 무를 볶는다.
7. 육수를 9부 정도 붓는다.
8. 콩나물을 넣는다.
9. 양념한 포북어를 넣는다.
10. 향채소류를 넣는다.
11. 소금 간을 한다.
12. 완성하여 담는다.

制作方法

1. 煮肉汤。
2. 将豆芽去尾。
3. 搅拌撕好的干明太鱼和鸡蛋。
4. 用盐、胡椒粉调味。
5. 将从肉汤中捞出来的萝卜切成大小0.5cm。
6. 用10g香油炒萝卜。
7. 加入90%左右的肉汤。
8. 加入豆芽。
9. 加入调好味的明太鱼。
10. 加入调味蔬菜。
11. 用盐调味。
12. 完成后盛菜。

How to cook

1. Boil stock.
2. Remove the ends of bean spouts.
3. Coat the torn dried pollack fillet with egg.
4. Season the ingredients with salt and ground black pepper.
5. Take the white radish out from the stock. Julienne the white radish into 0.5cm wide pieces.
6. Stir-fry the white radish with sesame oil.
7. Pour 90% of the meat stock.
8. Put the bean sprouts.
9. Put the seasoned dried pollack fillet.
10. Add condiment vegetable.
11. Season the ingredients with salt.
12. Finish.

TIP
建议

3. 달걀옷을 입혀서 포북어살을 부드럽게 만든다.
8. 콩나물 머리는 충분히 가열하면 비린내가 나지 않는다.
9. 포북어가 익을 때까지 젓지 않아야 국물이 맑다.

3. 搅拌鸡蛋和干明太鱼丝可以使明太鱼肉质更加柔软。
8. 充分加热豆芽头可以去除异味。
9. 在明太鱼熟透前不要搅拌才可以使汤的味道更清爽。

Step 3. Coat the dried pollack fillet with the egg to make the pollack fillet soft.
Step 8. Boil the head of the bean sprouts as much as possible to remove fishy smell.
Step 9. Do not stir the soup until the dried pollack fillet is fully cooked. This way will make the soup clean.

성게전복미역국

海胆鲍鱼海带汤

Sea urchin-abalone seaweed soup

재료 및 분량(2인분)
성게 60g, 전복(1마리) 60g, 마른 미역 10g, 표고버섯 1개, 마늘 3g, 국간장 3g, 소금 3g, 천연조미료 3g, 청주 5g, 참기름 10g
***육수** : 다시마 5g, 무 50g
***응용요리재료** : 홍합, 조개, 참치 등은 미역국 재료로 요리된다.

材料和份量(2人份)
海胆 60g, 鲍鱼(1只) 60g, 干海带 10g, 香菇 1个, 大蒜 3g, 汤用酱油 3g, 盐 3g, 天然调味料 3g, 白酒 5g, 香油 10g
***肉汤** : 海带 5g, 萝卜 50g
***应用料理材料** : 海虹, 扇贝, 金枪鱼等可用作海带汤的材料。

Ingredients and portion(2 portions)
60g sea urchin, 60g abalones, 10g dried seaweed, 1 shiitake mushroom(mushroom pyogo), 3g garlic, 3g soy sauce, 3g salt, 3g natural seasonings, 5g rice-wine, 10g sesame oil
***Meat stock** : 5g dried sea tangle, 50g white radish
***Note** : You can also cook sea urchin-abalone seaweed soup with mussel, clam and tuna.

육수 끓이기 煮肉汤 How to boil stoc

미역 손질
海带的处理方法
How to cook seaweed

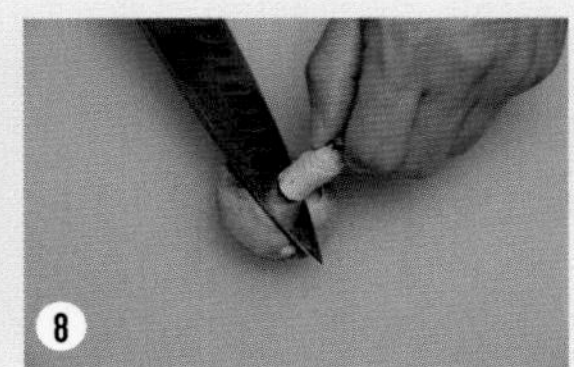

버섯 손질
蘑菇的处理方法
How to cook mushroom

만드는 방법

1. 무는 편 썬다.
2. 육수를 끓인다.
3. 끓으면 다시마를 건진다.
4. 육수를 체에 거른다.
5. 마른 미역을 불린다.
6. 불린 미역을 데친다.
7. 3x4cm로 채 썬다.
8. 표고기둥을 제거한다.
9. 밑둥을 다듬는다.
10. 0.2cm로 편 썰기
11. 끓는 물에 넣는다.
12. 살짝 데친다.

制作方法

1. 将萝卜切成片。
2. 煮肉汤。
3. 待汤煮开后捞出海带。
4. 将肉汤盛出来备用。
5. 用水泡开干裙带菜。
6. 把裙带菜用水焯一下。
7. 将裙带菜切成大小3x4cm。
8. 去除蘑菇杆。
9. 整理底部。
10. 将蘑菇切成大小0.2cm。
11. 将蘑菇放入烧开的水中。
12. 稍微焯一下。

How to cook

1. Cut the white radish into thin slices.
2. Boil meat stock.
3. When the meat stock boils, take the dried sea tangle out from the stock.
4. Sieve the meat stock.
5. Soak the dried seaweed in water.
6. Blanch the seaweed.
7. Julienne the sea weed into 3x4cm wide pieces.
8. Remove the base of the shiitake mushroom(mushroom pyogo).
9. Clean the base.
10. Cut the base into 0.2cm wide slices.
11. Put the slices into boiling water.
12. Blanch the slices slightly.

전복 손질
鮑鱼的处理方法
How to cook abalone

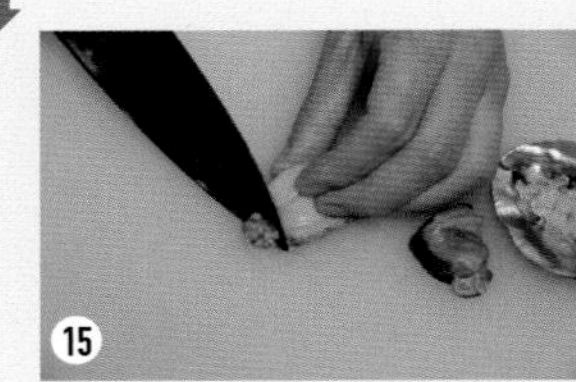

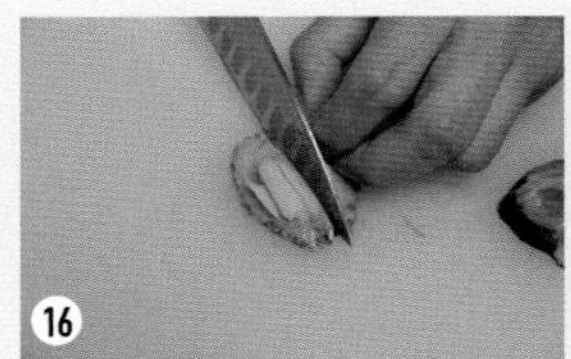

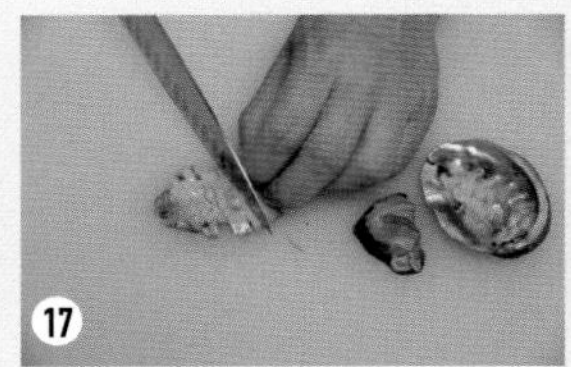

끓이기
煮
How to boil

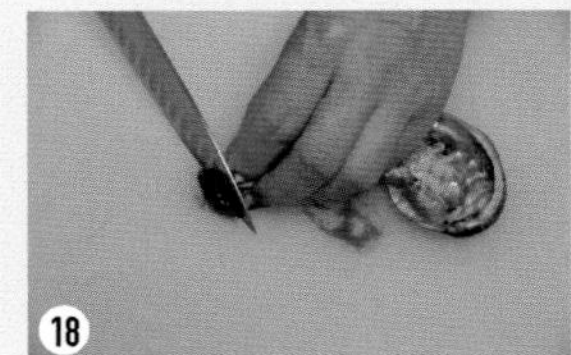

만드는 방법

13. 찬물에 담가 식힌다.
14. 전복, 내장, 껍질을 분리한다.
15. 입을 제거한다.
16. 0.2cm 간격으로 칼집을 낸다.
17. 0.2cm로 편 썬다.
18. 내장을 제거한다.
19. 참기름을 10g 넣는다.
20. 전복을 볶는다.
21. 미역을 볶는다.
22. 미역을 부드럽게 볶는다.
23. 육수를 붓는다.
24. 끓으면 성게를 넣는다.

制作方法

13. 放入凉水中冷却。
14. 分离鲍鱼，内脏和壳。
15. 去除嘴部。
16. 在鲍鱼上切出间隔0.2cm的刀口。
17. 切成大小为0.2cm的片状。
18. 去除内脏。
19. 加入香油10g。
20. 炒鲍鱼。
21. 炒裙带菜。
22. 炒海带，使其口感柔嫩。
23. 加入肉汤。
24. 待烧开后加入海胆。

How to cook

13. Cool the slices down in cold water.
14. Separate flesh, internal organs and shell.
15. Remove the mouth.
16. Make cuts at an interval of 0.2cm.
17. Slice the flesh into 0.2cm wide slices.
18. Remove the internal organs.
19. Pour 10g of sesame oil.
20. Fry the abalones.
21. Fry the seaweed.
22. Fry the seaweed softly.
23. Pour meat stock.
24. When the stock boils, put sea urchin.

만드는 방법

25. 표고버섯을 넣는다.
26. 국간장 3g을 넣어 색을 맞춘다.
27. 소금 3g으로 간을 맞춘다.
28. 천연조미료 3g을 넣는다.
29. 다진 마늘 3g을 넣는다.
30. 청주 5g을 넣는다.
31. 거품을 제거한다.
32. 따뜻하게 담는다.

制作方法

25. 加入香菇。
26. 加入汤用酱油3g调节颜色。
27. 加入盐3g调味。
28. 加入天然调味料3g。
29. 加入蒜泥3g。
30. 加入白酒5g。
31. 去除泡沫。
32. 盛放在容器里。

How to cook

25. Put shiitake mushroom(mushroom pyogo).
26. Adjust the color with soy sauce of 3g.
27. Season the ingredients with salt of 3g.
28. Add natural seasonings of 3g.
29. Put chopped garlic of 3g.
30. Add rice-wine of 5g.
31. Skim off the foam.
32. Serve the dish warm.

TIP
建议

5. 마른 미역은 충분히 불려 사용해야 결이 부드럽다.
19. 참기름이 타면 쓴맛의 원인이 되므로 주의하여 약불에서 볶는다.

5. 将干海带充分泡开后再使用，这样可以使口感柔嫩。
19. 如果香油烧焦了会有苦味儿，请注意此点并用微火炒。

Step 5. Soak the dried seaweed thoroughly to make the surface soft.
Step 19. Stir-fry the sesame oil over a low heat so that your dish will not taste bitter because of any burned sesame oil.

소고기미역국

海带牛肉汤

Beef-seaweed soup

재료 및 분량(2인분)
마른 미역 30g, 소고기(양지머리) 60g, 들깨가루 30g, 양파 5g, 까나리액젓 5g, 국간장 3g, 청주 5g, 마늘 5g, 소금 2g 참기름 5g
***육수** : 다시마 5g, 국멸치 2마리, 마른 새우 3g, 고추씨 2g, 양파 5g
***응용요리재료** : 마른 조갯살과 새우살 등을 미역국 재료로 활용할 수 있다.

材料和份量(2人份)
干海带 30g, 牛肉(颈部肉) 60g, 野芝麻粉 30g, 洋葱 5g, 玉筋鱼酱 5g, 汤用酱油 3g, 白酒 5g, 大蒜5g, 盐 2g 香油 5g
***肉汤** : 海带 5g, 汤用鳀鱼 2条, 干虾 3g, 辣椒籽 2g, 洋葱 5g
***应用料理材料**: 干扇贝肉和虾肉等可用作海带汤的材料。

Ingredients and portion(2 portions)
30g dried seaweed, 60g beef(brisket), 30g perilla seed flour, 5g onion, 5g sand eel sauce, 3g soy sauce, 5g rice-wine, 5g garlic, 2g salt, 5g sesame oil
***Meat stock** : 5g dried sea tangle, 2 anchovies, 3g dried shrimp, 2g pepper seed, 5g onion
***Note** : You can also use dried clam flesh and dried shrimp to cook soup.

만드는 방법

1. 불린 미역에 국간장 3g을 넣어 양념한다.
2. 까나리액젓 5g을 넣어 양념한다.
3. 들깨가루에 멸치육수 100g을 붓는다.
4. 청주 5g을 넣는다.
5. 마늘 5g을 넣는다.
6. 양파 5g을 넣는다.
7. 들깨즙을 완성한다.
8. 참기름 5g으로 소고기, 미역을 볶는다.
9. 들깨즙을 넣고 볶는다.
10. 육수를 붓는다.
11. 소금 2g으로 마무리한다.
12. 완성하여 담는다.

制作方法

1. 在泡好的裙带菜中加入3g汤用酱油来调味。
2. 加入5g玉筋鱼酱。
3. 在芝麻粉上倒入100g鳀鱼汤。
4. 加入白酒5g。
5. 加入大蒜5g。
6. 加入洋葱5g。
7. 芝麻汤制作完成。
8. 用5g香油炒牛肉和裙带菜。
9. 加入芝麻汤后一起炒。
10. 加入肉汤。
11. 加入2g盐。
12. 完成后盛菜。

How to cook

1. Soak seaweed. Season the seaweed with soy sauce of 3g.
2. Season the ingredients with sand eel sauce of 5g.
3. Pour 100g of anchovy stock to the perilla seed flour.
4. Put rice-wine of 5g.
5. Add garlic of 5g.
6. Put onion of 5g.
7. Finish the perilla juice.
8. Roast the beef and the seaweed with sesame oil of 5g.
9. Add perilla juice. Roast the ingredients all together.
10. Pour the meat stock.
11. Season the ingredients with salt of 2g.
12. Finish.

TIP
建议

들깨즙은 미역국 같은 맑은 국과 구수한 국물 맛을 내기 위해 많이 활용된다.

利用芝麻汤可以使海带汤的味道香酥可口。

Use the perilla juice to make the soup look clean and taste sweet.

수삼갈비탕

人参排骨汤

Fresh ginseng galbitang

재료 및 분량(2인분)
한우(소갈비) 400g, 소고기(양지머리) 400g, 당면 10g, 수삼 1뿌리, 대파 5g, 무 20g, 마늘 5g, 대추 2개, 생강 3g, 국간장 3g, 천연조미료 3g, 소금 3g
***응용요리재료** : 우거지갈비탕, 한방재료를 다양하게 끓이는 탕요리로 활용한다.

材料和份量(2人份)
韩牛(牛排骨) 400g, 牛肉(颈部肉) 400g, 粉条 10g, 水参 1根, 大葱 5g, 萝卜 20g, 大蒜 5g, 大枣 2个, 生姜 3g, 汤用酱油 3g, 天然调味料 3g, 盐 3g
***应用料理材料** : 可用于制作菜帮排骨汤和利用韩方材料的各种汤料理。

Ingredients and portion(2 portions)
400g Korea beef(beef ribs), 400g beef(brisket), 10g starch noodle, 1 fresh ginseng root, 5g leek, 20g white radish, 5g garlic, 2 jujubes, 3g ginger, 3g soy sauces, 3g natural seasonings, 3g salt
***Note** : You can also cook chinese cabbage galbitang with the ingredients of Fresh ginseng galbitang by adding some Chinese herbal ingredients.

재료 손질 材料处理方法 Clean ingredients

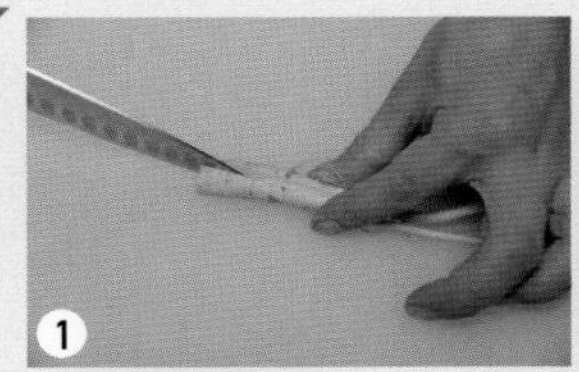

육수 끓이기
煮肉汤
How to boil meat stock

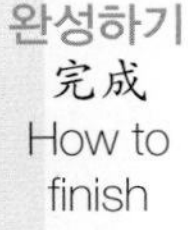

완성하기
完成
How to finish

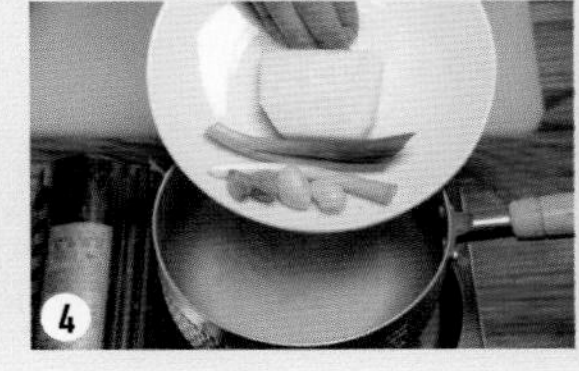

만드는 방법

1. 수삼을 손질한다.
2. 핏물 제거한 소갈비, 양지를 끓인다.
3. 고기를 건져 놓는다.
4. 육수에 향신채소를 끓여 제거한다.
6. 당면을 넣는다.
7. 따뜻하게 담는다.
8. 고명은 대추, 대파를 준비한다.
9. 갈비탕을 완성한다.

制作方法

1. 处理水参。
2. 煮牛排骨和颈部肉(无血水)。
3. 从锅里取出肉。
4. 用调味蔬菜煮肉汤，然后取出调味蔬菜。
6. 放入粉条。
7. 盛放在容器里。
8. 用大枣和大葱作为装饰菜。
9. 排骨汤制作完成。

How to cook

1. Clean fresh ginseng.
2. Remove any blood from the beef rib and the beef brisket. Boil them all together.
3. Take the beef out from the stock.
4. Put condiment vegetable to the stock. Remove the smell.
6. Put starch noodles.
7. Serve the dish warm.
8. Prepare jujube and green onion as a garnish.
9. Finish.

뚝배기에 끓이기 用石锅煮 How to boil in ttukbaegi pot

5-1. 불순물을 제거한다.

5-1. 清除汤中的杂质。

5-1. Remove alien substances.

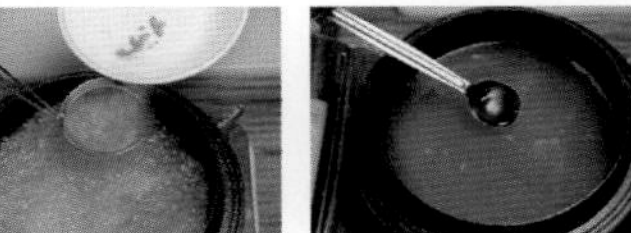

5-2. 국간장 3g과 소금 3g으로 간을 한다.

5-2. 加3g汤用酱油和3g盐

5-2. Season the ingredients with soy sauce of 3g and salt.

5-3. 천연조미료 3g을 넣는다.

5-3. 加入天然调味料3g。

5-3. Put natural seasonings of 3g.

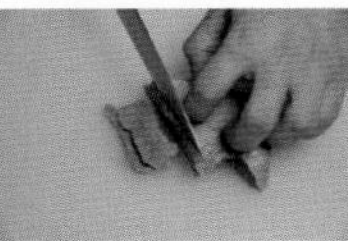

5-4. 양지머리를 편 썬다.

5-4. 将颈部肉切成细片。

5-4. Cut the beef brisket into thin slices.

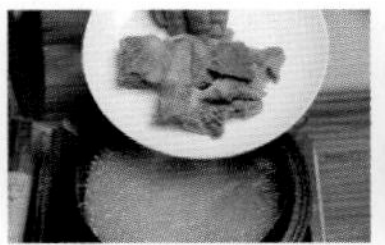

5-5. 소갈비, 양지편을 넣는다.

5-5. 加入牛排骨和颈部肉。

5-5. Put slices of the beef rib and the beef brisket.

5-6. 맑게 끓인다.

5-6. 多煮一段时间。

5-6. Boil the soup clean.

TIP 建议

2. 핏물을 완전히 제거하여 국물을 맑게 끓인다. (핏물 제거는 찬물에 담갔다가 헹구는 것이다.)

2. 干净的去除血水，保证汤水清新。（为了清除血水先把肉和骨泡在水里，然后进行多次清洗。）

Step 2. Remove the blood thoroughly to make the soup clean. (Remove the blood in cold water).

조개맑은탕

蛤蜊清汤

Clean clam soup

재료 및 분량(2인분)
조개(생합) 300g, 실파 5g, 홍고추 5g, 마늘 3g, 소금 3g, 청주 5g, 천연조미료 3g
***육수** : 다시마 5g, 통마늘 5g
***응용요리재료** : 홍합, 대합 등의 다양한 조개류를 재료로 활용하여 맑게 끓인다.

材料和份量(2人份)
蛤蜊(生蛤蜊) 300g, 小葱 5g, 红辣椒 5g, 大蒜 3g, 盐 3g, 白酒 5g, 天然调味料 3g
***肉汤** : 海带 5g, 大蒜 5g
***应用料理材料** : 利用海虹和大蛤等各种材料制作清汤。

Ingredients and portion(2 portions)
300g fresh clam, 5g green onion, 5g red chilli, 3g garlic, 3g salt, 5g rice-wine, 3g natural seasonings
***Meat stock** : 5g dried sea tangle, 5g garlic
***Note** : You can also use other shellfish as mussel and clam to make the soup clean.

재료 손질 材料处理方法 Clean ingredients

완성하기 完成 How to finish

만드는 방법
1. 생합을 해감한다.
2. 실파는 길이 4cm로 썬다.
3. 홍고추는 길이 4cm로 채 썬다.
4. 마늘은 곱게 다진다.
6. 다진 마늘 3g을 넣는다.
7. 청주 5g을 넣는다.
8. 천연조미료 3g을 넣는다.
9. 따뜻하게 담는다.

制作方法
1. 洗掉生蛤蜊的淤泥。
2. 将小葱切成长4cm的细丝。
3. 将红辣椒切成长4cm的丝细。
4. 将大蒜切碎。
6. 加入切碎的大蒜3g。
7. 加入白酒5g。
8. 加入天然调味料3g。
9. 盛放在容器里。

How to cook
1. Wash fresh clam thoroughly.
2. Cut the green onion into 4cm long pieces.
3. Julienne the red chilli 4cm long pieces.
4. Chop the garlic finely.
6. Put chopped garlic of 3g.
7. Pour rice-wine of 5g.
8. Add of natural seasonings of 3g.
9. Serve the dish warm.

육수 끓이기 制作肉汤 How to boil stock

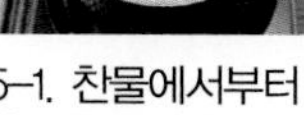

5-1. 찬물에서부터 육수를 끓인다.

5-1 .用凉水煮肉汤。

5-1. Boil meat stock in cold water.

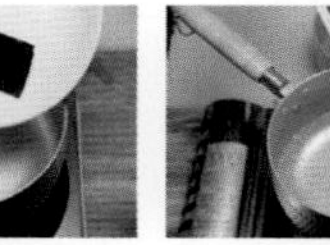

5-2. 끓는 물에서 다시마를 건진다.

5-2. 从烧开的水中捞出海带。

5-2. Take dried sea tangle out from boiling water.

5-3. 생합을 넣는다.

5-3. 加入生蛤蜊。

5-3. Put fresh clam.

5-4. 생합을 익힌다.

5-4. 将生蛤蜊煮熟。

5-4. Cook the fresh clam well.

5-5. 실파를 넣는다.

5-5. 加入小葱。

5-5. Put green onion.

5-6. 소금 3g으로 간하고, 홍고추를 넣는다.

5-6. 加入盐3g和红辣椒调味。

5-6. Put salt of 3g and red chilli of 3g.

TIP 建议

1. 생합의 해감은 소금을 3% 뿌리고, 빛이 없는 곳에 두면 깨끗하게 해감된다.
생합을 오래 끓이면 질겨지므로 주의한다.

1. 生蛤蜊淤泥处理方法是撒上3%的盐后放置在阴凉处淤泥就会被干净地去除。
注意如果生蛤蜊煮得时间太久就会变硬。

Step 1. When you wash the fresh clam, sprinkle salt of 3%. Leave the fresh clam in the shade. Be careful not to over-cook the fresh clam.
That will make the fresh clam tough.

콩나물냉국

豆芽凉汤

Cold bean sprouts soup

재료 및 분량(2인분)
콩나물 200g, 실파 5g, 양파 10g, 홍고추 5g, 풋고추 5g, 마늘 3g, 소금 3g, 통깨 2g
***육수** : 다시마 5g, 국멸치 2마리, 통마늘 2g, 대파 3g
***응용요리재료** : 미역콩나물냉국, 북어콩나물냉국 등으로 요리된다.

材料和份量(2人份)
豆芽 200g, 小葱 5g, 洋葱 10g, 红辣椒 5g, 青辣椒 5g, 大蒜 3g, 盐 3g, 芝麻 2g
***肉汤** : 海带 5g, 汤用鳀鱼 2条, 大蒜 2g, 大葱 3g
***应用料理材料** : 也可用作海带豆芽凉汤, 明太鱼豆芽凉汤的材料。

Ingredients and portion(2 portions)
200g bean sprouts, 5g green onion, 10g onion, 5g red chilli, 5g green chilli, 3g garlic, 2g salt, 2g sesame seed
***Meat stock** : 5g dried sea tangle, 2 anchovies, 2g garlic, 3g leek
***Note** : You can use the ingredients of cold bean sprouts soup to cook cold seaweed bean sprouts soup and cold dried pollack bean sprouts soup.

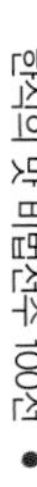

재료 손질 材料处理方法 Clean ingredients

냉국 만들기 制作凉汤 How to cook the cold soup

완성하기 完成 How to finish

만드는 방법

1. 콩나물 꼬리만 제거한다.
2. 콩나물과 육수재료를 함께 끓인다.
3. 콩나물에 맛을 들인다.
4. 콩나물을 걸러낸다.
5. 냉국에 소금 간을 한다.
6. 마늘을 곱게 다진다.
7. 다진 마늘을 3g 넣는다.
8. 실파는 길이 3cm로 채 썬다.
9. 홍고추, 풋고추를 길이 3cm로 채 썬다.
10. 양파는 3cm로 채 썬다.
11. 모든 재료를 혼합하여 통깨를 뿌린다.
12. 완성하여 담는다.

制作方法

1. 去掉豆芽尾。
2. 将豆芽和肉汤材料一起煮。
3. 使味道浸入豆芽。
4. 取出豆芽。
5. 在凉汤中加入盐调味。
6. 将大蒜切碎。
7. 加入3g切碎的大蒜。
8. 将小葱切成长3cm的丝细。
9. 将红辣椒和青辣椒切成长3cm的丝细。
10. 将洋葱切成长3cm的细丝。
11. 将所有材料混合后完成。
12. 完成后盛菜。

How to cook

1. Remove the ends of bean sprouts.
2. Boil the bean sprouts and the stock together.
3. Season the bean sprouts.
4. Take out the bean sprouts.
5. Season the cold soup with salt.
6. Finely chop the garlic up.
7. Add chopped garlic of 3g.
8. Julienne the green onion into 3cm long pieces.
9. Julienne the red chili and the green chili into 3cm long pieces.
10. Julienne the onion into 3cm long pieces.
11. Mix all the ingredients together. Garnish the dish with sesame seed.
12. Finish.

TIP
建议

3. 콩나물을 너무 오래 끓이면 아삭거리는 맛이 없어지므로 주의한다.
11. 냉장 보관 또는 얼음을 띄워서 시원하게 먹는다.

3. 请注意如果豆芽煮得时间太久了会失去原有的味道。
11. 将做好的凉汤放在冰箱里保管，或食用时加入冰块。

Step 3. Try not to over-cook the bean sprouts. You will lose the crispiness.
Step 11. Serve the dish cool. Keep the dish in the refrigerator before you serve or serve the dish with ice cubes.

연잎오리탕

莲叶鸭汤

Lotus leaf-duck soup

재료 및 분량(3인분)
생오리(1마리) 2kg, 연잎 1장, 엄나무 30g, 수삼 1뿌리, 황기 1뿌리, 당귀 2g, 가시오가피 10g, 구기자 3g, 대추 3개, 은행 3g, 대파 10g, 양파 15g, 밤 2개, 호박씨 3g, 통잣 3g, 통마늘 3g, 표고버섯 1장, 찹쌀 150g
***응용요리재료** : 토종닭과 생오리를 재료로 활용하여 한방백숙을 요리한다.

材料和份量(3人份)
鸭子(1只) 2kg, 莲叶 1张, 刺桐 30g, 水参 1根, 黄芪 1根, 当归 2g, 刺五加 10g, 枸杞子 3g, 大枣 3个, 银杏 3g, 大葱 10g, 洋葱 15g, 栗子 2个, 南瓜籽 3g, 松树籽 3g, 大蒜 3g, 香菇 1个, 粘米 150g
***应用料理材料** : 利用土鸡和鸭子制作韩方清蒸料理。

Ingredients and portion(3 portions)
2kg fresh duck, 1 lotus leaf, 30g kalopanax, 1 fresh ginseng root, 1 milk vetch root, 2g Korean angelica roots(dong guai), 10g Eleuthero, 3g chinese wolfberries (gugija), 3 jujubes, 3g ginkgo nut, 10g leek, 15g onion, 2 chestnuts, 3g pumpkin seeds, 3g pine nut, 1 shiitake mushroom(mushroom pyogo), 150g Glutinous rice
***Note** : Use a Korea chicken or a fresh duck to cook other Chinese herbal chicken soup.

재료 손질 材料处理方法 Clean ingredients

생오리 손질
鸭子的处理方法
How to cook the duck

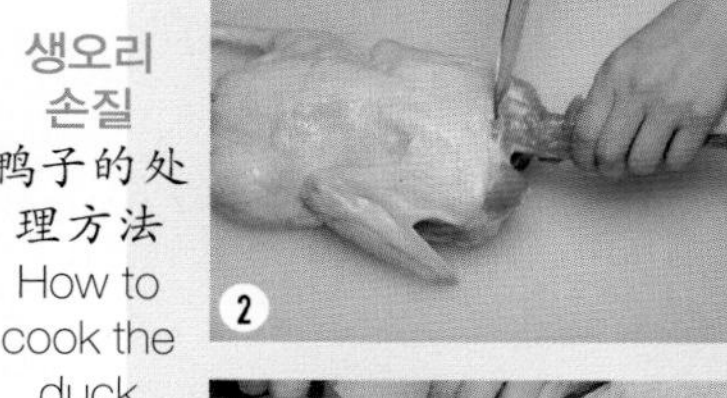

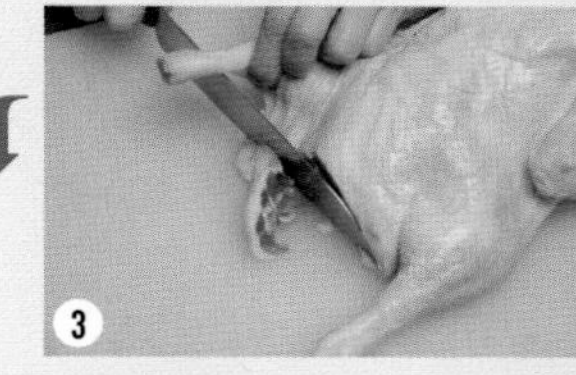

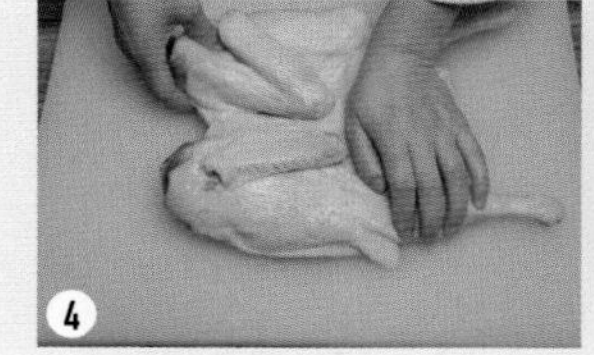

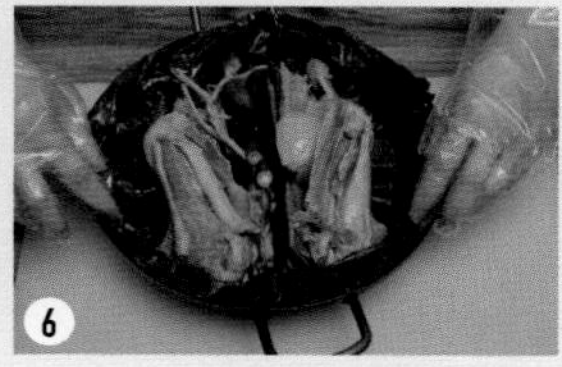

완성하기
完成
How to finish

만드는 방법

1. 불린 찹쌀과 부재료를 망에 넣는다.
2. 오리 목을 제거한다.
3. 엉덩이살을 제거한다.
4. 등쪽을 펴준다.
6. 삶은 오리를 건져 담는다.
7. 찹쌀을 붓는다.
8. 함께 담는다.
9. 국물을 붓는다.

制作方法

1. 将泡好的粘米和副材料放入网中。
2. 去除鸭子的脖子。
3. 去除鸭子的臀部肉。
4. 平铺背部。
6. 去除煮好的鸭肉并盛放在容器里。
7. 加入粘米。
8. 与鸭肉盛放在一起。
9. 加入肉汤。

How to cook

1. Soak Glutinous rice. Put the rice in a net with side ingredients.
2. Remove the neck of the duck.
3. Remove rump of the duck.
4. Make the back straightened up.
6. Take the boiled duck out.
7. Put the glutinous rice.
8. Put the ingredients all together.
9. Pour the soup.

끓이기
煮粥
How to boil

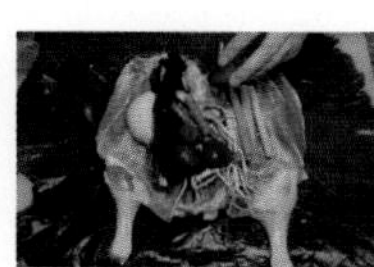

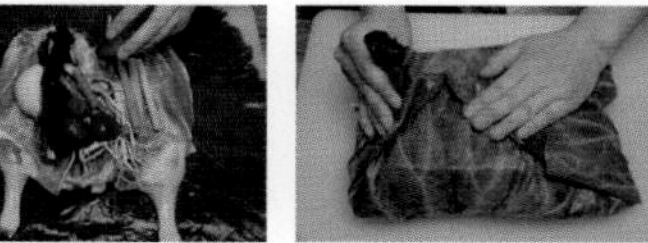

5-1. 한방재료를 넣는다.
5-1. 加入韩方中药材。
5-1. Put the Chinese herbs.

5-2. 연잎으로 감싸준다.
5-2. 用莲叶包好。
5-2. Cover the duck with the lotus leaf.

5-3. 압력솥에 담는다.
5-3. 放入压力锅中。
5-3. Put the duck into a pressure cooker.

5-4. 찬물을 1L 붓는다.
5-4. 加入1L凉水。
5-4. Pour cold water of 1L.

5-5. 준비된 ①을 넣는다.
5-5. 加入准备好的①。
5-5. Put the prepared ①.

5-6. 압력솥에 30분 삶는다.(불 조절)
5-6. 在压力锅中煮30分钟。(调整火候)
5-6. Boil the ingredients for 30 minutes in the pressure cooker (watch out for the heat).

TIP
建议

2~3. 백숙을 할 때 오리의 목과 엉덩이살을 제거하면 오리의 냄새를 제거할 수 있다.
5-2. 연잎은 오리냄새를 제거하는 데 많은 역할을 한다.
9. 남은 국물을 다시 끓여 따뜻하게 완성한다.

2~3. 这种处理方法是为了去除鸭肉的异味，主要用于清蒸。
5-2. 对去除鸭肉的异味起着很大的作用。
9. 加入肉汤后再继续煮，可以继续吃到热乎乎的饮食。

Step 2, Step 3. If you remove neck and rump of duck, you can remove the smell of the duck.
Step 5-2. You use the lotus leaf to remove the smell of the duck.
Step 9. Pour the soup and boil the dish again. Serve the dish warm.

홍초미역냉국

红醋海带凉汤

Cold red vinegar seaweed soup

재료 및 분량(4인분)
마른 미역 10g, 청오이 70g, 홍고추 15g, 풋고추 15g, 마늘 5g, 양파 30g, 실파 5g, 홍초 100g, 소금 3g, 올리고당 15g
***응용요리재료** : 홍초냉국 국물을 콩나물, 늙은 오이, 청포묵, 닭고기 등에 활용할 수 있다.

材料和份量(4人份)
干海带 10g, 青黄瓜 70g, 红辣椒 15g, 青辣椒 15g, 大蒜 5g, 洋葱 30g, 小葱 5g, 红醋 100g, 盐 3g, 低聚糖 15g
***应用料理材料** : 制作红醋凉汤可使用豆芽、老黄瓜、粉皮和鸡肉等。

Ingredients and portion(4 portions)
10g dried seaweed, 70g cucumber, 15g red chilli, 15g green chilli, 5g garlic, 30g onion, 5g spring onion, 100g red vinegar, 3g salt, 15g oligosaccharide
***Note** : You can use bean sprout, old cucumber, green lentil-jelly and chicken with cold red vinegar.

냉국 만들기 制作凉汤 How to cook the cold soup

만드는 방법

1. 생수 400g을 붓는다.
2. 홍초 100g을 붓는다.
3. 올리고당 15g을 넣는다.
4. 소금 3g을 붓는다.
5. 재료에 홍초 맛을 들인다.
6. 다진 마늘 3g을 넣는다.
7. 양파 0.2cm로 채 썬다.
8. 양파를 넣는다.
9. 홍초 맛이 밴 오이를 넣는다.
10. 홍초 맛이 밴 미역을 넣는다.
11. 다진 실파로 마무리한다.
12. 완성하여 담는다.

制作方法

1. 加入纯净水400g。
2. 加入红醋100g。
3. 加入低聚糖15g。
4. 加入盐3g。
5. 使材料浸入红醋的味道。
6. 加入切碎的大蒜3g。
7. 将洋葱切成大小0.2cm。
8. 加入洋葱。
9. 加入红醋味黄瓜。
10. 加入红醋裙带菜。
11. 加入切碎的小葱即完成。
12. 完成后盛菜。

How to cook

1. Pour cold water of 400g.
2. Pour red vinegar of 100g.
3. Put of oligosaccharide of 15g.
4. Sprinkle salt of 3g.
5. Season the ingredients with the red vinegar.
6. Put chopped garlic of 3g.
7. Julienne the onion into 0.2cm wide pieces.
8. Put the onion.
9. Put red vinegar-taste cucumber.
10. Put red vinegar and seaweed.
11. Garnish the dish with the chopped green onion.
12. Finish.

TIP
建议

2. 홍초 : 석류원액이 포함되어 있는 전통발효식초로 색상과 맛내기에 편리하다.

2. 红醋：含有石榴原液的传统发酵食醋，易于调节颜色和口味。

Step 2 : Red vinegar : Traditionally fermented vinegar with pomegranate juice included to make your dish even more colorful and taste better.

황기설렁탕

黄芪先浓汤

Milk vetch root seolleongtang

재료 및 분량((2인분)
한우뼈(사골) 800g, 소고기(양지머리) 900g, 황기 3g, 엄나무 5g, 구기자 2g, 감초 2g, 홍고추 10g, 대파 10g, 통마늘 5g, 생강 3g, 마른 표고버섯 1개, 사과 20g, 양파 20g
***응용요리재료** : 한우뼈(사골)를 사용하여 곰탕, 설렁탕, 해장국 등의 끓이는 탕류로 요리한다.

材料和份量(2人份)
韩牛骨(筛骨) 800g, 牛肉(颈部肉) 900g, 黄芪 3g, 刺桐 5g, 枸杞子 2g, 甘草 2g, 红辣椒 10g, 大葱 10g, 大蒜 5g, 生姜 3g, 干香菇 1个, 苹果 20g, 洋葱 20g
***应用料理材料** : 使用类似于牛杂碎汤的材料。

Ingredients and portion(2 portions)
800g Korea beef bone, 900g beef(brisket), 3g milk vetch root, 5g kalopanax, 2g chinese wolfberries(gugija), 2g licorice, 10g red chilli, 10g leek, 5g garlic, 3g ginger, 1 dried shiitake mushroom(dried mushroom pyogo), 20g apple, 20g onion
***Note** : You can also use Korean beef bone to cook the thick beef bone soup, solleongtang and Haejangguk.

만드는 방법

1. 사골 핏물을 제거한다.
2. 찬물 1.8L에 사골을 넣는다.
3. 한방재료, 향신채소, 소고기를 넣는다.
4. 압력솥에 약불로 5시간 정도 끓인다(양지는 1시간 후에 건져 낸다).
5. 기름기를 제거한다.
6. 양지머리는 0.2cm로 편 썰기
7. 대파를 송송 썬다.
8. 소금 간을 맞춘다.
9. 뚝배기에서 식지 않게 끓인다.
10. 따뜻하게 담는다.

制作方法

1. 去除筛骨的血水。
2. 将筛骨放入1.8L的凉水中。
3. 加入韩方中材料, 调味蔬菜和牛肉。
4. 放入压力锅中煮5个小时。(微火)
5. 去除油分。
6. 将颈部肉切成大小为0.2cm的片状。
7. 将大葱切成丝细。
8. 用盐调味。
9. 放入石锅中加热。
10. 盛放在容器里。

How to cook

1. Remove any blood from Korea beef bone.
2. Put the bone into 1.8L of cold water.
3. Put Chinese herbs, condiment vegetable and beef.
4. Boil the ingredients for 5 hours in a pressure cooker (At over a low heat. An hour later, take the brisket out.).
5. Remove the fatty residue.
6. Cut beef(brisket) into 0.2cm wide slices.
7. Chop up the leek.
8. Season the ingredients with salt.
9. Boil the ingredients warm in an earthen pot(ttukbaegi).
10. Serve the dish warm.

TIP
建议

1. 핏물을 충분히 제거하여 잡냄새를 없앤다.
4. 가마솥에서 12시간 정도 약불에 끓여야 국물의 제맛을 느낄 수 있다.

1. 充分地去除血水以去掉异味。
4. 只有在铁锅中用微火煮12个小时左右才能品尝到汤的真正味道。

Step 1. Remove blood thoroughly to eliminate any unpleasant smell.
Step 4. Boil the soup at least 12 hours in an iron pot over a low heat.

韩食美“味”秘方传授 100选 实践篇

Transmission of a Hundred Secrets of Korean Dishes Flavor 炖汤，炖锅

한식의 맛 비법전수 100선

실기편

찌개, 전골

김치돼지고기찌개

猪肉泡菜汤

Kimchi-pork jjigae

재료 및 분량(2인분)
김치 100g, 돼지고기(앞다리) 50g, 두부 30g, 감자 30g, 양파 20g, 애호박 15g, 새우젓 3g, 느타리버섯 10g, 표고버섯 1장, 홍고추 5g, 풋고추 5g, 대파 5g, 실파 3g, 생강 3g, 들기름 10g, 고추기름 10g, 들깨가루 5g, 청주 10g, 국간장 5g
***육수** : 다시마 5g, 국멸치 3마리, 통마늘 5g, 대파 5g
***응용요리재료** : 김치감자찌개, 김치두부찌개, 김치참치찌개로 다양하게 활용되는 찌개류다.

材料和份量(2人份)
泡菜 100g, 猪肉(前腿肉) 50g, 豆腐 30g, 土豆 30g, 洋葱 20g, 西葫芦 15g, 虾酱 3g, 平菇 10g, 香菇 1个, 红辣椒 5g, 青辣椒 5g, 大葱 5g, 小葱 3g, 生姜 3g, 芝麻油 10g, 辣椒油 10g, 野芝麻粉 5g, 白酒 10g, 汤用酱油 5g
***肉汤** : 海带 5g, 汤用鳀鱼 3条, 大蒜 5g, 大葱 5g
***应用料理材料** : 可用于制作泡菜土豆汤, 泡菜豆腐汤和泡菜金枪鱼汤等汤类。

Ingredients and portion(2 portions)
100g Kimchi, 50g pork(arm shoulder), 30g bean curd(tofu), 30g potato, 20g onion, 15g squash long, 3g salted shrimp, 10g oyster mushroom, 1 shiitake mushroom (mushroom pyogo), 5g red chilli, 5g green chilli, 5g leek, 3g green onion, 3g ginger, 10g perilla oil, 10g chili oil, 5g perilla seed flour, 10g rice-wine, 5g soy sauce
***Meat stock** : 5g dried sea tangle, 3 anchovies, 5g garlic, 5g leek
***Note** : You can use the ingredients of Kimchi-pork jjigae(stew) to cook other jjigae(stew) as Kimchi-potato jjigae, Kimchi-bean curd(tofu) jjigae and Kimchi-tuna jjigae.

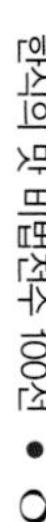

육수 끓이기 煮肉汤 How to cook the stock

재료 손질
材料处理方法
Clean ingredients

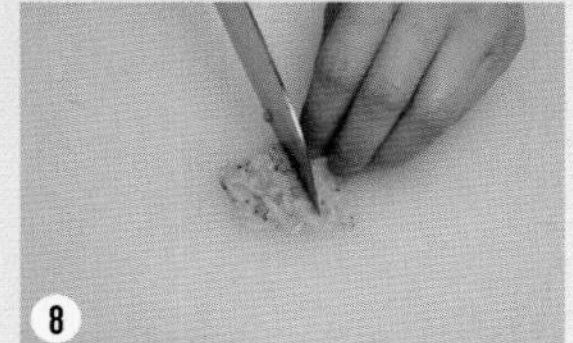

마무리하기
完成
How to finish

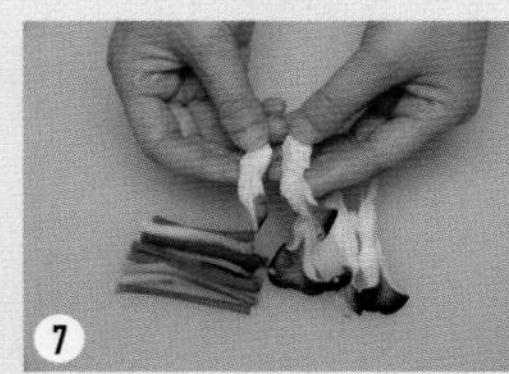

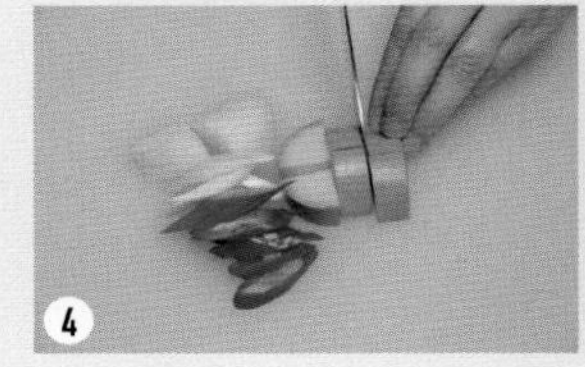

만드는 방법

1. 육수를 끓인다.
2. 두부를 3x4cm로 썬다.
3. 감자를 2x4cm로 썬다.
4. 호박, 양파는 깍둑썰기, 고추는 어슷썰기한다.
6. 육수와 들깨가루 5g을 넣는다.
7. 느타리버섯, 파를 넣는다.
8. 새우젓은 다진다.
9. 다진 새우젓 3g, 생강 3g으로 마무리한다.

制作方法

1. 煮肉汤。
2. 将豆腐切成大小3x4cm。
3. 将土豆切成大小2x4cm。
4. 将西葫芦、洋葱和辣椒切好备用。
6. 加入肉汤和野芝麻粉5g。
7. 加入平菇和葱。
8. 将虾酱捣碎。
9. 最后加入捣碎的虾酱3g和生姜3g后完成。

How to cook

1. Boil the meat stock.
2. Cut bean curd(tofu) into 3x4cm wide pieces.
3. Cut potato into 2x4cm wide pieces.
4. Cut squash long and onion into cubes. Cut the chilli diagonally.
6. Pour the meat stock and perilla seed flour of 5g.
7. Put oyster mushroom and leek.
8. Chop up salted shrimp.
9. Put of chopped salted shrimp 3g and of ginger 3g.

끓이기
煮粥
How to boil

5-1. 뚝배기에 들기름을 10g 넣는다.

5-1. 在石锅中加入芝麻油10g。

5-1. Put perilla oil of 10g into an earthen pot(ttukbaegi).

5-2. 고추기름을 10g 넣는다.

5-2. 加入辣椒油10g。

5-2. Put chili oil of 10g.

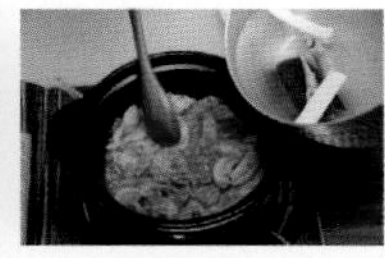

5-3. 김치 볶은 후 육수를 1/2 붓는다.

5-3. 将泡菜炒好后加入1/2的肉汤。

5-3. Stir-fry Kimchi. Pour 1/2 of the meat stock.

5-4. 돼지고기를 넣는다.

5-4. 加入猪肉。

5-4. Put pork.

5-5. 청주를 10g 넣는다.

5-5. 加入白酒10g。

5-5. Pour of rice-wine of 10g.

5-6. 국간장을 5g 넣는다.

5-6. 加入汤用酱油5g。

5-6. Add soy sauce of 5g.

TIP
建议

5-3. 김치는 익은 것을 사용하면 감칠맛이 상승되며, 충분히 볶아야 깊은 맛을 낼 수 있다.
8. 새우젓은 돼지고기의 지방분해 소화효소에 도움을 주며, 시원한 국물 맛을 낸다.

5-3. 如果使用腌熟的泡菜的话，泡菜的味道会更浓，只有充分地炒才能炒出味道来。
8. 虾酱有助于分解猪肉的脂肪和促进消化，并使汤的味道更爽快。

Step 5-3. Cook with fermented Kimchi to add a little more taste to the dish. Roast Kimchi as much as you can for more rich flavor.
Step 8. Salted shrimp works good for you as it removes fat that you gain by consuming pork. You can also use the salted shrimp to make your soup taste hearty.

동태찌개

明太鱼汤

Frozen pollack jjigae

재료 및 분량(3인분)
동태(1마리) 900g, 무 200g, 애호박 15g, 두부 15g, 표고버섯 10g, 새송이버섯 30g, 홍고추 10g, 풋고추 10g, 대파 10g, 마늘 3g, 쑥갓 5g, 실파 5g, 고춧가루 5g, 고추장 10g, 청주 15g, 국간장 5g, 된장 5g, 후춧가루 0.5g
***육수** : 다시마 5g, 멸치 3마리, 통마늘 2g, 생강 2g
***응용요리재료** : 생태, 대구, 조기 등의 다양한 생선재료를 활용할 수 있는 찌개류이다.

材料和份量(3人份)
明太鱼(1条) 900g, 萝卜 200g, 西葫芦 15g, 豆腐 15g, 香菇 10g, 杏鲍菇 30g, 红辣椒 10g, 青辣椒 10g, 大葱 10g, 大蒜 3g, 茼蒿 5g, 小葱 5g, 辣椒面儿 5g, 辣椒酱 10g, 白酒 15g, 汤用酱油 5g, 大酱 5g, 胡椒粉 0.5g
***肉汤** : 海带 5g, 鳀鱼 3条, 大蒜 2g, 生姜 2g
***应用料理材料** : 利用生明太鱼、鳕鱼和黄花鱼等各种鱼类材料制作汤类。

Ingredients and portion(3 portions)
900g frozen pollack fish, 200g white radish, 15g squash long, 15g bean curd(tofu), 10g shiitake mushroom(mushroom pyogo), 30g king oyster mushroom, 10g red chilli, 10g green chilli, 10g leek, 3g garlic, 5g edible chrysanthemum(ssukgot), 5g green onion, 5g chili powder, 10g red pepper paste(gochujang), 15g rice-wine, 5g soy sauce, 5g soybean paste, 0.5g ground black pepper
***Meat stock** : 5g dried sea tangle, 3 anchovies, 2g garlic, 2g ginger
***Note** : You can also use pollack, cod and white croakers fish to cook other fish jjigae(stew).

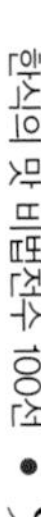

육수 끓이기 煮肉汤 How to cook stock

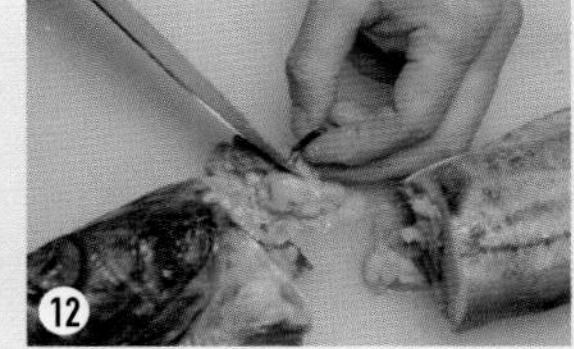

재료 손질
材料处理方法
Clean ingredients

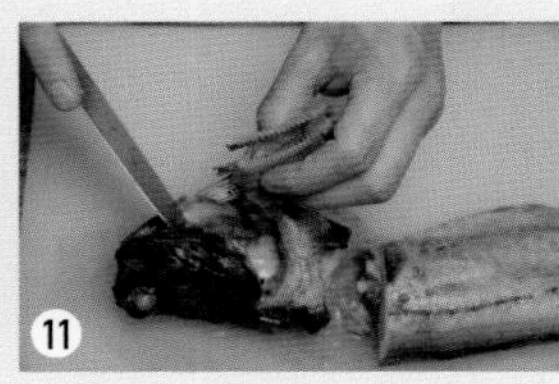

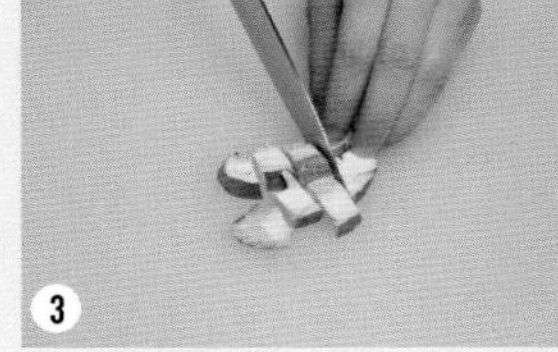

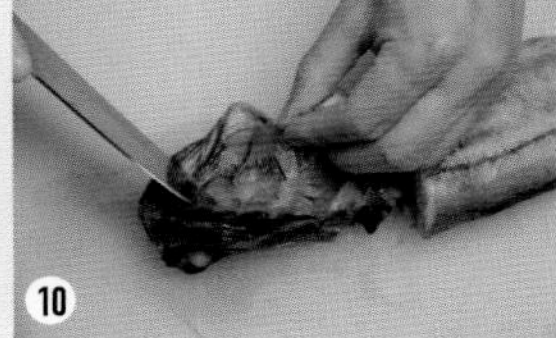

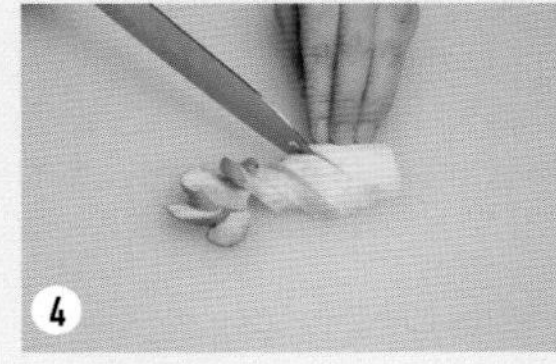

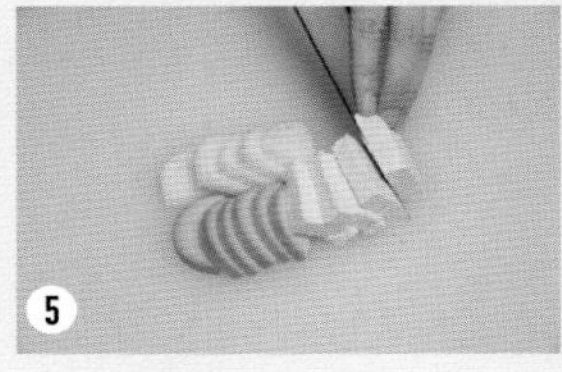

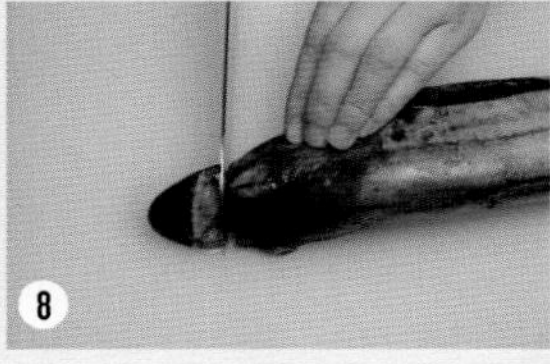

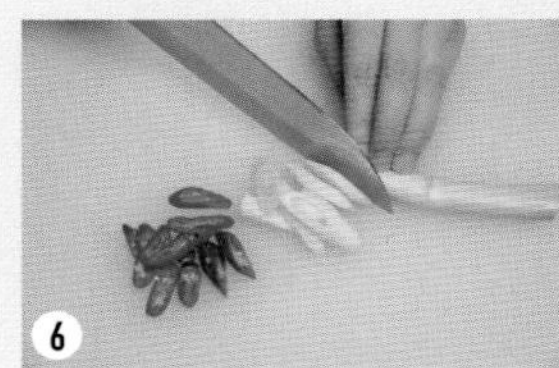

생선 손질
鱼的处理方法
How to prepare the fish

만드는 방법

1. 육수를 끓인다.
2. 쑥갓을 찬물에 담근다.
3. 표고버섯을 깍둑썰기한다.
4. 새송이는 어슷썰기한다.
5. 부재료 0.5cm로 편 썰기한다.
6. 고추, 대파 0.3cm로 어슷썰기
7. 지느러미를 제거한다.
8. 입을 제거한다.
9. 아가미를 벌린다.
10. 아가미에 칼집을 넣는다.
11. 아가미를 제거한다.
12. 쓸개를 제거한다.

制作方法

1. 煮肉汤。
2. 将茼蒿泡在凉水中。
3. 将香菇切成块。
4. 将杏鲍菇切成片。
5. 将副材料切成大小为0.5cm的片状。
6. 将辣椒和大葱切成大小0.3cm。
7. 去掉鱼鳍。
8. 去掉鱼嘴。
9. 打开鱼鳃。
10. 将刀插入鱼鳃。
11. 去除鱼鳃。
12. 去除苦胆。

How to cook

1. Boil meat stock.
2. Put edible chrysanthemum(ssukgot) in cold water.
3. Cut shiitake mushroom(mushroom pyogo) into cubes.
4. Cut king oyster mushroom diagonally.
5. Cut other ingredients into 0.5cm wide slices.
6. Cut diagonally the chili and the leek into 0.3cm wide pieces.
7. Remove a fin.
8. Remove a mouth.
9. Open a gill.
10. Make cuts in the gill.
11. Remove the gill.
12. Remove a gall bladder.

만드는 방법

13. 길이 5cm로 토막낸다.
14. 검은 막을 제거한다.
15. 내장은 소금으로 문지른다.
16. 소금물에 깨끗이 정리한다.
17. 먹는 내장만 다시 채운다.
18. 육수를 거른다.
19. 무와 고추장 10g을 푼다.
20. 국간장 5g을 넣는다.
21. 고춧가루 5g을 넣는다.
22. 표고버섯, 새송이버섯을 넣는다.
23. 된장을 5g 넣는다.
24. 후춧가루 0.5g을 넣는다.

制作方法

13. 切成长为5cm的块状。
14. 去除黑膜。
15. 用盐腌渍内脏。
16. 用盐水洗干净。
17. 仅将能食用的内脏重新放回去。
18. 过滤肉汤。
19. 加入萝卜和辣椒酱10g。
20. 加入汤用酱油5g。
21. 加入辣椒面儿5g。
22. 加入香菇和杏鲍菇。
23. 加入大酱5g。
24. 加入胡椒粉0.5g。

How to cook

13. Cut the fish into 5cm long pieces.
14. Remove the black film.
15. Rub edible internal organs only with salt.
16. Wash the ingredients again in salt water.
17. Stuff the fish with the edible internal organs only.
18. Sieve the stock.
19. Put the white radish and red pepper paste of 10g (gochujang).
20. Pour soy sauce of 5g.
21. Put chili powder of 5g.
22. Put shiitake mushroom and king oyster mushroom.
23. Put soybean paste of 5g.
24. Add ground black pepper of 0.5g.

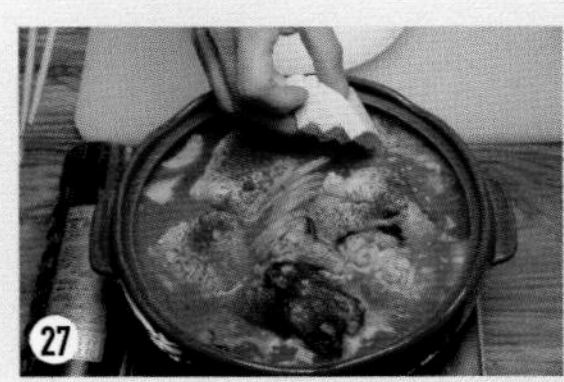

만드는 방법

25. 손질한 동태를 넣는다.
26. 애호박을 넣는다.
27. 두부를 넣는다.
28. 청주 15g을 넣는다.
29. 쑥갓으로 마무리한다.
30. 완성하여 담는다.

制作方法

25. 加入明太鱼。
26. 加入西葫芦。
27. 加入豆腐。
28. 加入白酒15g。
29. 最后加入茼蒿。
30. 完成后盛菜。

How to cook

25. Put the frozen pollack fish.
26. Put squash long.
27. Put bean curd(tofu).
28. Pour rice-wine of 15g.
29. Put edible chrysanthemum(ssukgot).
30. Finish.

TIP
建议

14. 검은 막을 제거하여 쓴맛을 없앤다.
25. 무가 완전히 익었을 때 동태를 넣으면 시원한 맛을 낸다.

14. 去掉黑膜可以去除苦味。
25. 若待萝卜完全熟透后再加入明太鱼的话，味道会更爽口。

Step 14. Remove the black film to eliminate bitter taste.
Step 25. Put the frozen pollack fish after the white radish is fully cooked. That will make your soup taste hearty.

돼지감자젓국찌개

土豆猪肉鲜味汤

Pork-potato-salted shrimp jjigae

재료 및 분량(2인분)
돼지고기(앞다리) 400g, 감자 150g, 두부 20g, 느타리 10g, 애호박 5g, 양파 5g, 홍고추 5g, 풋고추 5g, 대파 10g, 새우젓 3g, 된장 3g, 고추장 5g, 고춧가루 5g, 마늘 3g, 고추기름 5g, 국간장 5g
***육수** : 다시마 5g, 국멸치 3마리, 통마늘 3g, 마른 홍합 3g, 마른 새우 3g, 대파 5g, 마른 홍고추 5g, 풋고추 5g
***응용요리재료** : 돼지고기를 이용하여 감자, 김치, 해물 등에 다양하게 활용하는 찌개요리다.

材料和份量(2人份)
猪肉(前腿肉) 400g, 土豆 150g, 豆腐 20g, 平菇 10g, 西葫芦 5g, 洋葱 5g, 红辣椒 5g, 青辣椒 5g, 大葱 10g, 虾酱 3g, 大酱 3g, 辣椒酱 5g, 辣椒面儿 5g, 大蒜 3g, 辣椒油 5g, 汤用酱油 5g
***肉汤** : 海带 5g, 汤用鳀鱼 3条, 大蒜 3g, 干海虹 3g, 干虾 3g, 大葱 5g, 干红辣椒 5g, 青辣椒 5g
***应用料理材料** : 利用猪肉和土豆、泡菜及海鲜等制作各种汤类。

Ingredients and portion(2 portion)
400g pork(arm shoulder), 150g potato, 20g bean curd(tofu), 10g agaric, 5g squash long, 5g onion, 5g red chilli, 5g green chilli, 10g leek, 3g salted shrimp, 3g soybean paste, 5g pepper red paste(gochujang), 5g chili powder, 3g garlic, 5g chili oil, 5g soy sauce.
***Meat stock** : 5g dried sea tangle, 3 anchovies, 3g garlic, 3g dried mussel, 3g dried shrimp, 5g leek, 5g dried red chilli, 5g green chilli
***Note** : You can also add potato, Kimchi and seafood to Pork-potato-salted shrimp jjigae.

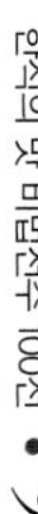

육수 끓이기 煮肉汤 How to cook stock

재료 손질
材料处理方法
Clean ingredients

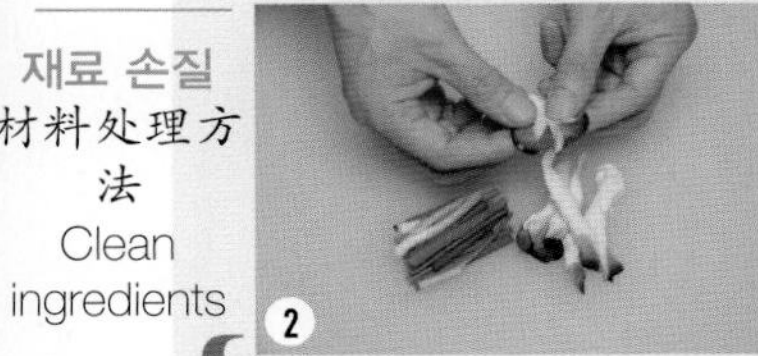

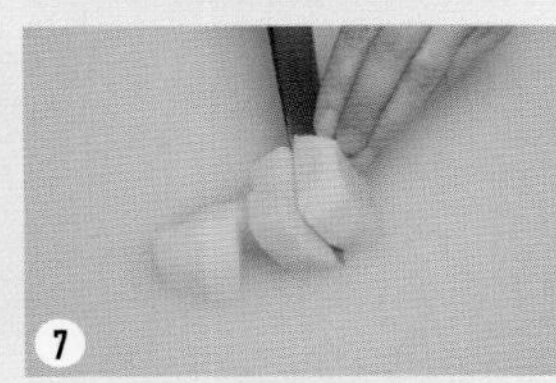

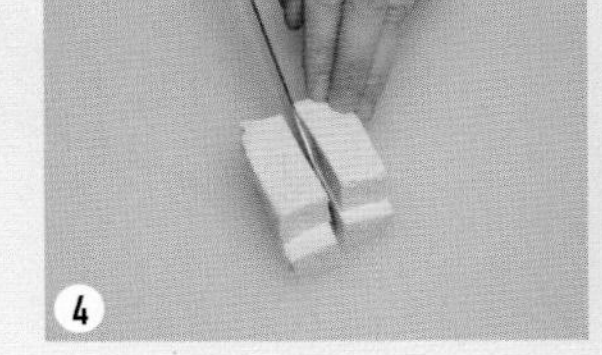

만드는 방법

1. 육수를 끓인다.
2. 실파와 느타리버섯을 손질한다.
3. 호박, 양파는 깍둑썰기, 고추, 대파는 어슷썰기한다.
4. 두부 2x3cm로 깍둑썰기한다.
6. 육수를 붓는다.
7. 감자 2x3cm로 깍둑썰기한다.
8. 재료를 끓인다.
9. 새우젓으로 마무리한다.

制作方法

1. 煮肉汤。
2. 将小葱和平菇整理好备用。
3. 将西葫芦、洋葱、辣椒和大葱切好备用。
4. 将豆腐切成大小为2x3的块状。
6. 加入肉汤。
7. 将土豆切成大小为2x3的块状。
8. 将材料一起煮。
9. 最后用虾酱调味。

How to cook

1. Cook stock.
2. Have small green onion and oyster mushroom neatly cut.
3. Cut green pumpkin and onion into cubes. Cut chili and spring onion diagonally.
4. Cut tofu into 2x3cm cubes.
6. Pour stock.
7. Cut potato into 2x3cm cubes.
8. Boil the content.
9. Finish it with salted shrimp.

끓이기
煮粥
How to boil

5-1. 뚝배기에 고추기름을 5g 넣는다.
5-1. 在石锅中加入辣椒油5g。
5-1. Put chili oil of 5g into an earthen pot(ttukbaegi).

5-2. 돼지고기를 두들긴다.
5-2. 拍猪肉。
5-2. Pound the pork.

5-3. 국간장을 5g 넣는다.
5-3. 加入汤用酱油5g。
5-3. Add soy sauce of 5g.

5-4. 된장을 3g 넣는다.
5-4. 加入大酱3g。
5-4. Put soybean paste of 3g.

5-5. 고추장을 5g 넣는다.
5-5. 加入辣椒酱5g。
5-5. Put red pepper paste(gochujang).

5-6. 고춧가루를 5g 넣고 볶는다.
5-6. 加入辣椒面儿5g后一起炒。
5-6. Stir-fry the ingredients with chili powder of 5g.

TIP 建议

5-4. 된장은 돼지고기의 잡냄새를 제거하는 역할을 한다.

5-4. 大酱起着去除猪肉异味的作用。

Step 5-4. Use soybean paste to remove bad smell of the pork.

된장찌개

大酱汤

Doenjang jjigae

재료 및 분량(2인분)
감자 30g, 두부 30g, 애호박 20g, 표고버섯 1개, 양파 15g, 홍고추 5g, 풋고추 5g, 대파 10g, 마늘 3g
***육수** : 다시마 5g, 무 30g, 멸치 5마리, 마른 새우 2g, 된장 30g, 고추씨 5g, 마른 홍고추 8g, 청량고추 5g, 감초 2g, 통마늘 2g, 표고버섯 5g
***응용요리재료** : 된장국의 기본으로 조개류, 꽃게 등의 해물을 추가하여 끓여도 시원한 국물 맛을 낸다.

材料和份量(2人份)
土豆 30g, 豆腐 30g, 西葫芦 20g, 香菇 1个, 洋葱 15g, 红辣椒 5g, 青辣椒 5g, 大葱 10g, 大蒜 3g
***肉汤** : 海带 5g, 萝卜 30g, 鳀鱼 5条, 干虾 2g, 大酱 30g, 辣椒籽 5g, 干红辣椒 8g, 青阳辣椒 5g, 甘草 2g, 大蒜 2g, 香菇 5g
***应用料理材料** : 若在大酱汤中加入蛤蜊类和螃蟹等海鲜的话，味道会更鲜美。

Ingredients and portion(2 portion)
30g potato, 30g bean curd(tofu), 20g squash long, 1 shiitake mushroom(mushroom pyogo), 15g onion, 5g red chilli, 5g green chilli, 10g leek, 3g of garlic
***Meat stock** : 5g dried sea tangle, 30g white radish, 5 anchovies, 2g dried shrimp, 30g of soybean paste, 5g pepper seed, 8g dried red chilli, 5g red hot chili peppers, 2g licorice, 2g garlic, 5g shiitake mushroom(mushroom pyogo)
***Note** : You can add shellfish or blue crab to make the jjigae(stew) taste hearty.

재료 손질 材料处理方法 Clean ingredients

만드는 방법

1. 양파는 1x2cm로 깍둑썰기한다.
2. 애호박은 반달썰기한다.
3. 두부는 2x3cm로 썰기한다.
4. 감자는 2x3cm로 썰기한다.
6. 준비된 채소를 넣는다.
7. 다진 마늘 3g을 넣는다.
8. 나머지 재료를 넣는다.
9. 따뜻하게 끓인다.

制作方法

1. 将洋葱切成大小为1x2cm的块状。
2. 将西葫芦切成半月状。
3. 将豆腐切成大小为2x3cm的块状。
4. 将土豆切成大小为2x3cm的块状。
6. 加入准备好的蔬菜。
7. 加入蒜泥3g。
8. 加入其它材料。
9. 将材料一起煮。

How to cook

1. Cut onion into 1x2cm wide cubes.
2. Cut squash long into a half-moon shape.
3. Cut bean curd(tofu) 2x3cm wide pieces.
4. Cut potato into 2x3cm wide pieces.
6. Put the vegetables.
7. Put chopped garlic of 3g.
8. Put the other ingredients.
9. Boil the ingredients warm.

육수 끓이기 煮粥 How to boil

5-1. 육수 재료를 준비한다.

5-1. 准备好肉汤 材料。

5-1. Prepare ingredients to cook stock.

5-2. 육수를 끓인다.

5-2. 煮肉汤。

5-2. Cook stock.

5-3. 된장을 풀어준다.

5-3. 用水冲开大酱。

5-3. Put soybean paste.

5-4. 약불에서 30분간 끓인다.

5-4. 用微火煮30分钟。

5-4. Boil the meat stock over a low heat for 30 minutes.

5-5. 체에 내린다.

5-5. 将肉汤取下来。

5-5. Sieve the stock.

5-6. 뚝배기에 육수를 담는다.

5-6. 将肉汤倒入石锅中。

5-6. Pour the stock to an earthen pot(ttukbaegi).

TIP 建议

5-4. 된장 풀어서 끓인 육수는 깊은 맛과 구수한 맛을 낸다.

5-4. 若将大酱用水冲开后再制作肉汤的话，汤的味道将更加美味和可口。

Step 5-4. Soybean paste makes meat stock taste sweeter.

모듬버섯전골

什锦蘑菇炖锅

All-sorts-of mushrooms jeongol

재료 및 분량(3인분)
자연산 버섯 100g, 느타리버섯 10g, 새송이버섯 5g, 양송이버섯 5g, 팽이버섯 5g, 목이버섯 5g, 콩나물 5g, 소고기(등심) 50g, 당근 10g, 쥬키니호박 10g, 청경채 5g, 애기배추 5g, 쑥갓 5g, 홍고추 5g, 풋고추 5g, 양파 5g, 마늘 5g, 국간장 5g, 까나리액젓 5g
***육수** : 다시마 5g, 국멸치 3마리, 감초 2g, 버섯 자투리(껍질) 5g, 통마늘 2g, 대파 5g
***응용요리재료** : 자연산 송이버섯을 올리면 향을 더욱 느낄 수 있으며, 낙지, 갈비 등의 재료를 다양하게 활용할 수 있는 전골류 요리다.

材料和份量(3人份)
天然野蘑菇 100g, 平菇 10g, 松茸 5g, 洋香菇 5g, 金针蘑 5g, 木耳 5g, 豆芽 5g, 牛肉(里脊) 50g, 胡萝卜 10g, 菜瓜 10g, 青菜 5g, 小白菜 5g, 茼蒿 5g, 红辣椒 5g, 青辣椒 5g, 洋葱 5g, 大蒜 5g, 汤用酱油 5g, 玉筋鱼酱 5g
***肉汤** : 海带 5g, 汤用鳀鱼 3条, 甘草 2g, 蘑菇碎片(皮) 5g, 大蒜 2g, 大葱 5g
***应用料理材料** : 放入天然野蘑菇，味道会更美味，利用八爪鱼和排骨等材料可以制成各种炖锅类料理。

Ingredients and portion(3 portions)
100g natural mushroom, 10g oyster mushroom, 5g king oyster mushroom, 5g button mushroom, 5g of mushroom winter, 5g black mushroom, 5g bean sprouts, 50g beef (sirloin), 10g carrot, 10g zucchini, 5g Bok choy, 5g baby chinese cabbage, 5g edible chrysanthemum(ssukgot), 5g red chilli, 5g green chilli, 5g onion, 5g garlic, 5g soy sauce, 5g sand eel sauce
***Meat stock** : 5g dried sea tangle, 3 anchovies, 2g licorice, 5g mushroom peels, 2g garlic, 5g leek
***Note** : If you want a deeper scent to you jeongol, add natural button mushroom. You can also use small octopus and beef rib.

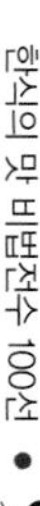

재료 손질 材料处理方法 Clean ingredients

완성하기
完成
How to finish

만드는 방법

1. 자연산 버섯은 소금으로 데쳐서 사용한다.
2. 소고기에 고춧가루를 10g 넣는다.
3. 국간장을 5g 넣는다.
4. 까나리액젓을 5g 넣는다.
6. 나머지 재료를 올린다.
7. 소고기양념장을 넣는다.
8. 육수를 붓는다.
9. 끓여 완성한다.

制作方法

1. 用盐水焯一下天然野蘑菇。
2. 在牛肉上撒上辣椒面儿10g。
3. 加入汤用酱油5g。
4. 加入玉筋鱼酱5g。
6. 将剩余的材料放上去。
7. 加入牛肉调味酱。
8. 加入肉汤。
9. 完成。

How to cook

1. Blanch natural mushrooms with salt.
2. Put chili powder to beef of 10g.
3. Put soy sauce of 5g.
4. Add sand eel sauce of 5g.
6. Put the rest of the ingredients on top.
7. Pour beef sauce.
8. Pour stock.
9. Boil.

끓이기
煮粥
How to boil

5-1. 육수를 끓인다.

5-1. 煮肉汤。

5-1. Boil stock.

5-2. 전골냄비에 콩나물을 담는다.

5-2. 在炖锅中放入豆芽。

5-2. Put bean sprouts to a jeongol pot.

5-3. 채소를 대칭으로 놓는다.

5-3. 将蔬菜对称放好。

5-3. Arrange the vegetables in a symmetrical manner.

5-4. 색스럽게 돌려 담는다.

5-4. 将锅里的材料美观地摆放好。

5-4. Put the ingredients beautifully in the pot.

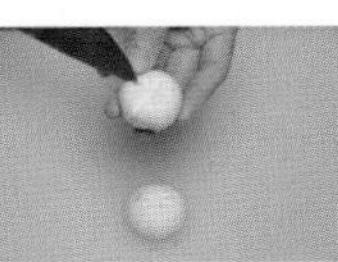

5-5. 양송이버섯에 칼집 모양을 낸다.

5-5. 用刀将洋香菇切出花纹。

5-5. Make cuts in button mushroom.

5-6. 전골형태를 갖춘다.

5-6. 使材料呈现炖锅模样。

5-6. Now, your jeongol actually looks like a jeongol.

TIP 建议

1. 자연산 버섯은 끓는 물에 소금을 넣고 살짝 데쳐서 흐르는 물에서 뜨거운 물기를 없애야 쫄깃하고 질감이 좋다.
5-1. 버섯전골은 버섯의 깊은 맛을 내기 위해 버섯 자투리로 육수를 끓인다.

1. 将天然野蘑菇放入盐水中稍微焯一下后，应将水份控干，这样蘑菇的口感才会劲道。
5-1. 为了体现蘑菇的味道，在蘑菇炖锅中加入用蘑菇碎片制成的汤。

Step 1. Blanch the natural mushrooms in boiling water seasoned with a pinch of salt. Wash the mushrooms in running water to make the mushrooms chewy.
Step 5-1. Cook the stock with mushroom remains to add a richer taste to the mushrooms jeongol(stew).

민물새우찌개

淡水虾汤

Freshwater shrimp jjigae

재료 및 분량(2인분)
민물새우 250g, 무 200g, 수제비 30g, 양파 20g, 느타리버섯 10g, 쑥갓 3g, 홍고추 5g, 풋고추 5g, 실파 3g, 부추 5g, 고추기름 10g, 고추장 15g, 국간장 10g, 까나리액젓 5g, 생강 3g
***육수** : 다시마 5g, 국멸치 2마리, 마른 홍합살 2g, 감초 2g, 표고버섯 20g, 통마늘 5g
***응용요리재료** : 꽃게찌개와 민물 생선류를 이용하여 끓이는 찌개류이다.

材料和份量(2人份)
淡水虾 250g, 萝卜 200g, 面疙瘩 30g, 洋葱 20g, 平菇 10g, 茼蒿 3g, 红辣椒 5g, 青辣椒 5g, 小葱 3g, 韭菜 5g, 辣椒油 10g, 辣椒酱 15g, 汤用酱油 10g, 玉筋鱼酱 5g, 生姜 3g
***肉汤** : 海带 5g, 汤用鳀鱼 2条, 干海虹肉 2g, 甘草 2g, 香菇 20g, 大蒜 5g
***应用料理材料** : 利用花蟹和淡水鱼类制成汤类。

Ingredients and portion(2 portion)
250g freshwater shrimp, 200g white radish, 30g sujebi(hand torn noodle soup), 20g onion, 10g oyster mushroom, 3g edible chrysanthemum(ssukgot), 5g red chilli, 5g green chilli, 3g green onion, 5g chives, 10g chili oil, 15g red pepper paste (gochujang), 10g soy sauce, 5g sand eel sauce, 3g ginger
***Meat stock** : 5g dried sea tangle, 2 anchovies, 2g dried mussel, 2g licorice, 20g shiitake mushroom(dried mushroom pyogo), 5g garlic
***Note** : You can also use blue crabs or other freshwater fish to cook seafood jjigae(stew).

육수 끓이기 煮肉汤 How to cook stock

재료 손질 材料处理方法 Clean ingredients

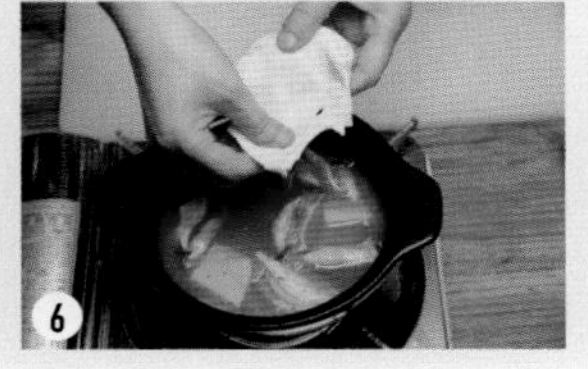

완성하기 完成 How to finish

만드는 방법

1. 육수를 끓인다.
2. 무는 0.3cm로 편 썬다.
3. 느타리버섯과 쑥갓 손질하기
4. 실파와 부추 5cm로 채 썬다.
6. 밀가루반죽을 떠서 넣는다.
7. 데친 민물새우를 넣는다.
8. 생강 3g을 넣는다.
9. 쑥갓으로 마무리한다.

制作方法

1. 煮肉汤。
2. 将萝卜切成大小为0.3cm的片状。
3. 将平菇和茼蒿整理好备用。
4. 将小葱和韭菜切成长5cm。
6. 加入面疙瘩。
7. 加入焯好的淡水虾。
8. 加入生姜3g。
9. 最后用茼蒿装饰。

How to cook

1. Boil stock.
2. Cut white radish into 0.3cm wide slices.
3. Clean oyster mushroom and edible chrysanthemum (ssukgot).
4. Julienne green onion and chives into 5cm long pieces.
6. Knead dough and put the dough.
7. Blanch freshwater shrimp. Put the shrimp.
8. Put ginger of 3g.
9. Put edible chrysanthemum(ssukgot).

끓이기 煮粥 How to boil

5-1. 무와 고추기름 10g을 볶은 후 고추장 15g을 넣는다.
5-1. 用10g辣椒油炒完萝卜后放入15g辣椒酱。
5-1. Stir-fry the white radish with chili oil of 10g. Put red pepper paste of 15g (gochujang).

5-2. 국간장 10g을 넣는다.
5-2. 加入汤用酱油10g。
5-2. Add soy sauce of 10g.

5-3. 까나리액젓 5g을 넣는다.
5-3. 加入玉筋鱼酱5g。
5-3. Put sand eel sauce og 5g.

5-4. 육수를 붓는다.
5-4. 倒入肉汤。
5-4. Pour stock.

5-5. 채 썬 양파를 넣는다.
5-5. 加入切好的洋葱。
5-5. Put julienne onion.

5-6. 느타리버섯을 넣는다.
5-6. 加入平菇。
5-6. Put oyster mushroom.

TIP 建议

5-1. 고추장은 민물새우의 냄새를 제거하는 역할을 한다.

5-1. 辣椒酱起着去除淡水虾异味的作用。

Step 5-1. Red pepper paste(gochujang) removes bad smell of freshwater shrimp.

불고기전골

烤肉炖锅

Bulgogi jeongol

재료 및 분량(3인분)

소고기(안심) 600g, 당면 40g, 새송이버섯 5g, 느타리버섯 5g, 표고버섯 1장, 양송이버섯 5g, 양파 5g, 당근 5g, 쥬키니호박 8g, 풋고추 5g, 홍고추 5g, 실파 5g, 대파 5g 쑥갓 3g, 배즙 20g
***육수** : 다시마 5g, 국멸치 3마리, 마른 새우 3g, 통마늘 2g, 감초 2g, 마른 홍고추 3g, 버섯 자투리 5g, 양파 2g, 당근 2g
***불고기소스** : 진간장 150g, 청주 150g, 다시마 5g, 국멸치 2마리, 마른 홍고추 5g, 대파 3g, 마늘 2g, 생강 3g, 양파 3g, 고춧가루 2g, 사과즙 30g, 설탕 15g, 올리고당 10g, 참기름 5g
***응용요리재료** : 불고기소스를 이용하여 기본적인 육류요리에 활용한다.

材料和份量(3人份)

牛肉(里脊) 600g, 粉条 40g, 杏鲍菇 5g, 平菇 5g, 香菇 1个, 洋香菇 5g, 洋葱 5g, 胡萝卜 5g, 菜瓜 8g, 青辣椒 5g, 红辣椒 5g, 小葱 5g, 大葱 5g, 茼蒿 3g, 梨汁 20g
***肉汤** : 海带 5g, 汤用鳀鱼 3条, 干虾 3g, 大蒜 2g, 甘草 2g, 干红辣椒 3g, 蘑菇碎片 5g, 洋葱 2g, 胡萝卜 2g
***烤肉调味料** : 浓酱油 150g, 白酒 150g, 海带 5g, 2汤用鳀鱼, 干红辣椒 5g, 大葱 3g, 大蒜 2g, 生姜 3g, 洋葱 3g, 辣椒面儿 2g, 苹果汁 30g, 白糖 15g, 低聚糖 10g, 香油 5g
***应用料理材料** : 利用烤肉调味料制作基本的肉类料理。

Ingredients and portion(3 portions)

600g beef(tenderloin), 40g starch noodle, 5g king oyster mushroom, 5g oyster mushroom, 1 shiitake mushroom(mushroom pyogo), 5g button mushroom, 5g onion, 5g carrot, 8g zucchini, 5g green chilli, 5g red chilli, 5g green onion, 5g leek, 3g chrysanthemum(ssukgot), 20g pear juice
***Meat stock** : 5g dried sea tangle, 3 anchovies, 3g dried shrimp, 2g garlic, 2g licorice, 3g dried red chili, 5g mushroom remains, 2g onion, 2g carrot
***Bulgogi sauce** : 150g thick soy sauce, 150g rice-win, 5g dried sea tangle, 2 anchovies, 5g dried red chilli, 3g green onion, 2g garlic, 3g ginger, 3g onion, 2g chili powder, 30g apple juice, 15g sugar, 10g oligosaccharide, 5g sesame oil
***Note** : You can use bulgogi sauce to cook other basic meat dishes.

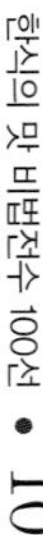

육수 끓이기 煮肉汤 How to cook stock

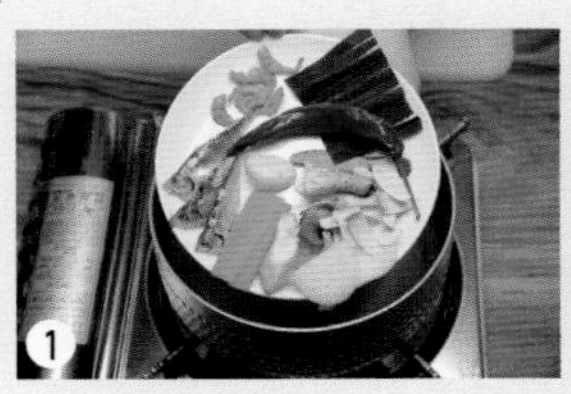

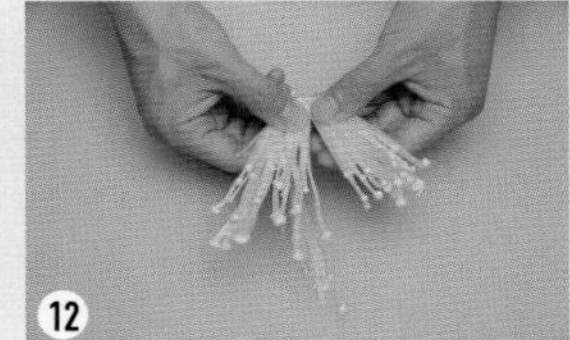

재료 손질
材料处理方法
Clean ingredients

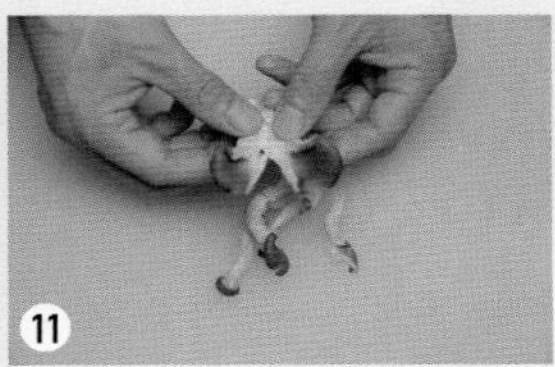

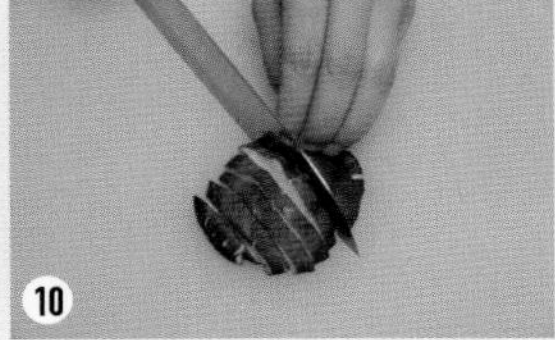

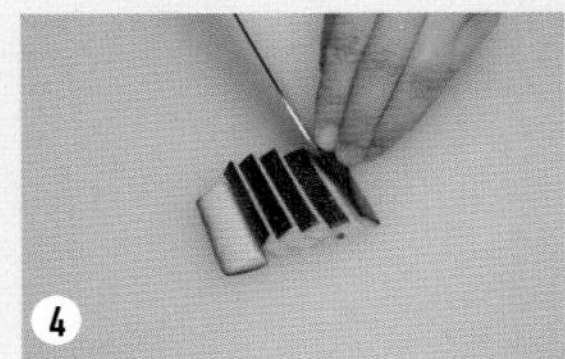

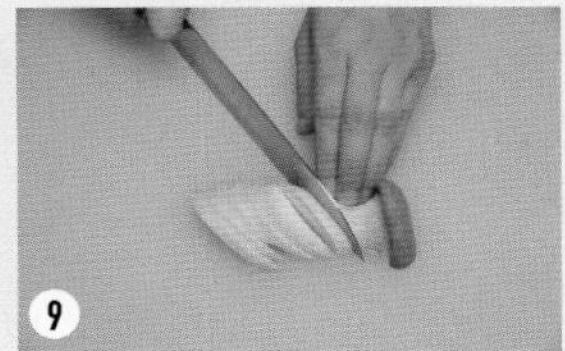

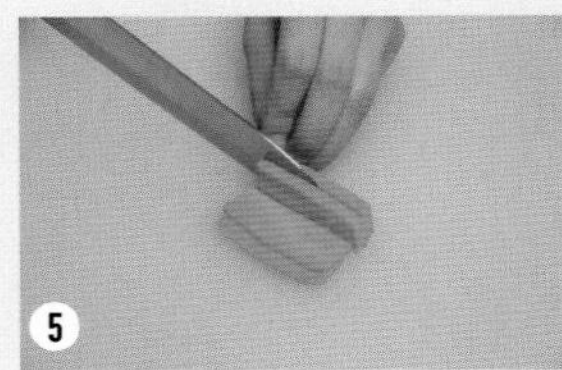

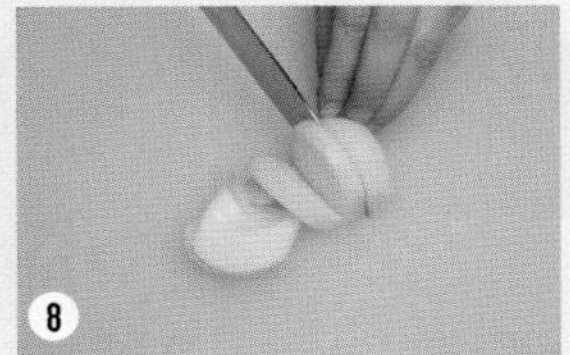

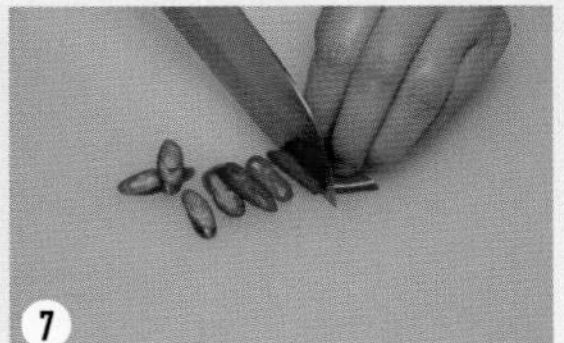

만드는 방법

1. 육수를 끓인다.
2. 불린 당면을 삶는다.
3. 소고기는 두들긴다.
4. 쥬키니호박은 골패형으로 썬다.
5. 당근은 골패형으로 썬다.
6. 대파는 7cm로 채 썬다.
7. 고추는 0.5cm로 어슷썰기한다.
8. 양파는 0.5cm로 채 썬다.
9. 새송이버섯은 0.5cm로 어슷썰기한다.
10. 표고버섯은 0.4cm로 채 썬다.
11. 느타리버섯을 찢는다.
12. 팽이버섯을 찢는다.

制作方法

1. 煮肉汤。
2. 将泡好的粉丝煮一下。
3. 敲打牛肉。
4. 将菜瓜切成骨牌形。
5. 将胡萝卜切成骨牌形。
6. 将大葱切成大小7cm。
7. 将辣椒切成大小0.5cm。
8. 将洋葱切成大小0.5cm。
9. 将杏鲍菇斜切成大小0.5cm。
10. 将香菇切成大小0.4cm。
11. 将平菇撕好备用。
12. 将金针蘑撕好备用。

How to cook

1. Boil stock.
2. Blanch starch noodle.
3. Percuss the beef.
4. Cut zucchini in a domino shape.
5. Cut carrot in a domino shape.
6. Julienne the leek into 7cm long slices.
7. Diagonally cut chili into 0.5cm wide pieces.
8. Julienne onion into 0.5cm wide slices.
9. Cut king oyster mushroom into 0.5cm wide pieces diagonally.
10. Julienne shiitake mushroom into 0.4cm wide pieces.
11. Tear oyster mushroom.
12. Tear mushroom winter.

불고기소스 만들기
制作烤肉调味料
How to make bulgogi sauce

전골 담기
制作炖锅
How to cook joengol

만드는 방법

13. 양송이버섯은 껍질을 제거한다.
14. 진간장 150g을 넣는다.
15. 청주 150g을 넣는다.
16. 마른 재료와 참기름 5g을 넣는다.
17. 약불에서 끓인다.
18. 설탕 15g을 넣는다.
19. 올리고당 10g을 넣는다.
20. 사과즙 30g을 넣는다.
21. 불고기소스를 체에 내린다.
22. 소고기에 배즙 20g을 갈아준다.
23. 불고기소스 60g에 버무린다.
24. 당면을 담는다.

制作方法

13. 将洋香菇的皮去掉。
14. 加入浓酱油150g。
15. 加入白酒150g。
16. 加入干材料和香油5g。
17. 用微火煮。
18. 加入白糖15g。
19. 加入低聚糖10g。
20. 加入苹果汁30g。
21. 取出烤肉调味料。
22. 在牛肉中加入梨汁20g。
23. 用烤肉调味料60g混合搅拌。
24. 放入粉丝。

How to cook

13. Peel off button mushroom.
14. Put thick soy sauce of 150g.
15. Put clean rice-wine of 150g.
16. Put dried ingredients and sesame oil of 5g.
17. Boil the ingredients over a low heat.
18. Add sugar of 15g.
19. Put oligosaccharide of 10g.
20. Add apple juice of 30g.
21. Sieve the bulgogi sauce.
22. Grind beef with pear juice of 20g.
23. Mix the beef with bulgogi sauce of 60g.
24. Put the starch noodle.

만드는 방법

25. 각종 버섯과 채소를 돌려 담는다.
26. 불고기소스 20g을 끼얹는다.
27. 육수를 7부 부어 끓인다.
28. 완성하여 담는다.

制作方法

25. 将各种蘑菇和蔬菜美观地放入锅中。
26. 加入烤肉调味料20g。
27. 加入肉汤70%左右后一起煮。
28. 完成后盛菜。

How to cook

25. Arrange the mushrooms and the vegetables in a circular manner.
26. Pour 20g of bulgogi sauce.
27. Pour 70% of the meat stock. Boil the ingredients.
28. Finish.

TIP
建议

21. 불고기소스는 냉장 보관하고, 하루 숙성하면 맛이 더욱 조화로워진다.
27. 육수는 7부만 붓고 끓여야 나중에 채소에서도 수분이 나와서 국물비율이 많아진다.

21. 若将烤肉调味料放在冰箱里保管一天的话，味道会更合口。
27. 肉汤只能倒入70%，因为最后从蔬菜里渗出水分后汤的量会增多。

Step 21. Keep bulgogi sauce in a refrigerator. If you ferment the sauce for a night, it will be good.
Step 27. Make sure that you pour 70% of meat stock only. This way, you will not get overwhelmed by how much water is after the meat stock is added with other water from vegetables.

순두부소고기찌개

嫩豆腐牛肉汤

Soft tofu-beef jjigae

재료 및 분량(2인분)
순두부 200g, 소고기(목심) 50g, 바지락 30g, 애호박 5g, 양파 3g, 느타리버섯 3g, 홍고추 3g, 풋고추 3g, 대파 3g, 고추기름 10g, 들기름 10g, 고춧가루 5g, 국간장 5g, 마늘 3g
***육수** : 다시마 5g, 국멸치 3마리, 마늘 3g
***응용요리재료** : 순두부를 이용하여 해물(꽃게, 굴) 등과 소고기를 함께 활용할 수 있는 찌개류다.

材料和份量(2人份)
软豆腐 200g, 牛肉(颈部肉) 50g , 蛤仔 30g, 西葫芦 5g, 洋葱 3g, 平菇 3g, 红辣椒 3g, 青辣椒 3g, 大葱 3g, 辣椒油 10g, 芝麻油 10g, 辣椒面儿 5g, 汤用酱油 5g, 大蒜 3g
***肉汤** : 海带 5g, 汤用鳀鱼 3条, 大蒜 3g
***应用料理材料** : 可利用软豆腐与海鲜(花蟹, 海蛎子)等及牛肉制成汤类。

Ingredients and portion(2 portions)
200g soft tofu(soft bean curd), 50g beef(chuck roll), 30g clam short neck live, 5g long squash, 3g onion, 3g oyster mushroom, 3g red chilli, 3g green chilli, 3g leek, 10g chili oil, 10g perilla oil, 5g chili powder, 5g soy sauce, 3g garlic
***Meat stock** : 5g dried sea tangle, 3 anchovies, 3g garlic
***Note** : You can also add seafood(blue crabs, oyster) and beef.

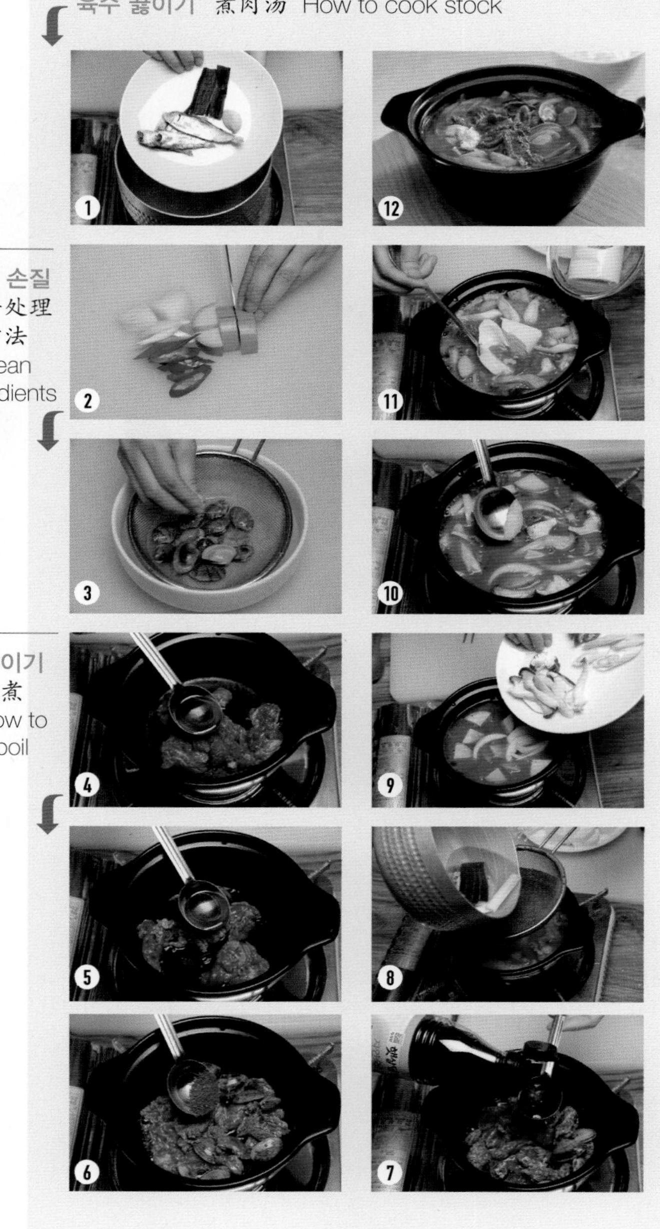

만드는 방법

1. 육수를 끓인다.
2. 부재료를 손질한다.
3. 바지락은 해감 처리한다.
4. 소고기와 고추기름 10g을 볶는다.
5. 들기름 10g을 넣고 볶는다.
6. 고춧가루 5g과 바지락을 볶는다.
7. 국간장 3g으로 간한다.
8. 육수를 붓는다.
9. 재료를 넣고 끓인다.
10. 다진 마늘 3g을 넣는다.
11. 순두부를 떠 넣는다.
12. 완성하여 담는다.

制作方法

1. 煮肉汤。
2. 将副材料整理好备用。
3. 对蛤仔的淤泥进行处理。
4. 将牛肉和辣椒油10g一起炒。
5. 然后再加入芝麻油10g一起炒。
6. 将5g辣椒面和蛤仔一起炒。
7. 加入汤用酱油3g调味。
8. 倒入肉汤。
9. 加入材料后一起煮。
10. 加入蒜泥3g。
11. 用勺子盛出软豆腐后加入其中。
12. 完成后盛菜。

How to cook

1. Boil stock.
2. Clean side ingredients.
3. Wash clam thoroughly.
4. Stir-fry beef with chili oil of 10g.
5. Stir-fry the ingredients with perilla oil of 10g.
6. Roast the clam with chili powder of 5g.
7. Season the ingredients with soy sauce of 3g.
8. Pour stock.
9. Put the ingredients and boil the ingredients.
10. Put chopped garlic of 3g.
11. Put soft tofu(soft bean curd).
12. Finish.

TIP
建议

5. 들기름으로 소고기의 냄새를 제거한다.
11. 순두부는 마지막에 넣어 덩어리 모양을 유지하도록 한다.

5. 用芝麻油可以去除牛肉的异味。
11. 软豆腐在最后放是为了保持软豆腐的块状模样。

Step 5. Perilla soil removes bad smell of the beef.
Step 11. Put the soft tofu(soft bean curd) at the finish so that you will not break the shape of the tofu.

육개장

辣牛肉汤

Yukgyejang

재료 및 분량(2인분)
소고기(양지머리) 400g, 토란대 20g, 고사리 20g, 숙주 15g, 대파 20g, 무 20g, 국간장 5g, 소금 2g, 후춧가루 0.5g
***고추기름** : 식용유 200g, 고춧가루 60g, 대파 5g, 생강 3g, 마늘 3g
***양념장** : 고추장 5g, 고춧가루 10g, 된장 3g, 다진 마늘 5g, 국간장 5g, 참기름 3g
***육수** : 다시마 5g, 국멸치 2마리, 마른 홍고추 1개, 통마늘 3g, 고추씨 5g, 감초 1g
***응용요리재료** : 해물재료를 넣은 해물육개장과 닭고기살을 넣은 닭개장으로 활용된다.

材料和份量(2人份)
牛肉(牛排) 400g, 芋头杆 20g, 蕨菜 20g, 绿豆芽 15g, 大葱 20g, 萝卜 20g, 汤用酱油 5g, 盐 2g, 胡椒粉 0.5g
***辣椒油** : 食用油 200g, 辣椒面儿 60g, 大葱 5g, 生姜 3g, 大蒜 3g
***调味酱** : 辣椒酱 5g, 辣椒面儿 10g, 大酱 3g, 蒜泥 5g, 汤用酱油 5g, 香油 3g
***肉汤**: 海带 5g, 汤用鳀鱼 2条, 干红辣椒 1个, 大蒜 3g, 辣椒籽 5g, 甘草 1g
***应用料理材料**: 可加入海鲜材料制成海鲜牛肉汤和加入鸡肉制成鸡肉汤。

Ingredients and portion(2 portions)
400g beef(brisket), 20g taro stem, 20g dried bracken(gosari), 15g mung Bean Sprout, 20g leek, 20g white radish, 5g soy sauce, 2g salt, 0.5g ground black pepper
*Chili oil : 200g salad oil, 60g chili powder, 5g leek, 3g ginger, 3g garlic
*Marinade : 5g gochujang, 10g chili powder, 3g soybean paste, 5g chopped garlic, 5g soy sauce, 3g sesame oil
*Meat stock : 5g ofdried sea tangles, 2 anchovies, 1 dried red chilli, 3g garlic, 5g red-pepper seeds, 1g licorice
*Note : If you add seafood ingredients, you have seafood yukgyejang. In the same context, if you put chicken, you have chicken yukgyejang.

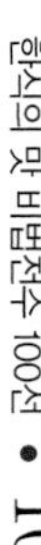

육수 끓이기 煮肉汤 How to cook stock

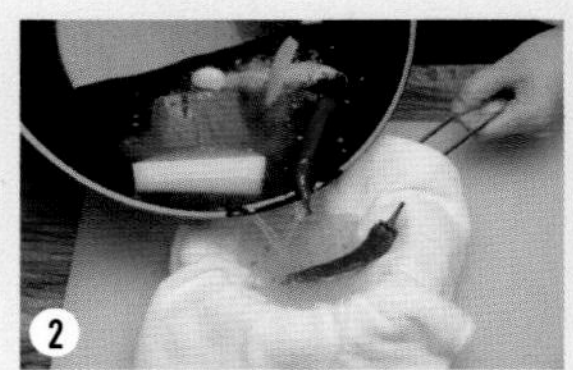

재료 손질
材料处理方法
Clean ingredients

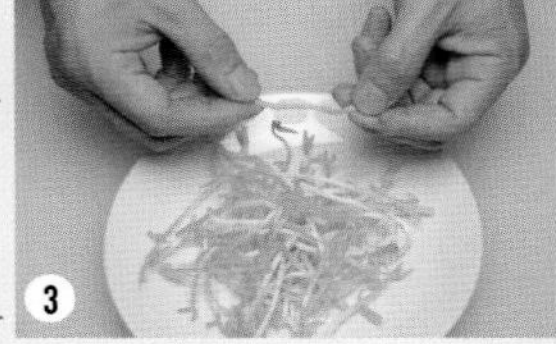

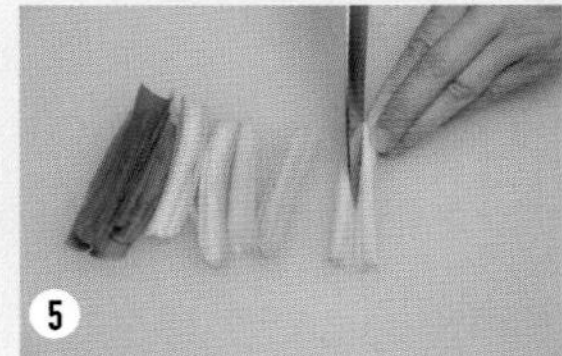

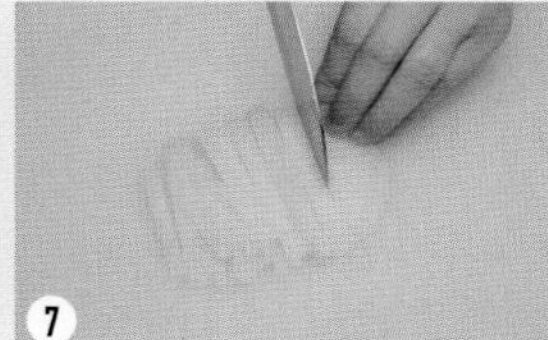

양념장 만들기
制作调味酱
How to make marinade

만드는 방법

1. 육수를 끓인다.
2. 체에 내린다.
3. 숙주는 꼬리를 다듬는다.
4. 끓는 물에 데친다.
5. 대파는 길게 채 썬다.
6. 끓는 물에 데친다.
7. 육수에서 건진 무는 0.3cm로 채 썬다.
8. 고사리는 불린다.
9. 끓는 물에 데친다.
10. 토란대도 데친다.
11. 마늘을 곱게 다진다.
12. 고추장 5g, 고춧가루 10g, 된장 3g, 다진 마늘 5g을 넣는다.

制作方法

1. 煮肉汤。
2. 用过滤网过滤肉汤。
3. 去掉绿豆芽尾。
4. 放入烧开的水中焯一下。
5. 将大葱切成细丝。
6. 放入烧开的水中焯一下。
7. 将从肉汤中捞出来的萝卜切成大小0.3cm的细丝。
8. 将蕨菜泡在水里。
9. 放入烧开的水中焯一下。
10. 将芋头杆焯一下。
11. 将大蒜捣碎。
12. 放入辣椒酱5g、辣椒面儿10g、大酱3g和蒜泥5g。

How to cook

1. Boil stock.
2. Sieve the stock.
3. Cut the ends of mung bean sprout.
4. Blanch the bean sprout green in boiling water.
5. Julienne the leek long.
6. Blanch the ingredients in boiling water.
7. Take white radish out from the meat stock. Julienne the white radish into 0.3cm wide slices.
8. Soak dried bracken(gosari).
9. Blanch dried bracken(gosari) in boiling water.
10. Blanch taro stem.
11. Chop garlic finely.
12. Put red pepper paste(gochujang) of 5g, chili powder of 10g, soybean paste of 3g and chopped garlic of 5g.

고추기름 만들기
制作辣椒油
How to make chili oil

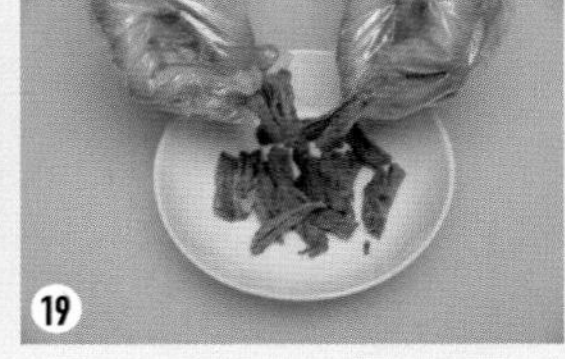

끓이기
煮制
How to boil

만드는 방법

13. 국간장을 5g 넣는다.
14. 참기름을 3g 넣는다.
15. 각종 나물과 재료에 양념한다.
16. 골고루 버무린다.
17. 식용유 200g에 고춧가루 60g, 향신채를 넣고 볶는다.
18. 체에 내려 사용한다.
19. 육수의 양지머리를 건져서 찢는다.
20. 냄비에 고추기름 20g을 넣는다.
21. 국간장 5g, 양지머리를 볶는다.
22. 재료를 볶는다.
23. 골고루 볶는다.
24. 육수를 붓는다.

制作方法

13. 放入汤用酱油5g。
14. 放入香油3g。
15. 将调味酱与各种蔬菜及材料混合。
16. 均匀混合搅拌。
17. 在食用油200g中放入辣椒面儿60g和调味蔬菜后用油炸。
18. 倒入过滤网中除去多余的油。
19. 从肉汤中捞出排骨撕成小块。
20. 在锅中倒入20g辣椒油。
21. 倒入5g酱油后炒排骨肉。
22. 倒入其他材料一起炒。
23. 均匀地炒。
24. 倒入肉汤。

How to cook

13. Put soy sauce of 5g.
14. Add sesame oil of 3g.
15. Mix all sorts of leaves and ingredients with the sauce.
16. Keep mixing them well.
17. Stir-fry chili powder of 60g and condiment vegetable with salad oil of 200g.
18. Sieve the ingredients.
19. Take beef(brisket) out from the stock. Tear the beef.
20. Put chili oil of 20g into a pot.
21. Stir-fry the beef(brisket) with soy sauce of 5g.
22. Roast the ingredients.
23. Keep roasting.
24. Pour stock.

만드는 방법

25. 함께 끓인다.
26. 소금 2g으로 간한다.
27. 후춧가루 0.5g을 넣는다.
28. 완성하여 담는다.

制作方法

25. 一起煮。
26. 加入2g盐。
27. 加0.5g胡椒粉。
28. 完成后盛菜。

How to cook

25. Boil the ingredients all together.
26. Season the ingredients with salt of 2g.
27. Put ground black pepper of 0.5g.
28. Finish and serve it.

TIP
建议

1. 소고기(양지머리) 육수를 진하게 약불에서 30분 정도 끓여야 육개장 맛이 깊어진다.

1. 只有将牛肉(排骨)汤用微火煮30分钟左右后汤才会又浓又香。

Step 1. Boil the beef(brisket) stock over a low heat for about 30 minutes. That will make your yukgyejang taste richer.

즉석불고기

铁板牛肉，石板牛肉

Quick bulgogi

재료 및 분량(3인분)

소고기(등심) 800g, 소갈비양념 소스 150g, 새송이버섯 5g, 느타리버섯 5g, 표고버섯 1개, 팽이버섯 3g, 양파 10g, 파프리카 8g, 피망 8g, 배 5g, 대파 5g, 마늘 3g, 찹쌀가루 15g, 멸치육수 200g

***불고기 소스** : 진간장 150g, 황설탕 30g, 다진 마늘 5g, 다시마 5g, 국멸치 2마리, 양파 3g, 대파 3g, 버섯 자투리 3g, 배즙 50g, 올리고당 15g

***응용요리재료** : 오징어불고기, 돼지고기불고기 등 메뉴에 활용이 가능한 즉석불고기 소스이다.

材料和份量(3人份)

牛肉(里脊) 800g, 牛排调味料 150g, 杏鲍菇 5g, 平菇 5g, 香菇 1个, 金针蘑 3g, 洋葱 10g, 大辣椒 8g, 青椒 8g, 梨 5g, 大葱 5g, 大蒜 3g, 粘米面儿 15g, 鳀鱼汤 200g

***烤肉调味料**: 浓酱油 150g, 黄糖 30g, 蒜泥 5g, 海带 5g, 汤用鳀鱼 2条, 洋葱 3g, 大葱 3g, 蘑菇碎片 3g, 梨汁 50g, 低聚糖 15g

***应用料理材料**: 可用于鱿鱼烤肉和烤猪肉等菜单的烤肉调味料。

Ingredients and portion(3 portions)

800g beef(sirloin), 150g beef rib barbecue sauce, 5g king oyster mushroom, 5g oyster mushroom, 1 shiitake mushroom(mushroom pyogo), 3g mushroom winter, 10g onion, 8g paprika, 8g bell pepper, 5g pear, 5g leek, 3g garlic, 15g Glutinous rice flour, 200g anchovy stock

***Bulgogi sauce** : 150g thick soy sauce, 30g brown sugar, 5g chopped garlic, 5g dried sea tangle, 2 anchovies, 3g onion, 3g leek, 3g mushroom remains, 50g pear juice, 15g oligosaccharide

***Note** : You can use bulgogi sauce to cook squid bulgogi and pork bulgogi.

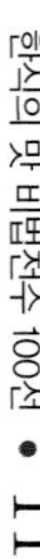

재료 손질 材料处理方法 Clean ingredients

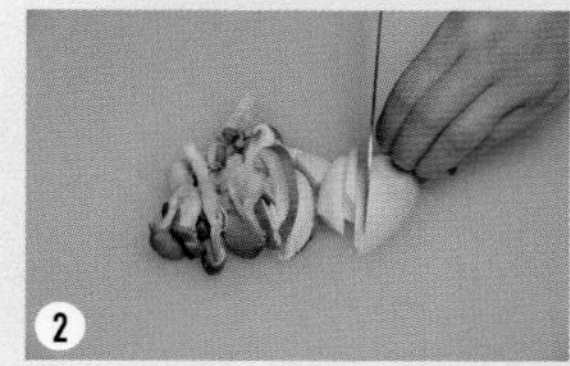

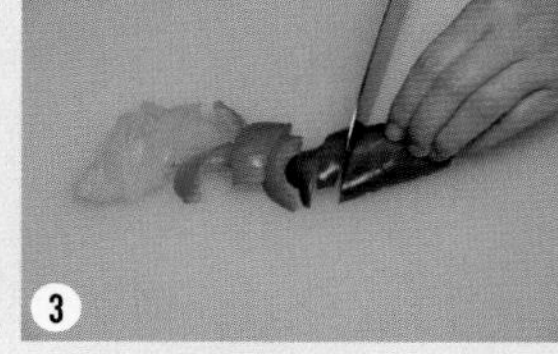

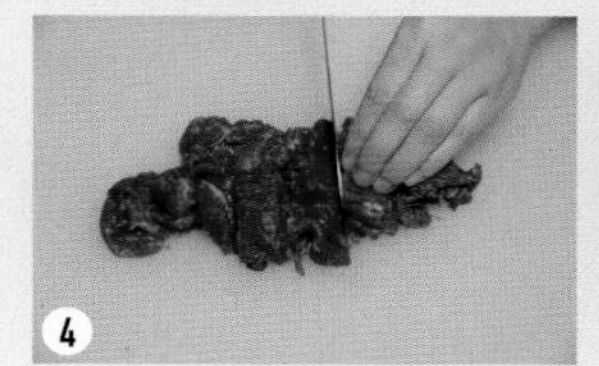

완성하기
完成
How to finish

만드는 방법

1. 양파는 0.5cm로 채 썬다.
2. 버섯은 0.3cm로 편 썬다.
3. 파프리카, 피망은 0.5cm로 채 썬다.
4. 소고기는 2x3cm로 썬다.
6. 소고기 소스로 버무린다.
7. 소고기를 볶다가 멸치육수 200g을 붓는다.
8. 채소와 찹쌀가루 15g을 넣고 농도를 잡는다.
9. 따뜻하게 담는다.

制作方法

1. 将洋葱切成宽0.5cm的丝。
2. 将蘑菇切成厚0.3cm的片。
3. 将大辣椒和青椒切成宽0.5cm的丝。
4. 牛肉切成2x3cm的块。
6. 搅拌牛肉调味料。
7. 炒好牛肉后加入200g鳀鱼汤。
8. 倒入蔬菜和15g糯米粉。
9. 盛放在容器里。

How to cook

1. Julienne onion into 0.5cm wide slices.
2. Julienne mushrooms into 0.3cm wide slices.
3. Julienne paprika and bell pepper into 0.5cm wide slices.
4. Cut beef into 2x3cm wide pieces.
6. Mix the ingredients with bulgogi sauce.
7. Stir-fry the beef. Pour anchovy stock of 200g.
8. Put vegetable and glutinous rice flour of 15g to adjust the consistency.
9. Serve the dish warm.

불고기 소스 끓이기
煮烤肉调味料
How to boil bulgogi sauce

5-1. 진간장 150g
5-1. 浓酱油150g。
5-1. Thick soy sauce of 150g.

5-2. 황설탕 30g
5-2. 黄糖30g。
5-2. Brown sugar of 30g.

5-3. 다진 마늘 5g
5-3. 蒜泥5g。
5-3. Chopped garlic of 5g.

5-4. 마른 재료를 넣는다.
5-4. 倒入干材料。
5-4. Put dried ingredients.

5-5. 배즙을 50g 넣는다.
5-5. 倒入50g梨汁。
5-5. Put pear juice of 50g.

5-6. 올리고당 15g을 넣는다.
5-6. 倒入15g低聚糖。
5-6. Put oligosaccharide of 15g.

TIP
建议

7. 소고기요리는 오래 끓이면 질겨지기 때문에 센 불에서 신속하게 익혀 완성한다.

7. 如果牛肉煮得时间太长，肉就会变硬，应快速煮熟。

Step 7. Do not over-cook the beef. That will make your beef tough.

주꾸미소고기전골

小章鱼牛肉炖锅

Baby octopus-beef jeongol

재료 및 분량(3인분)

주꾸미 200g, 소고기(목심) 200g, 중하 2마리, 새송이버섯 5g, 느타리버섯 3g, 양송이버섯 3g, 표고버섯 2개, 팽이버섯 3g, 청경채 5g, 겨자잎 3g, 쥬키니호박 10g, 당근 5g, 양파 5g, 쑥갓 3g, 고추장 50g, 마늘 10g, 생강 5g, 고춧가루 15g, 국간장 10g, 까나리액젓 8g, 후춧가루 0.5g, 청주 10g

***육수** : 다시마 5g, 국멸치 3마리, 마른 홍합 3g, 감초 2g

***응용요리재료** : 해물과 소고기가 어울리는 국물요리로 다양하게 활용하는 전골요리이다.

材料和份量(3人份)

小章鱼 200g, 牛肉(颈部肉) 200g, 中等蛤蜊 2只, 杏鲍菇 5g, 平菇 3g, 洋香菇 3g, 香菇 2个, 金针蘑 3g, 青菜 5g, 芥菜叶 3g, 菜瓜 10g, 胡萝卜 5g, 洋葱 5g, 茼蒿 3g, 辣椒酱 50g, 大蒜 10g, 生姜 5g, 辣椒面儿 15g, 汤用酱油 10g, 玉筋鱼酱 8g, 胡椒粉 0.5g, 白酒 10g

***肉汤**: 还嗲 5g, 汤用鳀鱼 3条, 干海虹 3g, 甘草 2g

***应用料理材料**: 利用适合海鲜和牛肉的汤类料理的各种炖锅料理。

Ingredients and portion(3 portions)

200g baby octopus, 200g beef(chuck roll), 2 medium shrimps, 5g king oyster mushroom, 5g oyster mushroom, 3g button mushroom, 2 shiitake mushroom (mushroom pyogo), 3g mushroom winter, 5g bok choy, 3g mustard leaf, 10g zucchini, 5g carrot, 5g onion, 3g edible chrysanthemum(ssukgot), 50g chili pepper paste, 10g garlic, 5g ginger, 15g chili powder, 10g soy sauce, 8g sand eel sauce, 0.5g ground black pepper, 10g rice-wine

***Meat stock** : 5g dried sea tangle, 3 anchovies, 3g dried mussel, 2g licorice

***Note** : You can also add other seafood ingredients and other beef meats.

양념장 만들기 制作调味料 How to make marinade

완성하기 完成 How to finish

재료 손질 材料处理方法 Clean ingredients

만드는 방법

1. 고추장 50g을 넣는다.
2. 다진 마늘 10g, 생강 5g, 고춧가루 15g을 넣는다.
3. 국간장 10g을 넣는다.
4. 까나리액젓 8g을 넣는다.
5. 후춧가루 0.5g을 넣는다.
6. 청주 10g을 넣는다.
7. 소고기와 주꾸미를 양념장으로 버무린다.
8. 살짝 볶는다.
9. 야채를 돌려 담는다.
10. 색스럽게 담는다.
11. 육수 부어 끓인다.
12. 완성하여 담는다.

制作方法

1. 加50g辣椒酱。
2. 加10g蒜泥, 5g生姜和15g辣椒面儿。
3. 加10g汤用酱油。
4. 加8g玉筋鱼酱。
5. 加入0.5g胡椒粉。
6. 加10g入白酒。
7. 将牛肉和小章鱼及调味料一起混合搅拌。
8. 稍微炒一下。
9. 将蔬菜放在四周。
10. 将蔬菜美观地摆放好。
11. 加入肉汤后一起煮。
12. 完成后盛菜。

How to cook

1. Put red pepper paste(gochujang) of 50g.
2. Put chopped garlic of 10g, ginger of 5g and chili powder of 15g.
3. Put soy sauce of 10g.
4. Put sand eel sauce of 8g.
5. Add ground black pepper of 0.5g.
6. Pour rice-wine of 10g.
7. Mix beef and baby octopus with the sauce
8. Roast the ingredients the above slightly.
9. Arrange the vegetables in a circular manner in a pot.
10. Decorate the dish even nicer.
11. Pour meat stock. Boil the ingredients.

TIP 建议

8. 소고기를 먼저 볶으면 생고기가 뭉치는 것을 막아준다.

8. 先炒牛肉可以防止生肉结块。

Step 8. Roast the beef in advance so that you will not have the beef crumpled into one big ball.

차돌박이청국장

雪花牛肉清国酱汤

Marbled beef cheonggukjang

재료 및 분량(2인분)
소고기(차돌박이) 150g, 청국장 70g, 두부 20g, 신김치 10g, 애호박 8g, 표고버섯 1개, 홍고추 5g, 풋고추 5g, 들깨가루 3g, 콩가루 3g, 멥쌀 30g, 고춧가루 3g, 마늘 3g, 들기름 10g
***육수** : 다시마 5g, 국멸치 3마리, 마른 새우 2g, 감초 2g, 통마늘 3g, 고추씨 3g, 마른 홍합 2g, 표고버섯 2g, 무 30g, 대파 5g, 마른 홍고추 3g
***응용요리재료** : 조개류도 함께 끓여서 활용할 수 있는 찌개류이다.

材料和份量(2人份)
牛肉(牛五花肉) 150g, 豆瓣酱 70g, 豆腐 20g, 新泡菜 10g, 西葫芦 8g, 香菇 1个, 红辣椒 5g, 青辣椒 5g, 野芝麻粉 3g, 豆粉 3g, 白米 30g, 辣椒面儿 3g, 大蒜 3g, 芝麻油 10g
***肉汤** : 海带 5g, 汤用鳀鱼 3条, 干虾 2g, 甘草 2g, 大蒜 3g, 辣椒籽 3g, 干海虹 2g, 香菇 2g, 萝卜 30g, 大葱 5g, 干海虹 3g
***应用料理材料**: 可以与蛤蜊类一起煮制的汤类。

Ingredients and portion(2 portions)
150g beef(marbled beef), 70g fermented soybeans, 20g bean curd(tofu), 10g over-fermented Kimchi, 8g long squash, 1 shiitake mushroom(mushroom pyogo), 5g red chilli, 5g green chilli, 3g perilla seed flour, 3g bean flour, 30g non-glutinous rice, 3g chili powder, 3g garlic, 10g perilla oil
***Meat stock** : 5g dried sea tangle, 3 anchovies, 2g dried shrimp, 2g licorice, 3g of garlic, 3g pepper seed, 2g dried mussel, 2g shiitake mushroom(mushroom pyogo), 30g white radish, 5g leek, 3g dried red chilli
***Note** : You can also add shellfish to cook jjigae(stew).

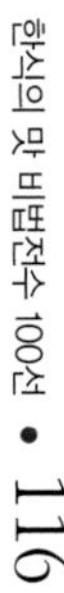

재료 손질
材料处理方法
Clean ingredients

만드는 방법

1. 육수를 끓인다.
2. 불린 쌀에서 쌀뜨물을 만든다.
3. 육수에 붓는다.
4. 재료를 손질한다.
5. 표고버섯은 깍둑썰기한다.
6. 두부는 깍둑썰기한다.
7. 소고기는 들기름 10g에 볶는다.
8. 콩가루 3g을 볶는다.
9. 육수를 부어 끓인다.
10. 청국장 70g을 넣는다.
11. 들깨가루와 1:1 비율로 고춧가루를 3g 넣는다.
12. 완성하여 담는다.

制作方法

1. 煮肉汤。
2. 从泡米中制成淘米水。
3. 倒入肉汤中。
4. 整理材料。
5. 把香菇切成块。
6. 把豆腐切成块。
7. 用10g芝麻油炒牛肉。
8. 再加3g豆粉。
9. 加入肉汤后一起煮。
10. 加清70g国酱。
11. 加3g辣椒粉（与芝麻粉成1:1的比率）。
12. 完成后盛菜。

How to cook

1. Boil the stock.
2. Make water from soaked rice.
3. Pour the rice water from Step 2 to the meat stock.
4. Clean ingredients.
5. Cut the shiitake mushroom(mushroom pyogo) into cubes.
6. Cut the bean curd(tofu) into cubes.
7. Stir-fry the beef with perilla oil of 10g.
8. Roast bean flour of 3g.
9. Boil the ingredients in the meat stock.
10. Putfermented soybeans of 70g.
11. Put perilla flour and chili powder of 20g at a ratio of 1 : 1.
12. Finish and serve it.

TIP
建议

2. 쌀뜨물을 사용하여 청국장찌개의 짠맛을 제거하고 구수한 국물 맛을 낸다.

2. 使用淘米水可以去除豆瓣酱汤的咸味，并使汤的味道香酥可口。

Step 2. The rice water removes salty taste of cheonggukjang jjigae(stew) making the dish taste sweet.

韩食美“味”秘方传授 100选 实践篇

Transmission of a Hundred Secrets of Korean Dishes Flavor 蒸菜，煮菜

한식의 맛 비법전수 100선

실기편

찜

매운 닭찜

韩式香辣炖鸡

Spicy chicken jjim

재료 및 분량(3인분)
닭(800g) 1마리, 감자 20g, 양파 10g, 당근 10g, 새송이버섯 5g, 표고버섯 1개, 브로콜리 10g, 은행 3g, 고추장 20g, 후춧가루 0.5g, 사과즙 10g, 매실즙 5g, 청주 5g, 대파 3g, 마늘 5g
***간장양념장** : 진간장 150g, 다시마 5g, 마른 홍고추 1개, 감초 1g, 통마늘 2g, 대파 3g, 고춧가루 15g, 설탕 20g, 청주 30g
***응용요리재료** : 돼지고기(아롱사태), 소고기(안심), 훈제오리 등으로 활용된다.

材料和份量(3人份)
鸡肉(800g) 1条, 土豆 20g, 洋葱 10g, 胡萝卜 10g, 杏鲍菇 5g, 香菇 1个, 西兰花 10g, 银杏 3g, 辣椒酱 20g, 胡椒粉 0.5g, 苹果汁 10g, 梅子汁 5g, 白酒 5g, 大葱 3g, 大蒜 5g
***酱油调味酱** : 浓酱油 150g, 海带 5g, 干红辣椒 1个, 甘草 1g, 大蒜 2g, 大葱 3g, 辣椒面儿 15g, 白糖 20g, 白酒 30g
***应用料理材料**: 可利用猪肉(胫骨肉)、牛肉(里脊)和熏制鸭肉等。

Ingredients and portion(3 portions)
800g chicken, 20g potato, 10g onion, 10g carrot, 5g king oyster mushroom, 1 shiitake mushroom(mushroom pyogo), 10g broccoli, 3g ginkgo nut, 20g red pepper paste(gochujang), 0.5g ground black pepper, 10g apple juice, 5g plum juice, 5g rice-wine, 3g leek, 5g garlic
***Soy sauce** : 150g thick soy sauce, 5g dried sea tangle, 1 dried red chilli, 1g licorice, 2g garlic, 3g leek, 15g chili powder, 20g sugar, 30g rice-wine
***Note** : You can cook another jjim dish with pork(center hell of shank), beef(tenderloin) and roasted duck.

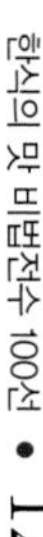

재료 준비 准备材料 Clean ingredients

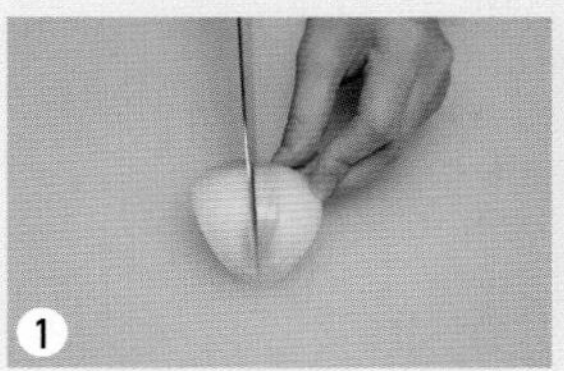

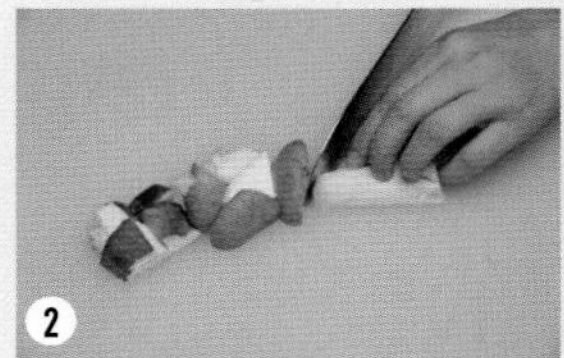

닭 손질 鸡肉的处理方法 How to prepare chicken

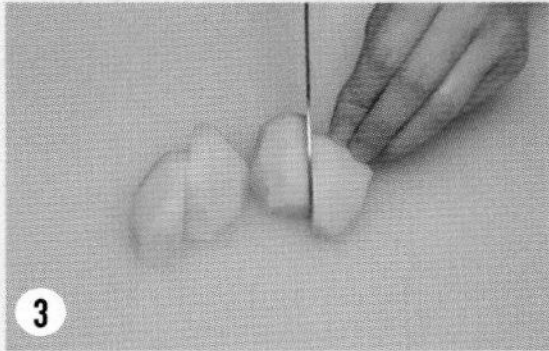

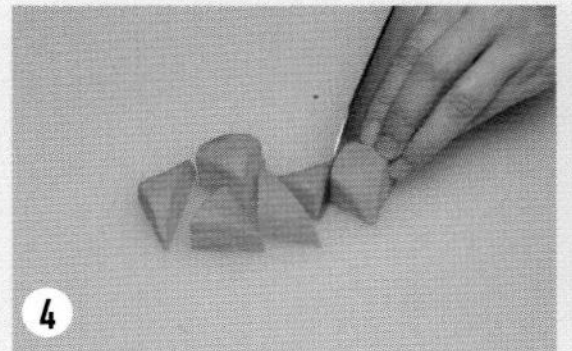

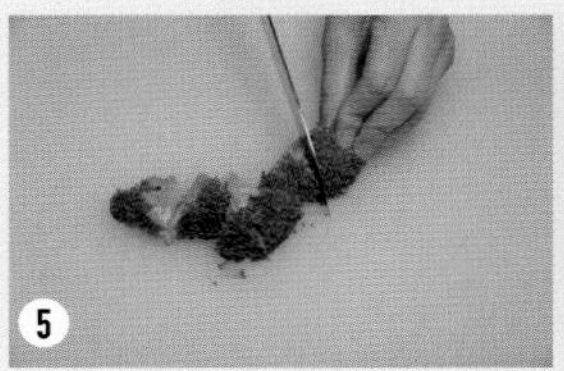

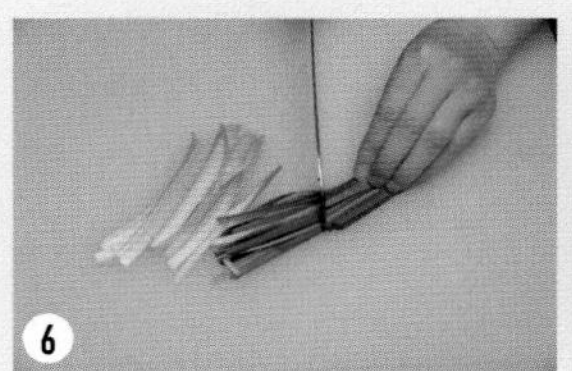

간장양념장 만들기 制作酱油调味酱 How to make soy sauce marinade

만드는 방법

1. 양파는 큼직하게 썬다.
2. 새송이버섯을 큼직하게 썬다.
3. 감자는 깍둑썰기한다.
4. 당근을 한입 크기로 썬다.
5. 브로콜리를 한입 크기로 썬다.
6. 대파는 길게 채 썰고 부추는 4cm로 썬다.
7. 진간장 150g(다시마, 홍고추, 감초, 통마늘, 대파)을 넣는다.
8. 고춧가루 15g을 넣는다.
9. 설탕 20g을 넣는다.
10. 청주 30g을 넣는다.
11. 끓는 물에 닭을 데친다.
12. 사과즙 10g을 넣는다.

制作方法

1. 将大葱切成大块。
2. 将杏鲍菇切成大块。
3. 将土豆切成方块。
4. 胡萝卜切成合适的大小。
5. 西兰花切成合适的大小。
6. 将大葱切成丝细，韭菜切成大小4cm。
7. 加150g酱油(海带、红辣椒、甘草、大蒜和大葱)。
8. 加15g辣椒面儿。
9. 加20g白糖。
10. 加30g白酒。
11. 将鸡肉在烧开的水中焯一下。
12. 加10g苹果汁。

How to cook

1. Cut the onion into chunks.
2. Cut the king oyster mushroom into chunks.
3. Cut the potato into cubes.
4. Cut the carrot into bite-size pieces.
5. Cut the broccoli into bite-size pieces.
6. Julienne the leek long. Cut the chives into 4cm long pieces.
7. Put thick soy sauce of 150g(add dried sea tangle, red chili, licorice, garlic and leek).
8. Put chili powder of 15g.
9. Put sugar of 20g.
10. Add rice-wine of 30g.
11. Blanch the chicken in boiling water.
12. Pour apple juice of 10g.

끓이기
煮制
How to boil

만드는 방법

13. 후춧가루 0.5g을 뿌린다.
14. 고추장 20g을 넣는다.
15. 골고루 버무린다.
16. 냄비에 담는다.
17. 양념장 100g을 넣는다.
18. 물 200g을 넣는다.
19. 매실즙 5g을 넣는다.
20. 청주 5g을 넣는다.
21. 함께 조린다.
22. 대파를 넣는다.

制作方法

13. 加0.5g胡椒粉。
14. 加20g辣椒酱。
15. 均匀混合搅拌。
16. 将材料放入锅中。
17. 放100g调味酱。
18. 加200g水。
19. 加5g梅子汁。
20. 加5g白酒。
21. 一起煮。
22. 放大葱。

How to cook

13. Sprinkle ground black pepper of 0.5g.
14. Put red pepper paste(gochujang) of 20g.
15. Mix the ingredients all together.
16. Put the ingredients all into a pot.
17. Put sauce of 100g.
18. Pour water of 200g.
19. Put plum juice of 5g.
20. Add rice-wine of 5g.
21. Boil the ingredients down.
22. Put the leek.

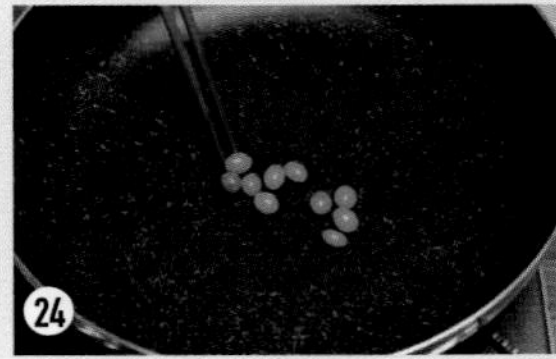

만드는 방법

23. 다진 마늘 5g을 넣는다.
24. 은행을 볶아서 고명으로 사용한다.
25. 완성하여 담는다.

制作方法

23. 放入5g蒜泥。
24. 将银杏炒一下用作装饰菜。
25. 完成后盛放在容器里。

How to cook

23. Put chopped garlic of 5g.
24. Stir-fry ginkgo nuts. Use the nuts as a garnish.
25. Serve the dish well.

TIP
建议

15. 닭 양념을 1시간 정도 재워 놓으면 더욱 깊은 맛을 낸다.
간장양념장은 다양한 조림으로 응용할 수 있는 소스이다.

15. 若将鸡肉调味料腌制1个小时左右，味道将渗入其中。
酱油调味酱可用于各种煮制料理。

Step 15. Leave the chicken with the sauce for 1 hour to add a richer taste to the chicken.
You can cook a variety of dishes with soy sauce.

도가니찜

炖牛膝骨

Cow knee cartilage jjim

재료 및 분량(2인분)
삶은 도가니 500g, 찹쌀가루 10g, 목이버섯 3g, 대파 3g, 홍고추 3g, 풋고추 3g, 참기름 3g, 마늘 2g, 통깨 2g, 국간장 5g, 설탕 5g
***응용요리재료** : 도가니무침, 도가니수육을 만들 때 활용 가능한 요리이다.

材料和份量(2人份)
蒸好的牛膝骨肉 500g, 粘米面儿 10g, 木耳 3g, 大葱 3g, 红辣椒 3g, 青辣椒 3g, 香油 3g, 大蒜 2g, 芝麻 2g, 汤用酱油 5g, 白糖 5g
***应用料理材料**: 可用于制作牛膝骨肉拌菜, 牛膝骨汤的料理。

Ingredients and portion(2 portions)
500g boiled Knee cartilage of cow, 10g Glutinous rice flour, 3g black mushroom, 3g leek, 2g red chilli, 3g green chilli, 3g sesame oil, 2g garlic, 2g sesame seed, 5g soy sauce, 5g sugar
***Note** : You can use the recipe also to cook seasoning cow knee cartilage and boiling cow knee cartilage.

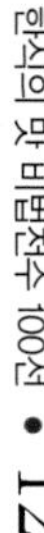

만드는 방법

1. 물 50g에 국간장을 5g 넣는다.
2. 설탕 5g을 넣는다.
3. 다진 마늘을 2g 넣는다.
4. 삶은 도가니를 2x3cm로 자른다.
5. 팬에 볶는다.
6. 조린다.
7. 부재료를 넣는다.
8. 섞어준다.
9. 찹쌀가루를 1:1 비율로 물과 섞는다.
10. 찹쌀물을 끼얹어준다.
11. 약불에서 마무리한다.
12. 완성한다.

制作方法

1. 在50g水中加5g汤用酱油。
2. 加5g白糖。
3. 加2g蒜泥。
4. 将煮熟的牛膝骨肉切成2x3cm的块。
5. 放入平底锅中炒。
6. 炖。
7. 加入副材料。
8. 混合搅拌。
9. 将粘米面儿和水以1:1的比例混合。
10. 加入粘米面儿水。
11. 用微火炖。
12. 完成。

How to cook

1. Pour soy sauce of 5g to water of 50g.
2. Put sugar of 5g.
3. Add chopped garlic of 2g.
4. Cut boiled knee cartilage of cow into 2x3cm wide pieces.
5. Pan-fry the ingredients.
6. Boil the ingredients down.
7. Put side ingredients.
8. Mix the ingredients well.
9. Mix glutinous rice flour and water at a ratio of 1:1.
10. Pour glutinous rice flour water.
11. Finish cooking over a low heat.
12. Finish.

TIP 建议

4. 소 무릎의 도가니는 향신채를 넣고 2시간 정도 삶아서 사용한다.

4. 在牛膝骨中放入调味蔬菜，然后煮2个小时后使用。

Step 4. Boil knee cartilage of cow with condiment vegetable for about 2 hours before you use the knee cartilage of cow.

돼지등뼈찜

炖猪里脊

Pork chine jjim

재료 및 분량(3인분)
돼지등뼈 1kg, 토란대 30g, 시래기 50g, 양파 30g, 팽이버섯 5g, 느타리버섯 5g, 고추장 10g, 된장 20g, 고춧가루 5g, 들깨가루 5g, 실파 3g, 대파 5g, 홍고추 5g, 풋고추 5g, 마늘 3g, 생강 3g, 매운 갈비 소스 80g, 국간장 10g, 찹쌀가루 20g
***육수** : 다시마 5g, 마른 홍고추 3g, 통마늘 3g
***응용요리재료** : 돼지고기 부위를 알맞게 손질하여 찜을 할 수 있는 요리이다.

材料和份量(3人份)
猪里脊骨 1kg, 芋头杆 30g, 干菜叶 50g, 洋葱 30g, 金针蘑 5g, 平菇 5g, 辣椒酱 10g, 大酱 20g, 辣椒面儿 5g, 野芝麻粉 5g, 小葱 3g, 大葱 5g, 红辣椒 5g, 青辣椒 5g, 大蒜 3g, 生姜 3g, 辣味排骨调味料 80g, 汤用酱油 10g, 粘米面儿 20g
***肉汤** : 海带 5g, 干红辣椒 3g, 大蒜 3g
***应用料理材料**: 用猪肉的适当部位制作的蒸炖类料理。

Ingredients and portion(3 portions)
1kg pork backbone, 30g taro stem, 50g dried radish leaves, 30g onion, 5g mushroom winter, 5g oyster mushroom, 10g pepper red paste(gochujang), 20g soybean paste, 5g chili powder, 5g perilla seed flour, 3g green onion, 5g leek, 5g red chilli, 5g green chilli, 3g garlic, 3g ginger, 80g spicy galbi sauce, 10g soy sauce, 20g Glutinous rice flour
***Meat stock** : 5g dried sea tangle, 3g dried red chilli, 3g garlic
***Note** : You can cook other jjim dishes with other pork meats.

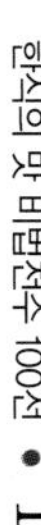

만드는 방법

1. 찬물에서 핏물을 제거한다.
2. 등뼈를 조린다.
3. 매운 갈비 소스 80g을 넣는다.
4. 다시마육수 800g을 붓는다.
5. 양념에 조린다.
6. 토란대, 시래기나물은 국간장 10g로 간한다.
7. 나물을 올린다.
8. 부재료를 넣고 익힌다.
9. 찹쌀가루 20g으로 농도를 맞춘다.
10. 섞어준다.
11. 따뜻하게 담는다.
12. 완성한다.

制作方法

1. 用水去除血水。
2. 炖里脊骨。
3. 加80g辣味排骨调味料。
4. 倒入800g海带汤。
5. 用调味料调味。
6. 在芋头杆和干菜叶中加入10g汤用酱油。
7. 放上蔬菜。
8. 放入副材料后炖熟。
9. 加20g糯米粉调整浓度。
10. 混合搅拌。
11. 盛放在容器里。
12. 完成。

How to cook

1. Remove blood in cold water.
2. Boil the backbone down.
3. Put spicy galbi sauce of 80g.
4. Pour dried sea tangle stock of 800g.
5. Boil the ingredients down with sauce.
6. Season the taro stem and dried radish leaves with soy sauce of 10g.
7. Place the leaves on top.
8. Put the other ingredients. Boil the ingredients.
9. Adjust the consistency with glutinous rice flour of 20g.
10. Mix the ingredients well.
11. Serve the dish warm.
12. Finish.

TIP
建议

2. 등뼈 냄새는 끓는 물에 된장 20g을 풀어서 1시간 삶으면 제거된다.
5. 양념 소스에 20분 이상 조려준다.

2. 为去除里脊骨的异味，在烧开的水中加入20g大酱后煮1个小时。
5. 在调味料中炖20分钟以上。

Step 2. Put soybean paste of 20g in boiling water to remove smell of backbone.
Step 5. Boil the ingredients down for 20 minutes more.

묵은지찜

炖陈泡菜

Over-fermented jjim

재료 및 분량(2인분)
묵은 김치 1kg, 돼지고기(삼겹살) 1kg, 느타리버섯 20g, 양송이버섯 3g, 양파 20g, 당근 5g, 대파 5g, 홍고추 3g, 풋고추 3g, 대파 5g, 마늘 3g, 생강 3g, 된장 30g
***찜 소스** : 고추장 20g, 고춧가루 5g, 국간장 15g, 매실즙 15g, 후춧가루 0.5g, 올리고당 10g
***육수** : 다시마 5g, 국멸치 3마리, 양파 3g, 대파 3g, 당근 2g, 표고버섯 3g
***응용요리재료** : 묵은지닭도리탕, 묵은지고등어찜, 묵은지갈비찜 등에 활용된다.

材料和份量(2人份)
陈泡菜 1kg, 猪肉(五花肉) 1kg, 平菇 20g, 洋香菇 3g, 洋葱 20g, 胡萝卜 5g, 大葱 5g, 红辣椒 3g, 青辣椒 3g, 大葱 5g, 大蒜 3g, 生姜 3g, 大酱 30g
***炖锅用调味料**: 辣椒酱 20g, 辣椒面儿 5g, 汤用酱油 15g, 梅子汁 15g, 胡椒粉 0.5g, 低聚糖 10g
***肉汤**: 海带 5g, 汤用鳀鱼 3条, 洋葱 3g, 大葱 3g, 胡萝卜 2g, 香菇 3g
***应用料理材料**: 可用于制作陈泡菜鸡肉炖土豆、陈泡菜炖青花鱼和陈泡菜炖排骨等。

Ingredients and portion(2 portions)
1kg over-fermented Kimchi, 1kg pork belly, 20g oyster mushroom, 3g button mushroom, 20g onion, 5g carrot, 5g leek, 3g red chilli, 3g green chilli, 5g leek, 3g garlic, 3g ginger, 30g soybean paste
***Jjim sauce** : 20g red pepper paste(gochujang), 5g chili powder, 15g soy sauce, 15g plum juice, 0.5g ground black pepper, 10g oligosaccharide
***Meat stock** : 5g dried sea tangle, 3 anchovies, 3g onion, 3g leek, 2g carrot, 3g shiitake mushroom(mushroom pyogo)
***Note** : You can use the recipe to cook over-fermented Kimchi simmered chicken, over-fermented Kimchi simmered mackerel and over-fermented Kimchi braised beef ribs.

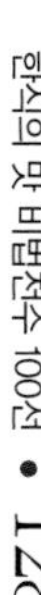

고기 손질 肉的处理方法 How to prepare the meat

찜 소스 만들기 制作炖锅用调味料 How to make jjim sauce

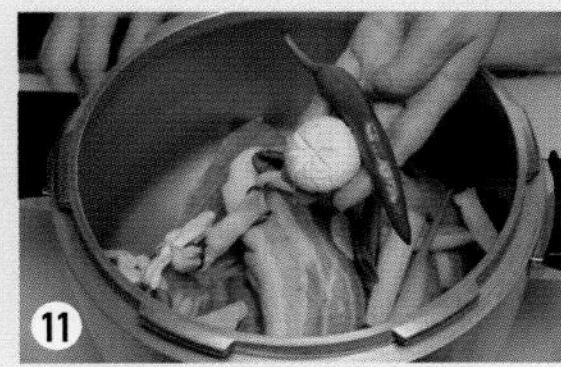

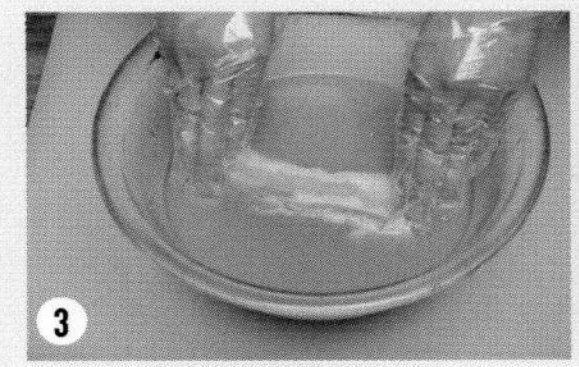

김치 손질 泡菜的处理方法 How to prepare Kimchi

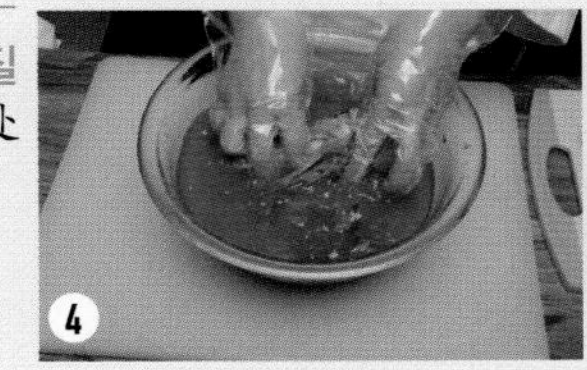

끓이기 煮 How to boil

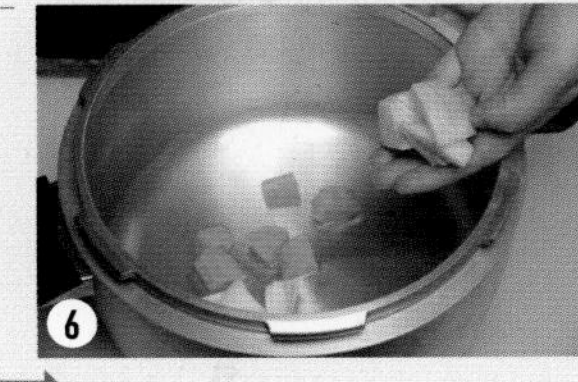

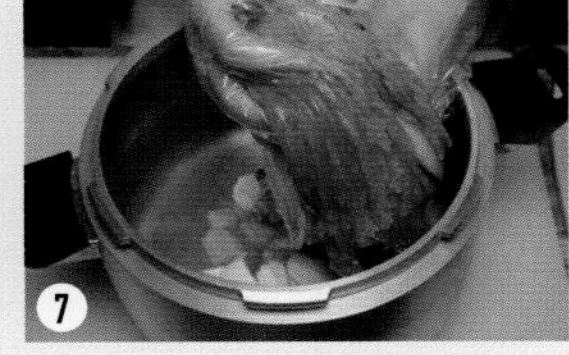

만드는 방법

1. 찬물에 된장 30g을 넣는다.
2. 삼겹살덩어리를 된장물에 넣는다.
3. 냄새를 제거한다.
4. 묵은 김치를 헹군다.
5. 물기를 제거한다.
6. 냄비에 향신채를 담는다.
7. 묵은 김치를 바닥에 깐다.
8. 삼겹살을 올린다.
9. 향신채를 넣는다.
10. 느타리버섯을 넣는다.
11. 양송이버섯과 풋고추를 넣는다.
12. 고추장 20g을 넣는다.

制作方法

1. 在凉水中加30g大酱。
2. 将五花肉放入大酱水中。
3. 去除异味。
4. 用水冲洗陈泡菜。
5. 控除水分。
6. 在锅里放入调味蔬菜。
7. 将陈泡菜放在最下面。
8. 泡菜上放五花肉。
9. 放入调味蔬菜。
10. 加入平菇。
11. 加入洋香菇和青辣椒。
12. 加20g辣椒酱。

How to cook

1. Put soybean paste of 30g in cold water.
2. Put the pork belly into the soybean water.
3. Remove smell.
4. Wash over-fermented Kimchi.
5. Remove water from Kimchi.
6. Put condiment vegetable into a pot.
7. Place the over-fermented Kimchi on the bottom.
8. Put pork belly on top.
9. Put condiment vegetable.
10. Put oyster mushroom.
11. Add button mushroom and green chilli.
12. Put red pepper paste(gochujang) of 20g.

완성하기
完成
How to finish

만드는 방법

13. 고춧가루 5g을 넣는다.
14. 국간장 15g을 넣는다.
15. 매실즙 15g을 넣는다.
16. 후춧가루 0.5g을 넣는다.
17. 올리고당 10g을 넣는다.
18. 준비된 육수 300g을 붓는다.
19. 김치찜 소스를 완성한다.
20. 찜 소스를 붓는다.
21. 국자로 골고루 붓는다.
22. 천천히 부어준다.
23. 압력솥에 20분 정도 삶는다.
24. 완성된 통삼겹살 편 썰기

制作方法

13. 加5g辣椒面儿。
14. 加15g汤用酱油。
15. 加15g梅子汁。
16. 加0.5g胡椒粉。
17. 加10g低聚糖。
18. 倒入300g肉汤。
19. 完成泡菜调味料的制作。
20. 加入炖锅用调味料。
21. 用锅铲搅拌均匀。
22. 慢慢地倒入。
23. 放在压力锅中炖20分钟左右。
24. 将做好的整块五花肉切成片状。

How to cook

13. Put chili powder of 5g.
14. Add soy sauce of 15g.
15. Put plum juice of 15g.
16. Put ground black pepper of 0.5g.
17. Add oligosaccharide of 10g.
18. Pour meat stock of 300g.
19. Finish with cooking Kimchi jjim sauce.
20. Pour the jjim sauce.
21. Pour the sauce with a ladle.
22. Keep pouring the sauce slowly.
23. Boil the ingredients in a pressure cooker for about 20 minutes.
24. Cut the cooked pork belly into thin slices.

완성하기
完成
How to finish

만드는 방법
25. 묵은 김치와 함께 담는다.
26. 곁들여 담는다.
27. 완성한다.

制作方法
25. 与陈泡菜盛放在一起。
26. 整理使其看起来美观。
27. 完成。

How to cook
25. Serve the dish with over-fermented Kimchi.
26. Garnish the dish with other ingredients as you like.
27. Finish.

TIP
建议

3. 된장을 풀어 삼겹살의 냄새를 제거한다.
4. 묵은 김치는 찬물에 헹궈 양념을 씻어 사용해야 빛깔이 곱게 완성된다.
23. 압력솥을 이용하여 30분 정도 삶는다. (불 조절 필요)

3. 用大酱水可以去除五花肉的异味。
4. 陈泡菜只有用凉水冲洗才能保持原有的光泽。
23. 利用压力锅炖30分钟左右。(需要调整火候)

Step 3. Soybean paste removes bad smell of pork belly.
Step 4. Wash over-fermented Kimchi in cold water so that you have a beautiful color on your dish.
Step 23. Boil the ingredients in a pressure cooker for about 30 minutes(watch over the heat).

삼겹살보쌈

菜包五花肉

Pork belly bossam

재료 및 분량(2인분)
돼지고기(삼겹살) 600g, 배 30g, 양파 15g, 사과 10g, 마른 홍고추 3g, 대파 5g, 진간장 100g, 구기자 3g, 감초 2g, 가시오가피 5g, 당귀 2g, 생강 2g, 마늘 2g, 황설탕 15g
***응용요리재료** : 돼지고기(목살), 족발보쌈에 활용되는 요리이다.

材料和份量(2人份)
猪肉(五花肉) 600g, 梨 30g, 洋葱 15g, 苹果 10g, 干红辣椒 3g, 大葱 5g, 浓酱油 100g, 枸杞子 3g, 甘草 2g, 刺五加 5g, 当归 2g, 生姜 2g, 大蒜 2g, 黄糖 15g
***应用料理材料**: 可用于制作猪肉(颈部肉)和猪蹄菜包的料理。

Ingredients and portion(2 portions)
600g pork belly, 30g pear, 15g onion, 10g apple, 3g dried red chilli, 5g leek, 100g thick soy sauce, 3g chinese wolfberries(gugija), 2g licorice, 5g Eleuthero, 2g Korean angelica roots(dong guai), 2g ginger, 2g garlic, 15g brown sugar
***Note** : You can also use the recipe to cook a dish of pork(neck), pork ankle and bossam.

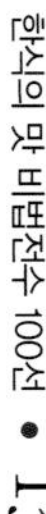

고기 손질 肉的处理方法 How to prepare the meat

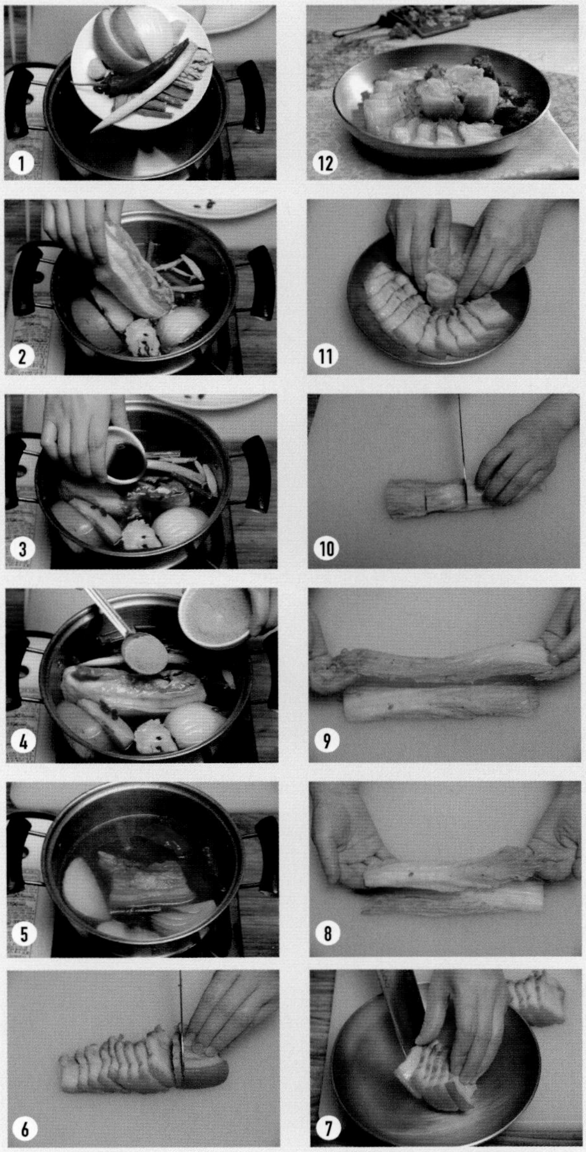

김치 손질 泡菜的处理方法 How to prepare Kimchi

만드는 방법

1. 찬물 1.5L에 한방재료와 고기를 넣는다.
2. 끓으면 고기를 넣는다.
3. 진간장 100g을 넣는다.
4. 황설탕 15g을 넣는다.
5. 한 시간 삶는다.
6. 식은 후 편 썰어준다.
7. 완성그릇에 펼쳐 담는다.
8. 물에 씻은 김치는 한 장씩 포갠다.
9. 서로 반대로 쌓는다.
10. 동그랗게 잡아 길이 5cm로 썬다.
11. 보쌈을 완성하여 담는다.
12. 완성한다.

制作方法

1. 在1.5L的凉水中放入肉和韩方中药材料。
2. 待水烧开后放入肉。
3. 加100g浓酱油。
4. 加15g黄糖。
5. 煮1个小时。
6. 冷却后切成片状。
7. 盛放在盘子里。
8. 将用水冲洗过的泡菜一张一张地展开。
9. 彼此按照相反的方向堆放。
10. 使模样呈现圆形，切成长度5cm。
11. 菜包完成后盛放在盘子上。
12. 完成后盛菜。

How to cook

1. Put Chinese herbal ingredients into cold water of 1.5L.
2. When the water boils, put the meat.
3. Put thick soy sauce of 100g.
4. Put brown sugar of 15g.
5. Boil the ingredients for 1 hour.
6. Leave the ingredients to cool down. Cut the meat into thin slices.
7. Serve the dish on a wide plate.
8. Wash Kimchi in water. Pile Kimchi with one sheet on another.
9. Make sure that you pile the Kimchi sheets zigzag.
10. Roll the sheets. Cut the sheets into 5cm long pieces.
11. Finish. Serve the dish on a plate.
12. Finish.

TIP 建议

10. 고기를 쌈할 수 있도록 크기를 조절하여 김치를 썬다.

10. 切泡菜时须使泡菜的大小可以包住肉。

Step 10. Cut Kimchi big enough so that you can wrap the meat with it.

송이갈비찜

松茸炖排骨

Button mushroom-galbi jjim

재료 및 분량(2인분)
한우(소갈비) 400g, 자연산 송이버섯 50g, 대추 3개, 양파 30g, 배 20g, 밤 10g, 마늘 3g, 은행 3g, 대파 10g, 진간장 50g, 올리고당 30g, 배즙 30g, 양파즙 20g, 청주 10g
***응용요리재료** : 소등갈비찜, 매콤갈비찜, 전복갈비찜 등의 다양한 갈비찜으로 이용된다.

材料和份量(2人份)
韩牛(牛排骨) 400g, 天然松茸 50g, 大枣 3个, 洋葱 30g, 梨 20g, 栗子 10g, 大蒜 3g, 银杏 3g, 大葱 10g, 浓酱油 50g, 低聚糖 30g, 梨汁 30g, 洋葱汁 20g, 清酒 10g
***应用料理材料**: 可用于制作炖牛排骨、辣味炖排骨和鲍鱼炖排骨等各种炖排骨类料理。

Ingredients and portion(2 portions)
400g Korea beef(beef ribs), 50g natural button mushroom, 3 jujubes, 30g onion, 20g pear, 10g chestnuts, 3g garlic, 3g ginkgo nut, 10g leek, 50g thick soy sauce, 30g oligosaccharide, 30g pear juice, 20g onion juice, 10g rice-wine
***Note** : You can use the recipe also to cook other galbi jjim dishes as beef back-galbi jjim, spicy galbi jjim and abalones galbi jjim.

송이버섯 손질 松茸的处理方法 How to prepare button mushroom

완성하기
完成
How to finish

만드는 방법

1. 송이버섯 뿌리를 다듬는다.
2. 2등분한다.
3. 뿌리 쪽을 손질한다.
4. 진간장과 1:1 비율의 물에 조린다.
6. 소갈비를 건진다.
7. 소갈비를 손질하여 완성한다.
8. 진간장과 1:1 비율의 물 소스에 조린다.
9. 올리고당, 배즙, 양파즙을 넣고 조린다.

制作方法

1. 整理松茸的根。
2. 2等分。
3. 整理根部。
4. 将酱油与水以1:1的比例混合。
6. 取出牛排骨。
7. 牛排骨处理完成。
8. 加入酱油与水1:1的调味料一起炖。
9. 加入低聚糖、梨汁和洋葱汁。

How to cook

1. Clean the ends of the button mushroom.
2. Cut the base in half.
3. Trim the ends
4. Boil the button mushroom down with water and thick soy bean at a ratio of 1 : 1.
6. Take the beef rib(galbi) out from the ginger water.
7. Finish with cleaning the beef rib(galbi).
8. Boil the beef rib(galbi) down with soy bean and water at a ratio of 1 : 1.
9. Add oligosaccharide, pear juice and onion juice. Boil the ingredients down.

고기 손질
肉的处理方法
How to prepare the meat

5-1. 핏물 제거
5-1. 去除血水。
5-1. Remove blood.

5-2. 소갈비 막 제거
5-2. 去除牛排骨膜。
5-2. Remove the film of the beef rib(galbi).

5-3. 기름기 제거
5-3. 去除油分。
5-3. Remove fatty residue.

5-4. 넓게 펼쳐 손질한다.
5-4. 铺开后处理。
5-4. Spread the meat. Clean the meat.

5-5. 생강물을 만든다.
5-5. 制作生姜水。
5-5. Make ginger water.

5-6. 청주를 넣고 소갈비를 데친다.
5-6. 用生姜水焯一下牛排骨。
5-6. Blanch the beef rib(galbi) in ginger water.

TIP 建议

9. 갈비가 완전히 익었을 때 송이버섯이 첨가되어야 향을 진하게 느낄 수 있다.
양파, 배, 황설탕은 사용하지 않을 수 있다.

9. 只有在排骨完全熟透时加入松茸才能使料理的香味浓郁。
可以不使用洋葱、梨、黄糖

Step 9. Put button mushroom after the beef rib(galbi) is fully cooked. That will get a richer flavor to your dish.
You might not use onion, pear, or brown sugar.

아귀찜

炖鮟鱇鱼

Monkfish jjim

재료 및 분량(2인분)
아귀 1kg, 찜용 콩나물 600g, 미나리 80g, 양파 20g, 홍고추 10g, 풋고추 10g, 느타리버섯 5g, 대파 5g, 홍합살 40g, 된장 15g, 소금 5g, 설탕 5g, 찹쌀가루 30g
***양념장** : 고춧가루 30g, 국간장 10g, 후춧가루 0.5g, 올리고당 15g, 청주 15g, 마늘 10g, 생강 5g, 홍합살 30g, 새우살 30g, 천연조미료 5g
***육수** : 다시마 5g, 국멸치 3마리. 홍고추 5g, 풋고추 5g, 대파 2g
***응용요리재료** : 해물재료를 활용하여 매콤한 찜요리를 만들 수 있다.

材料和份量(2人份)
鮟鱇鱼 1kg, 豆芽 600g, 水芹菜 80g, 洋葱 20g, 红辣椒 10g, 青辣椒 10g, 平菇 5g, 大葱 5g, 海虹肉 40g, 大酱 15g, 盐 5g, 白糖 5g, 粘米面儿 30g
***调味料:** 辣椒面儿 30g, 汤用酱油 10g, 胡椒粉 0.5g, 低聚糖 15g, 白酒 15g, 大蒜 10g, 生姜 5g, 海虹肉 30g, 虾 30g, 天然条调味料 5g
***肉汤:** 海带 5g, 汤用鳀鱼 3条, 红辣椒 5g, 青辣椒 5g, 大葱 2g
***应用料理材料** : 可利用海鲜材料制作辣味炖锅类料理。

Ingredients and portion(2 portions)
1kg omonkfish, 800g thick bean sprouts, 80g watercress(Minari), 20g onion, 10g red chilli, 10g green chilli, 5g oyster mushroom, 5g leek, 40g mussel, 15g soybean paste, 5g salt, 5g sugar, 30g glutinous rice flour
***Sauce** : 30g chili powder, 10g soy sauce, 0.5g ground black pepper, 15g oligosaccharide, 15g rice-wine, 10g garlic, 5g ginger, 30g mussel, 30g shrimp, 5g natural seasonings
***Stock** : 5g dried sea tangle, 3 anchovies, 5g red chilli, 5g green chilli, 2g leek
***Note** : You can cook spicy jjim dish with seafood ingredients.

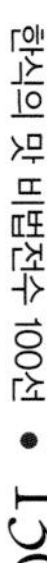

재료 손질 材料处理方法 Clean ingredients

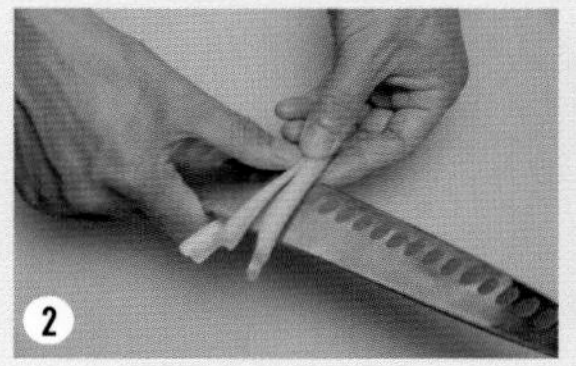

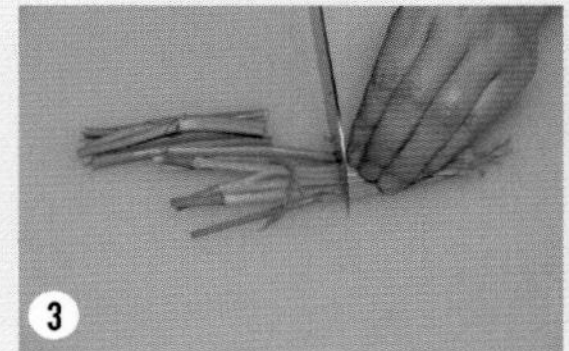

찜양념장 만들기 制作炖锅调味料 How to make jjim marinade

찜용 콩나물 손질 豆芽的处理方法 How to prepare bean sprouts exclusively for jjim recipe

아귀 손질 鮟鱇鱼的处理方法 How to prepare monkfish

만드는 방법

1. 양파 0.3cm로 채 썬다.
2. 대파를 가늘게 채 썬다.
3. 미나리는 길이 6cm로 채 썬다.
4. 고추 0.3cm로 어슷썰기한다.
5. 찜용 콩나물 꼬리만 다듬어 찐다.
6. 찬물에 식힌다.
7. 찬물에 담가 놓는다.
8. 아귀는 된장 15g의 물에 데친다.
9. 고춧가루 30g, 국간장 10g을 넣는다.
10. 후춧가루를 0.5g 넣는다.
11. 올리고당을 15g 넣는다.
12. 청주를 15g 넣는다.

制作方法

1. 将洋葱切成宽0.3cm丝。
2. 将大葱切成细丝。
3. 将水芹菜切成长6cm。
4. 将辣椒切成宽0.3cm。
5. 将豆芽去尾。
6. 放入凉水中冷却。
7. 放在凉水中。
8. 用加15g大酱的水焯一下鮟鱇鱼。
9. 加30g辣椒面儿和10g汤用酱油。
10. 加0.5g胡椒粉。
11. 加15g低聚糖。
12. 加15g白酒。

How to cook

1. Julienne the onion into 0.3cm wide slices.
2. Julienne the leek thin and long.
3. Julienne watercress(minari) into 6cm long slices.
4. Cut the chili into 0.3cm wide pieces diagonally.
5. Remove the ends of the bean sprouts. Boil the bean sprouts.
6. Cool the bean sprouts down in cold water.
7. Leave the bean sprouts in cold water for a while.
8. Blanch the monkfish in water with soybean paste of 15g.
9. Put chili powder of 30g and soy sauce of 10g.
10. Put ground black pepper of 0.5g.
11. Add oligosaccharide of 15g.
12. Pour rice-wine of 15g.

끓이기
煮
How to boil

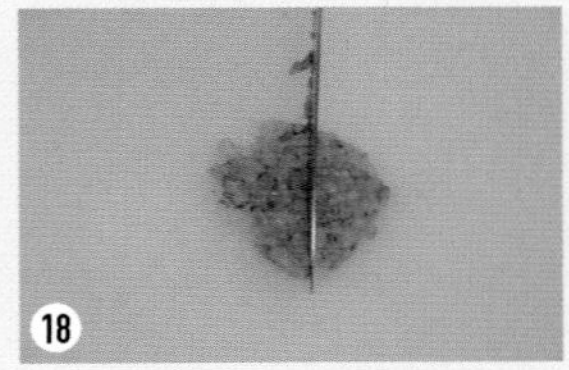

만드는 방법

13. 다진 마늘을 10g 넣는다.
14. 다진 생강을 5g 넣는다.
15. 천연조미료를 5g 넣는다.
16. 홍합수염을 제거한다.
17. 홍합살을 다진다.
18. 새우살을 다진다.
19. 다진 홍합살을 넣는다.
20. 다진 새우살을 넣는다.
21. 양념을 잘 섞는다.
22. 멸치육수 700g에 찜양념장 150g을 넣는다.
23. 데친 아귀를 넣는다.
24. 채소, 홍합살을 넣는다.

制作方法

13. 加10g蒜泥。
14. 加5g捣碎的生姜。
15. 加5g天然调味料。
16. 去除海虹须。
17. 将海虹肉捣碎。
18. 绞碎虾。
19. 加入捣碎的海虹肉。
20. 放入绞碎的虾。
21. 将所有材料混合。
22. 在700g鳀鱼汤中加150g炖锅调味料。
23. 加入焯好的鮟鱇鱼。
24. 加入蔬菜和海虹肉。

How to cook

13. Put chopped garlic of 10g.
14. Put chopped ginger of 5g.
15. Add natural seasonings of 5g.
16. Wash mussel thoroughly.
17. Chop the mussel up.
18. Chop the shrimp up.
19. Put the chopped mussel.
20. Put the chopped shrimp.
21. Mix the ingredients together.
22. Put 150g of jjim sauce to 700g of anchovy stock.
23. Put the blanched monkfish.
24. Put the vegetables and mussel.

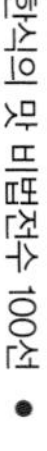

완성하기
完成
How to finish

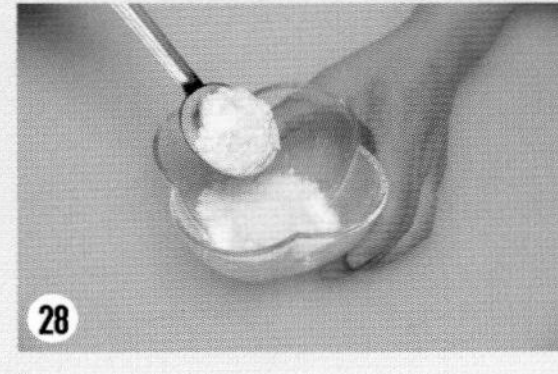

만드는 방법

25. 소금 5g을 넣는다.
26. 설탕 5g을 넣는다.
27. 찜콩나물, 미나리를 넣는다.
28. 찹쌀가루 30g을 넣는다.
29. 물 30g을 넣는다.
30. 농도를 맞춘다.
31. 센 불에서 섞는다.
32. 따뜻하게 담는다.

制作方法

25. 加5g盐。
26. 加5g白糖。
27. 加入豆芽和水芹菜。
28. 加30g粘米面儿。
29. 加30g水。
30. 调整浓度。
31. 用大火搅拌。
32. 盛放在容器里。

How to cook

25. Put salt of 5g.
26. Put sugar of 5g.
27. Add boiled thick bean sprouts and watercress(minari).
28. Put glutinous rice flour of 30g.
29. Pour water of 30g.
30. Adjust the consistency.
31. Mix the ingredients over a high heat.
32. Serve the dish warm.

TIP
建议

7. 콩나물을 아삭하게 하기 위해 찬물에 담가 냉장고에 보관해서 사용한다.
21. 양념장을 만들어 하루 정도 숙성하여 사용하면 더욱 깊은 맛을 낸다.

7. 为了使豆芽香脆，放入凉水中后保管在冰箱里备用。
21. 若用海鲜制成的调味料放置一天后使用的话，味道会更浓厚。

Step 7. Keep the thick bean sprouts cold in water and place them in a refrigerator. This will make the mung bean sprouts crispy.
Step 21. Leave the sauce for another night. That will get a richer flavor to the sauce.

전복갈비찜

鲍鱼炖排骨

Abalone galib jjim

재료 및 분량(3인분)
한우(소찜갈비) 400g, 전복 3마리, 배 20g, 양파 20g, 대파 20g, 대추 3개, 밤 10g, 마늘 5g, 은행 3g, 진간장 100g, 올리고당 50g, 배즙 30g, 양파즙 20g, 청주 20g
***응용요리재료** : 낙지갈비찜, 매운 갈비찜 등의 요리에 응용된다.

材料和份量(3人份)
韩牛(牛排骨) 400g, 鲍鱼 3只, 梨 20g, 洋葱 20g, 大葱 20g, 打造 3个, 栗子 10g, 大蒜 5g, 银杏 3g, 浓酱油 100g, 低聚糖 50g, 梨汁 30g, 洋葱汁 20g, 清酒 20g
***应用料理材料**: 可用于制作八爪鱼炖排骨和辣味炖排骨的菜单。

Ingredients and portion(3 portions)
400g Korea beef(beef ribs), 3 abalones, 20g pear, 20g onion, 20g leek, 3 jujubes, 10g chestnuts, 5g garlic, 3g ginkgo nut, 100g thick soy sauce, 50g oligosaccharide, 30g pear juice, 20g onion juice, 20g rice-wine.
***Note** : You can use the recipe to cook small octopus galib jjim and spicy galbi jjjim.

고기 손질 肉的处理方法 How to prepare the meat

전복 손질 鲍鱼的处理方法 How to prepare abalone

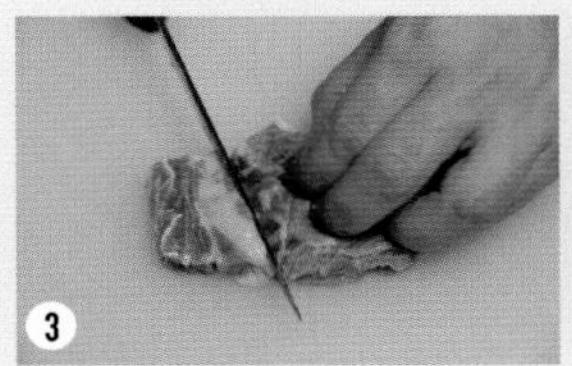

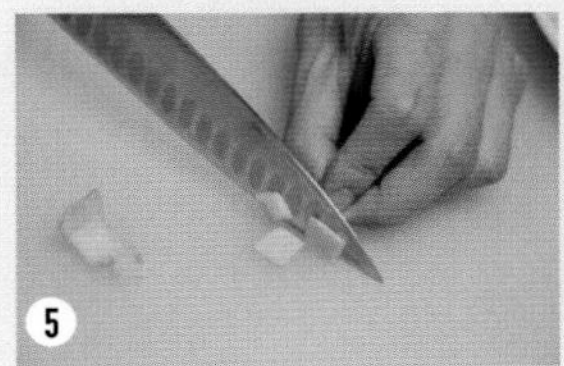

만드는 방법

1. 핏물을 제거한다.
2. 소갈비 막을 제거한다.
3. 기름기를 제거한다.
4. 넓게 펼쳐 준비한다.
5. 생강은 편 썰기한다.
6. 끓는 물에 생강을 넣는다.
7. 소갈비살을 넣는다.
8. 청주를 넣고 데쳐낸다.
9. 소갈비를 건져낸다.
10. 소갈비 준비를 완성한다.
11. 전복을 소금으로 문질러준다.
12. 전복살과 내장을 떼어낸다.

制作方法

1. 去除血水。
2. 去除牛排骨膜。
3. 去除油分。
4. 铺平备用。
5. 将生姜切成片状。
6. 在烧开的水中放入生姜。
7. 放入牛排骨。
8. 焯一下。
9. 取出牛排骨。
10. 牛排骨准备完成。
11. 用盐处理鲍鱼。
12. 取出鲍鱼的肉和内脏。

How to cook

1. Remove the blood.
2. Remove the film of the beef rib(galbi).
3. Remove fatty residue.
4. Spread the meat out.
5. Cut the ginger into slices.
6. Put the ginger slices in boiling water.
7. Put the beef rib(galbi).
8. Blanch the ingredients with rice-wine.
9. Take the beef rib(galbi) out the water.
10. Now, you are all ready to cook galbi.
11. Rub abalones with salt.
12. Separate abalone flesh and internal organs.

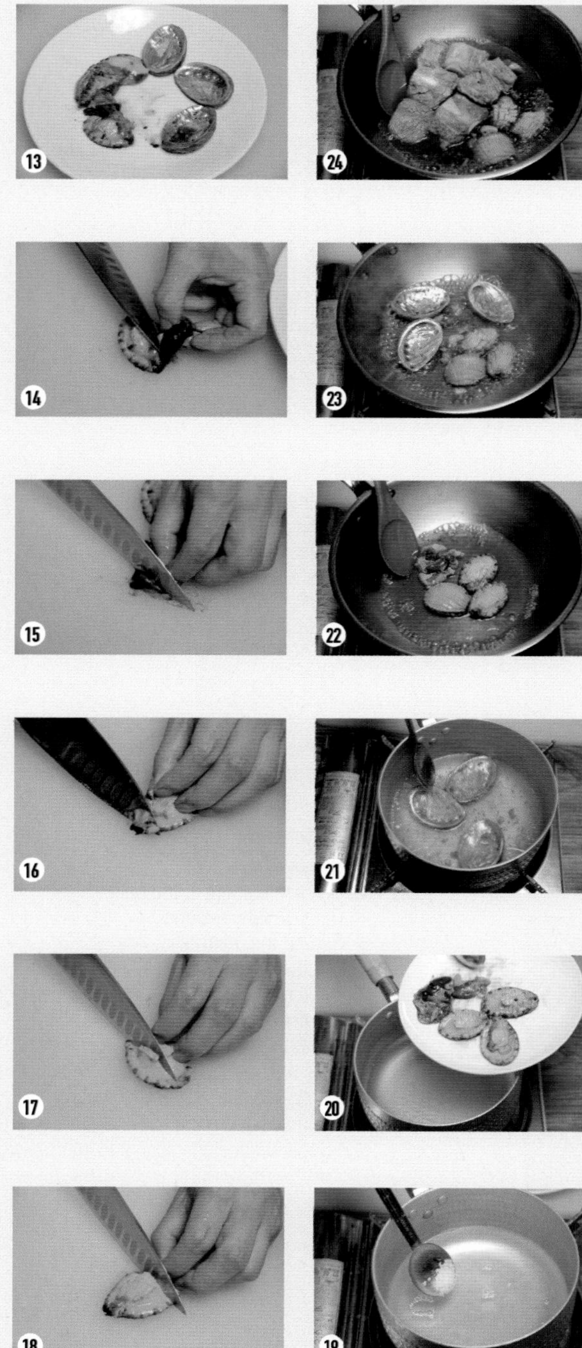

완성하기
完成
How to finish

만드는 방법

13. 껍질과 전복살을 분리한다.
14. 전복 내장을 자른다.
15. 쓸개를 제거한다.
16. 입을 제거한다.
17. 살에 칼집을 넣는다.
18. 0.2cm로 편 썰기한다.
19. 끓는 물에 소금을 넣는다.
20. 데친다.
21. 껍질을 데친다.
22. 진간장과 1:1 비율로 물에 조린다.
23. 전복과 껍질도 조린다.
24. 소갈비와 함께 조린다.

制作方法

13. 将壳和鲍鱼肉分离。
14. 切鲍鱼内脏。
15. 切成小块。
16. 去除嘴部。
17. 在鲍鱼肉上开口。
18. 切成大小为0.2cm的片状。
19. 在烧开的水中加入盐。
20. 焯一下。
21. 焯一下壳。
22. 在酱油与水1:1的混合物中炖。
23. 鲍鱼和壳一起炖。
24. 与牛排骨一起炖。

How to cook

13. Separate the shell from the flesh.
14. Cut the internal organs.
15. Remove a gall bladder.
16. Remove a mouth.
17. Make cuts in the abalones.
18. Cut the abalones into 0.2cm wide slices.
19. Put salt in boiling water.
20. Blanch the abalones.
21. Blanch the shell.
22. Boil the ingredients down with thick soy sauce and water at a ratio of 1:1.
23. Boil the abalones and the shell down.
24. Boil the ingredients down with beef rib(galbi).

만드는 방법

25. 간장양념장에 배즙, 양파즙, 올리고당을 넣는다.
26. 윤기 나게 조려서 완성한다.
27. 완성품을 담는다.

制作方法

25. 在酱油调味料中加入梨汁、洋葱汁和低聚糖。
26. 炖出光泽后即完成。
27. 将制成的料理盛放在容器里。

How to cook

25. Put pear juice, onion juice and oligosaccharide to soy sauce.
26. Boil the ingredients down until they get shiny.
27. Serve the dish on a plate.

TIP
建议

18. 전복의 칼집을 깊고 일정하게 내야 꽃처럼 핀다.
22. 전복은 오래 조리면 크기가 줄어들면서 질겨지므로 갈비를 마무리할 때 첨가하여 완성한다.
간장양념장에 배, 양파를 생략할 수 있다.

18. 只有鲍鱼的开口深度一致才会像花一样绽开。
22. 因为如果鲍鱼炖的时间太久，鲍鱼就会越来越小，口感就会越来越硬，所以应该在排骨炖得差不多的时候加入鲍鱼。
酱油调味汁中可省略梨、洋葱

Step 18. Make sure that you make regular cuts in abalones. This way, your abalone will make like a flower.
Step 22. Abalones will get hard and reduced if you overcook them. Put the abalones when the beef rib(galbi) is nearly cooked.
You might not use pear or onion to soy sauce.

채소달걀찜

蔬菜鸡蛋糕
Vegetable-egg jjim

재료 및 분량(2인분)
달걀 2개, 실파 3g, 표고버섯 1개, 당근 3g, 소금 3g, 뚝배기 1개
***응용요리재료** : 명란젓달걀찜, 꽃게살달걀찜, 새우젓달걀찜 등의 요리에 응용된다.

材料和份量(2人份)
鸡蛋 2个, 小葱 3g, 香菇 1个, 胡萝卜 3g, 盐 3g, 石锅 1个
***应用料理材料**: 可用于制作明太鱼籽鸡蛋糕和虾酱鸡蛋糕等的料理。

Ingredients and portion(2 portions)
2 eggs, 3g green onion, 1 shiitake mushroom(mushroom pyogo), 3g carrot, 3g salt, earthen pot(ttukbaegi)
***Note** : You can use the recipe to cook salted pollack roe-egg jjim, crab-egg jjim and salted shrimp-egg jjim.

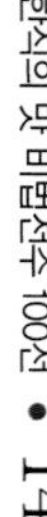

재료 손질 材料处理方法 Clean ingredients

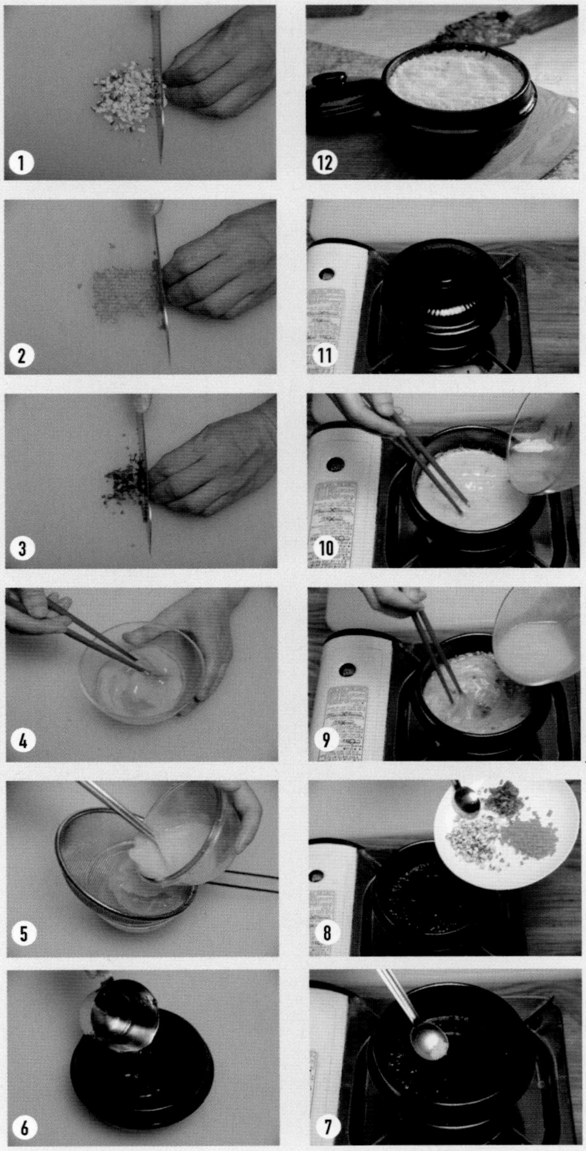

1 2 3 4 5 6 7 8 9 10 11 12

완성하기
完成
How to finish

만드는 방법

1. 표고버섯을 0.3cm로 다진다.
2. 당근을 0.3cm로 다진다.
3. 실파를 0.3cm로 송송 썬다.
4. 달걀을 풀어준다.
5. 체에 내려 끈을 제거한다.
6. 뚝배기에 달걀 2배의 물을 붓는다.
7. 소금 3g으로 간을 맞춘다.
8. 물이 끓으면 다진 재료를 넣는다.
9. 온도 100℃에서 달걀을 풀어준다.
10. 달걀을 혼합한 후 재빨리 저어준다.
11. 센 불에서 뚜껑 덮고 10초 후에 불을 끈다.
12. 완성하여 담는다.

制作方法

1. 将香菇切成大小0.3cm的细丁。
2. 将胡萝卜切成大小0.3cm的细丁。
3. 将小葱切成0.3cm葱沫。
4. 搅拌鸡蛋。
5. 用筛子过滤。
6. 在石锅中加入相当于鸡蛋2倍的水。
7. 加入3g盐。
8. 水烧开后放入切好的材料。
9. 在温度100℃时放入鸡蛋。
10. 待鸡蛋混合后马上快速地搅拌。
11. 盖上盖子用大火煮10秒后熄火。
12. 完成后盛菜。

How to cook

1. Chop shiitake mushroom(mushroom pyogo) into 0.3cm wide pieces.
2. Chop the carrot into 0.3cm wide pieces.
3. Chop the green onion into 0.3cm wide pieces.
4. Whip the eggs.
5. Sieve the eggs. Remove strings.
6. Pour water twice as much as the eggs into an earthen pot(ttukbaegi).
7. Season the ingredients with salt of 3g.
8. When water boil, put the chopped ingredients.
9. Whip the eggs at a temperature of 100℃.
10. Quickly whip the eggs.
11. Boil the ingredients over a high heat with a lid on for 10 seconds. Turn off the heat.
12. Finish.

TIP 建议

10. 달걀을 재빨리 저어주면서 온도를 떨어뜨리지 않고 바로 뚜껑을 덮어 센 불에서 10초 후에 불을 끄고 뜸 들인다. (달걀찜을 많이 부풀게 할 때는 약간의 소다를 사용한다.)

10. 迅速地搅拌鸡蛋，温度不会降低，盖上盖子后用大火煮10秒后熄火，然后再焖一会。(当鸡蛋糕溢出太多时，可使用少量的苏打。)

Step 10. Whip the eggs quickly because you would not want to cool down the heat. Put a lid on immediately and cook the eggs over a high heat for 10 seconds. Turn off the heat and leave them cool down for a while.(Use a pinch of soda to make the egg risen even more.)

해물찜

韩式炖海鲜

Seafood jjim

재료 및 분량(3인분)

꽃게 300g, 아귀 300g, 낙지 1마리, 중하 4마리, 생합 3개, 찜용 콩나물 600g, 미나리 80g, 양파 10g, 풋고추 10g, 홍고추 10g, 느타리버섯 5g, 마늘 5g, 생강 3g, 소금 5g, 찹쌀가루 30g

***찜용 양념장** : 고춧가루 30g, 국간장 10g, 후춧가루 0.5g, 올리고당, 10g, 청주 10g, 마늘 10g, 생강 5g, 천연조미료 5g

***육수** : 다시마 5g, 국멸치 3마리, 느타리버섯 3g, 풋고추 3g, 홍고추 3g, 통마늘 2g, 생강 2g, 대파 3g, 꽃게살 10g, 새우살 8g

***응용요리재료** : 낙지찜, 코다리찜, 대하찜, 꽃게찜 등에 활용되는 요리이다.

材料和份量(3人份)

花蟹 300g, 鮟鱇鱼 300g, 八爪鱼 1条, 中等蛤蜊 4只, 生蛤 3个, 豆芽 800g, 水芹菜 80g, 洋葱 10g, 青辣椒 10g, 红辣椒 10g, 平菇 5g, 大蒜 5g, 生姜 3g, 盐 5g, 粘米面儿 30g

***炖锅用调味料**: 辣椒面儿 30g, 汤用酱油 10g, 胡椒粉 0.5g, 低聚糖 10g, 白酒 10g, 大蒜 10g, 生姜 5g, 天然调味料 5g

***肉汤**: 海带 5g, 汤用鳀鱼 3条, 平菇 3g, 青辣椒 3g, 红辣椒 3g, 大蒜 2g, 生姜 2g, 大葱 3g, 花蟹肉 10g, 虾 8g

***应用料理材料** : 可用于制作炖八爪鱼、炖鳕鱼片、炖大蛤和炖花蟹等的料理。

Ingredients and portion(3 portions)

300g blue crabs, 300g monkfish, 1 small octopus, 4 medium shrimp, 3 mussel, 600g thick bean sprouts, 80g watercress(minari), 10g onion, 10g green chilli, 10g red chilli, 5g oyster mushroom, 5g garlic, 3g ginger, 5g salt, 30g glutinous rice flour

***Sauce for steaming** : 30g chili powder, 10g soy sauce, 0.5g ground black pepper, 10g oligosaccharide, 10g rice-wine, 10g garlic, 5g ginger, 5g natural seasonings

***Stock** : 5g dried sea tangle, 3 anchovies, 3g oyster mushroom, 3g green chilli, 3g red chilli, 2g garlic, 2g ginger, 3g leek, 10g blue crab, 8g shrimp

***Note** : You can use the recipe to cook small octopus jjim, dried pollack jjim, big shrimp jjim and blue crab jjim.

육수 끓이기 煮肉汤 How to make stock

재료 손질 材料处理方法 Clean ingredients

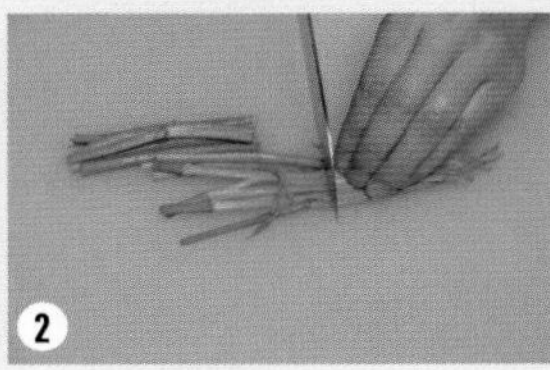

해물 손질 海鲜的处理方法 How to prepare seafood

만드는 방법

1. 육수를 끓인다.
2. 미나리는 길이 6cm로 썬다.
3. 찜용 콩나물은 꼬리만 다듬어 찐다.
4. 찬물에 헹군다.
5. 찬물에 담가 놓는다.
6. 생합을 해감처리한다.
7. 내장을 제거한다.
8. 꽃게를 손질한다.
9. 된장을 풀어 준비한다.
10. 해물을 헹궈 비린내를 제거한다.
11. 낙지는 밀가루로 비린내를 제거한다.
12. 머리 쪽 내장을 제거한다.

制作方法

1. 煮肉汤。
2. 将水芹菜切成长6cm。
3. 将豆芽去尾。
4. 用凉水冲洗。
5. 放在凉水中备用。
6. 对生哈进行淤泥处理。
7. 去除内脏。
8. 对花蟹进行处理。
9. 将大酱用水冲开备用。
10. 用大酱水冲洗海鲜，以去除异味。
11. 用面粉为八爪鱼去除腥味。
12. 去除头部的内脏。

How to cook

1. Boil meat stock.
2. Cut watercress(minari) into 6cm long pieces.
3. Cut the ends of the thick bean sprouts. and boil the thick bean sprouts.
4. Wash the ingredients in cold water.
5. Leave the ingredients in cold water.
6. Wash mussel thoroughly.
7. Remove internal organs.
8. Clean blue crab.
9. Put soybean paste.
10. Wash seafood to reduce fishy smell.
11. Use flour to reduce fishy smell of small octopus.
12. Remove internal organs in the head.

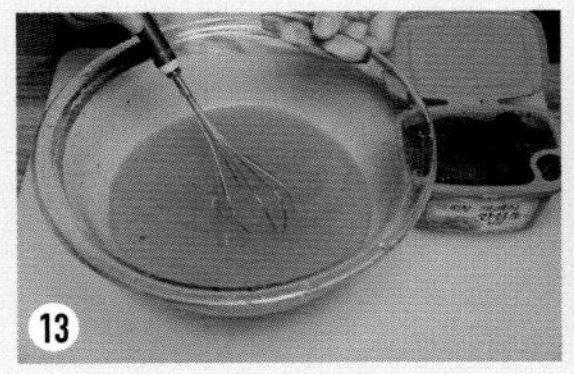

끓이기
煮
How to boil

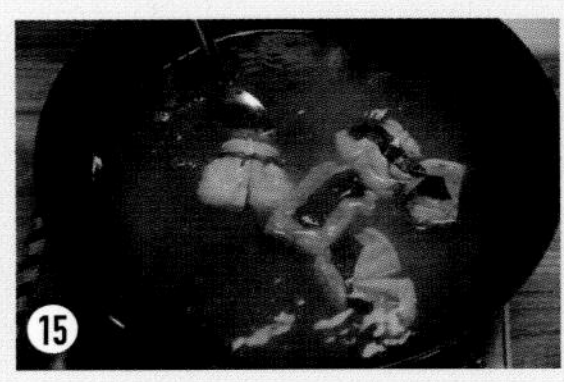

찜양념장 만들기
制作炖锅用调味料
How to make jjim marinade

만드는 방법

13. 된장을 풀어 준비한다.
14. 아귀의 비린내를 제거한다.
15. 끓는 물에 데친다.
16. 고춧가루 30g, 국간장 10g을 넣는다.
17. 후춧가루 0.5g을 넣는다.
18. 올리고당 10g을 넣는다.
19. 청주 10g을 넣는다.
20. 다진 마늘 10g을 넣는다.
21. 다진 생강 5g을 넣는다.
22. 천연조미료 5g을 넣는다.
23. 육수를 붓는다.
24. 양념장 150g을 풀어준다.

制作方法

13. 将大酱用水冲开备用。
14. 去除鮟鱇鱼的腥味。
15. 用烧开的水焯一下。
16. 加30g辣椒面儿和10g汤用酱油。
17. 加0.5g胡椒粉。
18. 加10g低聚糖。
19. 加10g白酒。
20. 加10g蒜泥。
21. 加5g捣碎的生姜。
22. 加5g天然调味料。
23. 倒入肉汤。
24. 调好150g的调味料。

How to cook

13. Put soybean paste.
14. Remove the fishy smell of the monkfish.
15. Blanch the monkfish in boiling water.
16. Put 30g of chili powder and 10g of soy sauce.
17. Put 0.5g of pepper.
18. Add 10g of oligosaccharide
19. Pour 10g of rice-wine.
20. Put 10g of chopped garlic.
21. Put 5g of chopped ginger.
22. Put 5g of natural seasonings.
23. Pour stock.
24. Put 150g of sauce.

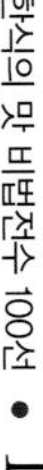

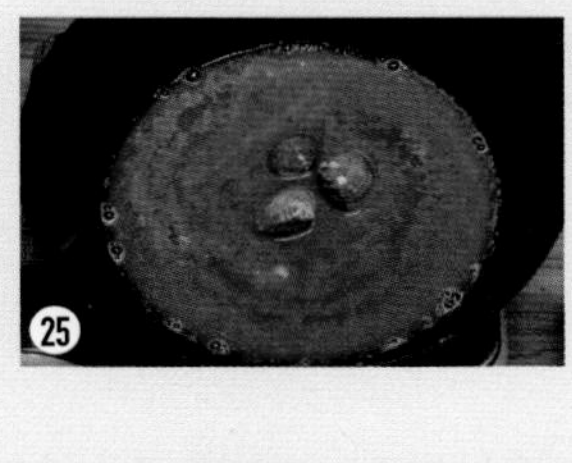

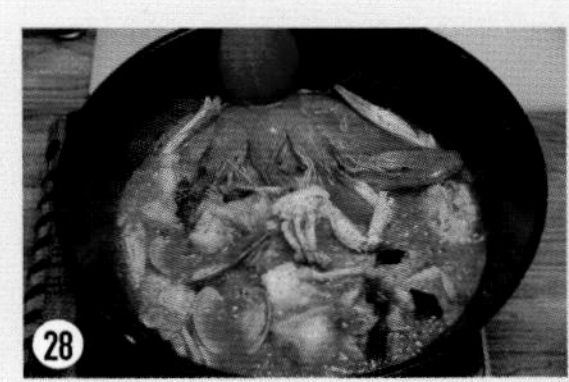

완성하기
完成
How to finish

만드는 방법

25. 생합을 먼저 끓인다.
26. 아귀를 넣는다.
27. 소금 5g을 넣는다.
28. 해물을 넣고 조린다.
29. 낙지, 찜콩나물, 미나리를 넣는다.
30. 찹쌀가루를 풀어준다.
31. 센 불에서 버무려준다.
32. 따뜻하게 담는다.
33. 완성하여 담는다.

制作方法

25. 先煮生哈。
26. 放入鮟鱇鱼。
27. 加5g盐。
28. 放入海鲜后一起炖。
29. 放入八爪鱼、豆芽和水芹菜。
30. 用水冲开粘米面儿。
31. 用大火炖。
32. 盛放在容器里。
33. 完成后盛菜。

How to cook

25. Boil the mussel first.
26. Put the monkfish.
27. Add salt of 5g.
28. Put seafood and boil the ingredients down.
29. Put small octopus, thick bean sprouts and watercress(minari).
30. Put the glutinous rice flour.
31. Mix the ingredients over a high heat.
32. Serve the dish warm.
33. Finish. Serve the dish on a plate.

TIP
建议

3. 찜용 콩나물은 짧은 시간에 쪄야 아삭거린다.
5. 찜용 콩나물은 찬물에 담가 냉장고에 보관하여 사용한다. (하루 정도 지나면 아삭거리는 맛이 떨어진다.)

3. 蒸豆芽的时间短才能保证豆芽的香脆。
5. 将豆芽放在凉水中后保管在冰箱里备用。(过了一天左右，香脆的味道会逐渐消失。)

Step 3. Boil the bean sprouts as quickly as possible because you will lose the crispiness of the bean sprouts.
Step 5. Put the thick bean sprouts in cold water and keep them in a refrigerator (Or the bean sprouts will lose crispiness after a day).

韩食美“味”秘方传授 100选 **实践篇**

Transmission of a Hundred Secrets of Korean Dishes Flavor 饼，油炸类

실기편

한 식 의 맛 비 법 전 수 1 0 0 선

전, 튀김

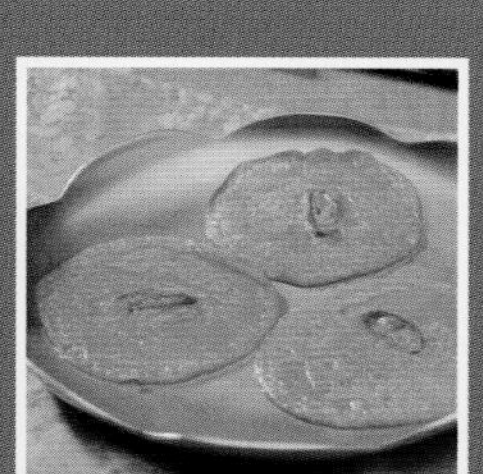

감자전

土豆饼

Potato jeon(pancake)

재료 및 분량(2인분)
감자(1개) 100g, 밀가루(중력분) 150g, 찹쌀가루 30g, 홍고추 10g, 풋고추 10g, 쑥갓 3g, 마늘 15g, 설탕 3g, 소금 3g, 식용유 5g
***응용요리재료** : 단호박전, 부추전, 파프리카전, 채소를 갈아서 색스럽게 만드는 전요리이다.

材料和份量(2人份)
土豆(1个) 100g, 面粉(中力粉) 150g, 粘米面儿 30g, 红辣椒 10g, 青辣椒 10g, 茼蒿 3g, 大蒜 15g, 白糖 3g, 盐 3g, 食用油 5g
***应用料理材料**: 将南瓜、韭菜、大辣椒和蔬菜搅碎后制成的饼料理。

Ingredients and portion(2 portions)
100g potato, 150g medium flour, 30g glutinous rice flour, 10g red chilli, 10g green chilli, 3g edible chrysanthemum(ssukgot), 15g garlic, 3g sugar, 3g salt, 5g salad oil
***Note** : You can also cook colorful jeon with a sweet pumpkin, chives, paprika and other vegetable.

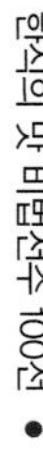

재료 손질 材料处理方法 Clean ingredients

완성하기 完成 How to finish

만드는 방법

1. 감자를 넣는다.
2. 마늘 15g을 넣는다.
3. 설탕 3g을 넣는다.
4. 찹쌀가루 20g을 넣는다.
5. 소금 3g을 넣는다.
6. 밀가루 150g을 넣는다.
7. 물을 80~100ml 넣는다.
8. 곱게 갈아준다.
9. 팬에 반죽을 붓는다.
10. 어슷썬 홍, 풋고추 고명을 올린다.
11. 완전히 익으면 쑥갓을 올린다.
12. 뒤집어서 불을 끄고 익힌다.

制作方法

1. 放入土豆。
2. 放入15g大蒜。
3. 放入3g白糖。
4. 放20g糯米粉。
5. 放入3g盐。
6. 放入150g面粉。
7. 加入80~100ml的水。
8. 均匀地搅碎。
9. 在平底锅上放上和好的面。
10. 放上切好的红、青辣椒和装饰菜。
11. 待完全熟透后放上茼蒿。
12. 翻面熄火后用余温烙熟。

How to cook

1. Put a potato.
2. Put garlic of 15g.
3. Put sugar of 3g.
4. Put glutinous rice flour of 20g.
5. Put salt of 3g.
6. Add flour of 150g.
7. Pour water of 80ml~100ml .
8. Grind the ingredients finely.
9. Put the dough onto a pan.
10. Cut red chilli and green chilli diagonally. Garnish the dish with the chili.
11. When the dough is fully cooked, put edible chrysanthemum(ssukgot) on top.
12. Flip the dough. Turn off the heat. Wait until the jeon is fully cooked.

TIP 建议

3. 설탕을 소량 사용하여 감자갈변을 막아준다.

3. 使用少量白糖可以防止土豆变成褐色。

Step 3. Use a pinch of sugar to prevent the potato from turning brown.

김치전

泡菜饼
Kimchi jeon

재료 및 분량(2인분)
김치 400g, 밀가루(중력분) 300g, 양파 30g, 실파 20g, 고추장 10g, 된장 20g, 찹쌀가루 30g, 올리브유 10g
***응용요리재료** : 장떡, 메밀김치전, 굴김치전, 두부참치김치전 등으로 활용할 수 있는 요리이다.

材料和份量(2人份)
泡菜 400g, 面粉(中力粉) 300g, 洋葱 30g, 小葱 20g, 辣椒酱 10g, 大酱 20g, 糯米粉 30g, 橄榄油 10g
***应用料理材料**: 可制作酱煎饼、荞麦泡菜饼、海蛎子泡菜饼和豆腐金枪鱼泡菜饼的料理。

Ingredients and portion(2 portions)
500g Kimchi, 300g medium flour, 30g onion, 20g spring onion, 20g pepper red paste(gochujang), 30g soybean paste, 30g glutinous rice flour, 10g olive oil
***Note** : You can use the recipe to cook jangtteok, buckwheat-kimchi jeon and tofu-tuna-Kimchi jeon.

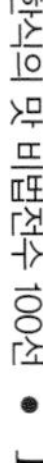

재료 손질 材料处理方法 Clean ingredients

완성하기
完成
How to finish

만드는 방법

1. 양파는 0.3cm로 채 썬다.
2. 실파는 길이 5cm로 썬다.
3. 물에 헹군 김치는 1cm로 채 썬다.
4. 밀가루 200g에 고추장 10g, 된장 20g을 넣는다.
5. 물을 150ml 넣고 풀어준다.
6. 양파, 실파를 넣어준다.
7. 김치를 넣는다.
8. 골고루 섞는다.
9. 올리브유를 두른다.
10. 반죽을 얇게 펼친다.
11. 뒤집어 익힌다.
12. 완성한다.

制作方法

1. 将洋葱切成宽0.3cm的丝。
2. 将小葱切成长5cm。
3. 用水冲洗泡菜后将其切成大小1cm的条。
4. 在面粉200g中加10g辣椒酱和20g大酱。
5. 加入150ml水后冲开。
6. 放入洋葱和小葱。
7. 放入泡菜。
8. 搅拌均匀。
9. 放入橄榄油。
10. 将和好的面薄薄地铺在平底锅上。
11. 翻面烙熟。
12. 完成。

How to cook

1. Slice the onion into 0.3cm.
2. Cut the green onion into 5cm long pieces.
3. Wash Kimchi in water. Julienne Kimchi into 1cm wide slices.
4. Put red pepper paste(gochujang) of 10g, soybean paste of 20g and flour of 200g.
5. Pour water of 150ml. Stir the ingredients.
6. Put onion and green onion.
7. Put Kimchi.
8. Mix the ingredients well.
9. Put olive oil into a pan.
10. Spread the dough thin.
11. Flip the dough. Get the dough turn brown.
12. Finish.

TIP
建议

3. 김치전은 신김치로 전을 만들어야 감칠맛이 살아난다.
5. 반죽을 계속 저어주면 글루텐이 나와 탄력이 생긴다.

3. 泡菜饼只有用新泡菜做才有泡菜味儿。
5. 若持续地和面的话，面中会出现面筋，从而面会更加劲道。

Step 3. Use over-fermented Kimchi for Kimchi jeon to add a savorier taste to the jeon.
Step 5. Keep stirring the dough. The gluten will increase the flexibility of the dough.

동태전

明太鱼饼

Pollack jeon

재료 및 분량(2인분)
동태(살) 300g, 달걀 2개, 밀가루(박력분) 150g, 홍고추 10g, 풋고추 10g, 소금 5g, 후춧가루 0.5g, 식용유 10g
***응용요리재료** : 대구전, 도미전, 생선전류 등으로 다양하게 활용할 수 있는 요리이다.

材料和份量(2人份)
明太鱼(肉) 300g, 鸡蛋 2个, 面粉(薄力粉) 150g, 红辣椒 10g, 青辣椒 10g, 盐 5g, 胡椒粉 0.5g, 食用油 10g
***应用料理材料**: 可制作鳕鱼饼和各种鱼饼等的料理。

Ingredients and portion(2 portions)
300g frozen pollack fish, 2 eggs, 150g flour(weak flour), 10g red chilli, 10g green chilli, 5g salt, 0.5g ground black pepper, 10g salad oil
***Note** : You can use cod, snapper or ther fish to the recipe to cook different jeons.

만드는 방법

1. 홍, 풋고추를 0.2cm로 어슷썰기한다.
2. 씨를 제거한다.
3. 동태살에 소금, 후춧가루를 뿌린다.
4. 밀가루를 입힌다.
5. 노른자옷을 입힌다.
6. 식용유를 두른다.
7. 코팅처리한다.
8. 동태전을 익힌다.
9. 약불에서 익힌다.
10. 고명을 올린다.
11. 앞 · 뒤 골고루 익힌다.
12. 완성하여 담는다.

制作方法

1. 将红、青辣椒切成大小0.2cm。
2. 去除籽。
3. 在明太鱼肉上撒上盐和胡椒粉。
4. 用面粉包住。
5. 用鸡蛋黄包住。
6. 放入食用油。
7. 用油煎出一层外皮。
8. 使明太鱼饼烙熟。
9. 用微火烙熟。
10. 放上装饰菜。
11. 前后均匀烙熟。
12. 完成后盛菜。

How to cook

1. Cut red chilli and green chilli into 0.2cm wide pieces diagonally.
2. Remove seeds.
3. Sprinkle salt and ground black pepper the pollack fish.
4. Coat the fish with flour.
5. Coat the fish with the egg yolk.
6. Put salad oil.
7. Coat the pan with the salad oil.
8. Cook pollack jeon.
9. Cook the jeon over a low heat.
10. Put a garnish.
11. Make sure that you have the jeon turned brown on both sides.
12. Finish.

TIP
建议

3. 동태의 수분을 완전히 제거해야 달걀옷이 안 떨어진다.

3. 只有完全控除明太鱼的水分后才能使外层的鸡蛋黄不会脱落。

Step 3. Remove water completely from the pollack before you coat the fish with the egg yolk.

두부튀김

炸豆腐

Fried tofu

재료 및 분량(2인분)
두부 150g, 전분가루 30g, 튀김기름 200g
***튀김간장 소스** : 멸치육수 200g, 무즙 30g, 생강즙 5g, 진간장 8g, 고춧가루 10g, 실파 3g, 레몬 3g
***멸치육수** : 다시마 5g, 국멸치 5마리, 새송이버섯 5g, 표고버섯 5g, 느타리버섯 5g, 팽이버섯 5g, 대파 3g

材料和份量(2人份)
豆腐 150g, 淀粉 30g, 油炸粉 200g
***炸类料理用酱油调味料** : 鳀鱼汤 200g, 萝卜汁 30g, 生姜汁 5g, 浓酱油 8g, 辣鸡面儿 10g, 小葱 3g, 柠檬 3g
***鳀鱼汤** : 海带 5g, 汤用鳀鱼 5条, 杏鲍菇 5g, 香菇 5g, 平菇 5g, 金针蘑 5g, 大葱 3g

Ingredients and portion(2 portions)
150g bean curd(tofu), 30g starch flour, 200g frying oil
***Frying soy sauce** : 200g anchovy stock, 20g white radish juice, 5g ginger juice, 8g thick soy sauce, 10g chili powder, 3g green onion, 3g lemon
***Anchovy stock** : 5g dried sea tangle, 5 anchovies, 5g king oyster mushroom, 5g shiitake mushroom(mushroom pyogo), 5g oyster mushroom, 5g mushroom winter, 3g leek

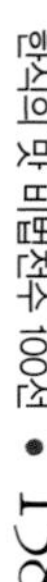

만드는 방법

1. 두부는 도톰하게 썬다.
2. 전분가루를 뿌린다.
3. 160℃에 튀긴다. 방울토마토를 살짝 튀긴다.
4. 튀김두부를 담는다.
5. 멸치육수를 200g 붓는다.
6. 무를 갈아준다.
7. 무즙을 30g 넣는다.
8. 생강즙을 5g 넣는다.
9. 진간장을 8g 넣는다.
10. 고춧가루를 10g 넣는다.
11. 송송 썬 실파와 레몬을 띄운다.
12. 완성하여 담는다.

制作方法

1. 将豆腐切成大块。
2. 撒上淀粉。
3. 在160℃温度下油炸。将小西红柿也稍微炸一下。
4. 放入油炸豆腐。
5. 倒入200g鳀鱼汤。
6. 将萝卜碾碎。
7. 加30g萝卜汁。
8. 加5g生姜汁。
9. 加8g浓酱油。
10. 加10g辣椒面儿。
11. 放上切成丝细的小葱和柠檬。
12. 完成后盛菜。

How to cook

1. Cut bean curd(tofu) into chunks.
2. Sprinkle glutinous rice flour over the bean curd(tofu).
3. Fry the bean curd(tofu) at a temperature of 160℃. Stir-fry bell cherry tomato.
4. Put the fried tofu on a plate.
5. Pour anchovy stock of 200g.
6. Grind the white radish.
7. Pour radish juice of 30g.
8. Put ginger juice of 5g.
9. Add thick soy sauce of 8g.
10. Put chili powder of 10g.
11. Put chopped green onion and lemon slices.
12. Finish.

TIP
建议

2. 두부는 수분이 많으므로 전분가루를 뿌려서 10분 정도 준비한다.
3. 바삭하게 두 번 튀겨준다.

2. 因为豆腐水分较多，所以先撒上淀粉，放置10分钟左右备用。
3. 为使炸出来的豆腐香脆，建议炸两遍。

Step 2. Tofu is moist. So sprinkle flour over tofu and leave the tofu for 10 minutes.
Step 3. Fry twice so that the tofu will get crispier.

모둠채소튀김

韩式炸蔬菜

Fried vegetables

재료 및 분량(3인분)
고구마 60g, 가지 60g, 양파 80g, 깻잎 2장, 풋고추 2개, 수삼 40g, 표고버섯 2개, 새송이버섯 60g, 느타리버섯 40g, 팽이버섯 50g, 브로콜리 40g, 달걀 1개, 박력분 400g, 얼음 50g
***튀김간장소스** : 실파 10g, 진간장 30g, 고춧가루 10g, 무 10g, 레몬 5g, 생강 3g
***응용요리재료** : 단호박, 더덕, 도라지 등의 다양한 채소류에 이용된다.

材料和份量(3人份)
地瓜 60g, 茄子 60g, 洋葱 80g, 苏子叶 2张, 青辣椒 2个, 水参 40g, 香菇 2个, 杏鲍菇 60g, 平菇 40g, 金针蘑 50g, 西兰花 40g, 鸡蛋 1个, 薄力粉 400g, 冰块 50g
***油炸类料理用酱油调料** : 小葱 10g, 浓酱油 30g, 辣椒粉 10g, 萝卜 10g, 柠檬 5g, 生姜 3g
***应用料理材料** : 用于南瓜, 沙参, 桔梗等蔬菜类。

Ingredients and portion(3 portions)
60g sweet potato, 60g eggplant, 80g onion, 2 sesame leaves, 2 green chilli, 40g fresh ginseng, 2 shiitake mushroom(mushroom pyogo), 60g king oyster mushroom, 40g oyster mushroom, 50g mushroom winter, 40g broccoli, 1 egg, 400g weak flour, 50g ice cube
***Frying thick soy sauce** : 10g green onion, 30g thick soy sauce, 10g chili power, 10g white radish, 5g lemon, 3g ginger
***Note** : You can add sweet pumpkin, todok(Codonopsis lanceolata) and bellflower roots.

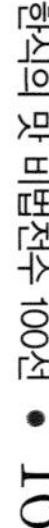

수삼 손질 水参的处理方法 How to prepare fresh ginseng

1 12 2 11 3 10

버섯 손질 蘑菇的处理方法 How to prepare mushroom

4 9 5 8 6 7

채소 손질 蔬菜的处理方法 How to prepare vegetables

만드는 방법

1. 수삼은 머리를 제거한다.
2. 2등분하여 손질한다.
3. 칼집을 넣는다.
4. 팽이버섯을 찢는다.
5. 느타리버섯을 찢는다.
6. 새송이버섯은 편 썬다.
7. 표고버섯의 칼집을 낸다.
8. 양파는 1cm로 편 썬다.
9. 양파는 고정시킨다.
10. 고구마를 손질한다.
11. 0.5cm로 편 썬다.
12. 풋고추를 2등분한다.

制作方法

1. 切掉水参的头。
2. 将水参切成2等分。
3. 切口子。
4. 将金针蘑撕好备用。
5. 将平菇撕好备用。
6. 将杏鲍菇切成片状。
7. 将香菇用到刻出纹样。
8. 将洋葱切成大小为1cm的片状。
9. 将洋葱固定。
10. 处理地瓜。
11. 将地瓜切成大小为0.5cm的片状。
12. 将青辣椒2等分。

How to cook

1. Remove the head of the fresh ginseng.
2. Cut the fresh ginseng in half.
3. Make cuts in the fresh ginseng.
4. Tear mushroom winter.
5. Tear oyster mushroom.
6. Slice king oyster mushroom.
7. Make cuts in shiitake mushroom(mushroom pyogo).
8. Slice the onion into 1cm wide pieces.
9. Have the onion fixed.
10. Clean sweet potato.
11. Cut the sweet potato into 0.5cm wide pieces.
12. Cut green chili in half.

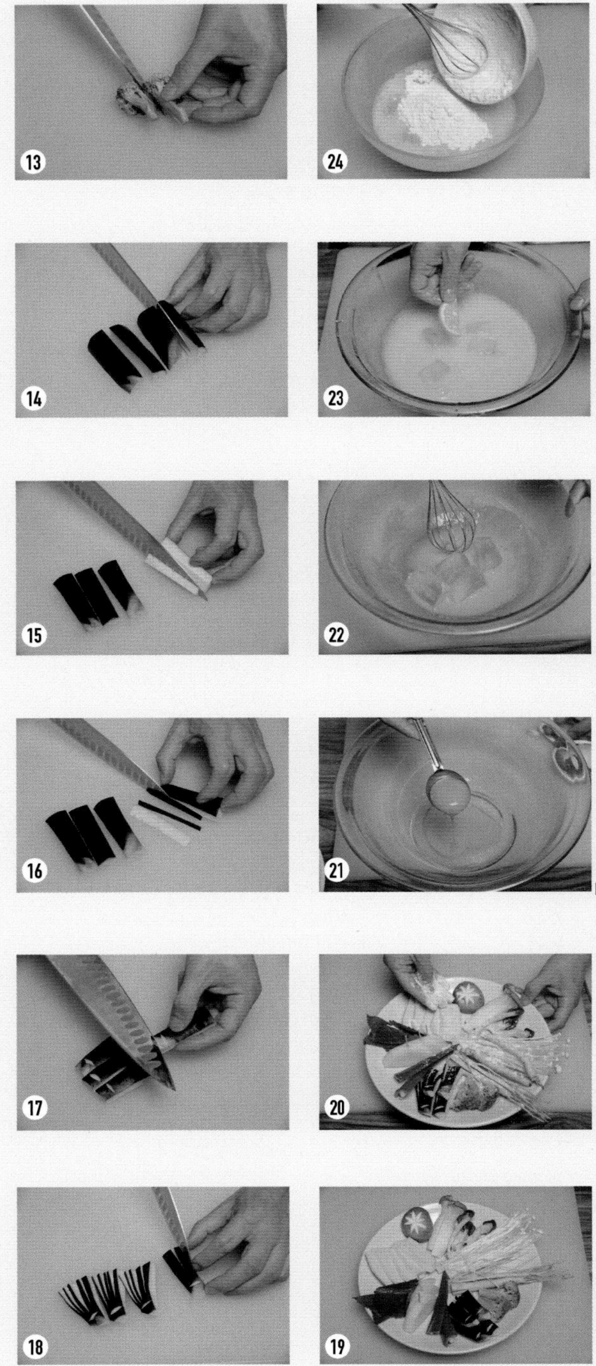

튀김옷 만들기
制作油炸外层
How to make frying cover

만드는 방법

13. 브로콜리를 2등분한다.
14. 가지를 4등분한다.
15. 가지를 스틱모양으로 썬다.
16. 모서리를 제거한다.
17. 칼집을 낸다.
18. 부채모양을 잡는다.
19. 모든 재료를 손질한다.
20. 박력분을 덧가루로 뿌린다.
21. 달걀 노른자를 2개 넣는다.
22. 얼음을 넣는다.
23. 찬물 1L와 레몬을 넣는다.
24. 박력분을 넣는다.

制作方法

13. 将西兰花2等分。
14. 将茄子4等分。
15. 将茄子切成杆状。
16. 去除边角。
17. 用刀刻出模样。
18. 刻出扇子形。
19. 处理所有材料。
20. 撒上薄力粉。
21. 放入2个蛋黄。
22. 放入冰块。
23. 放入凉水1L和柠檬。
24. 放入薄力粉。

How to cook

13. Cut the broccoli in half.
14. Cut the eggplant in quarters.
15. Cut the eggplant into a stick-shape.
16. Remove the corners.
17. Make cuts.
18. Arrange the eggplants in a fan shape.
19. Clean all the other ingredients.
20. Sprinkle weak flour.
21. Put two egg yolks.
22. Put ice cubes.
23. Put cold water of 1L and a lemon.
24. Put weak flour.

튀기기
油炸
How to fry

만드는 방법

25. 섞어준다.
26. 반죽을 완성한다.
27. 160℃에서 튀긴다.
28. 온도를 유지하면서 튀긴다.
29. 바삭하게 튀겨낸다.
30. 여유옷을 떨어뜨린다.
31. 두 번씩 튀겨낸다.
32. 완성한다.

制作方法

25. 混合搅拌。
26. 油炸用面糊完成。
27. 在160℃高温下油炸。
28. 油炸时保持温度。
29. 使油炸料理口味香脆。
30. 去掉多余的外层。
31. 炸两次。
32. 用纸去除油分。

How to cook

25. Mix the ingredients all together.
26. Make dough.
27. Fry the ingredients at a temperature of 160℃.
28. Keep the temperature.
29. Fry the ingredients until they get crispy.
30. Drop the frying cover remains.
31. Fry twice.
32. Put the ingredients on a sheet of paper to remove oil.

TIP
建议

튀김반죽은 차가운 얼음물과 밀가루(박력분)를 1:1 비율로 섞고, 튀김온도는 180℃에서 바삭하게 튀겨낸다.

冰水和面粉（薄力粉）以1:1的比率进行搅拌，搅拌至无颗粒状面糊。沾好面糊的蔬菜放进在180℃的油锅里进行油炸。

For the dough, mix flour(weak flour) with icy-cold water at a ratio of 1:1. Fry the ingredients at a temperature of 180℃ until they get crispy.

부추전

韭菜饼
Chives jeon

재료 및 분량(2인분)
부추 30g, 밀가루(부침가루) 200g, 찹쌀가루 10g, 마늘 5g, 소금 2g, 식용유 10g

材料和份量(2人份)
韭菜 30g, 面粉(油炸粉) 200g, 粘米面儿 10g, 大蒜 5g, 盐 2g, 食用油 10g

Ingredients and portion(2 portions)
30g chives, 200g flour(wheat flour), 10g glutinous rice flour, 5g garlic, 2g salt, 10g salad oil

반죽 만들기 和面 How to knead the dough

완성하기
完成
How to finish

만드는 방법

1. 부추는 길이 5cm로 썬다.
2. 부추를 믹서기에 넣는다.
3. 마늘 5g을 넣는다.
4. 찹쌀가루 10g을 넣는다.
5. 부침가루 200g을 넣는다.
6. 물 130~150g을 넣는다.
7. 소금 2g을 넣는다.
8. 믹서에서 곱게 갈아준다.
9. 부추 색을 살린다.
10. 팬에 붓는다.
11. 일정한 크기로 조절한다.
12. 고명을 올려 완성한다.

制作方法

1. 将韭菜切成大小5cm。
2. 将韭菜放入搅拌机里。
3. 放入5g大蒜。
4. 加10g糯米粉。
5. 放入200g油炸粉。
6. 加130~150g水。
7. 加2g盐。
8. 用搅拌机绞碎。
9. 保持韭菜的颜色。
10. 倒入平底锅中。
11. 调整大小。
12. 放上装饰菜完成。

How to cook

1. Cut chives into 5cm long pieces.
2. Put chives in a blender.
3. Put garlic of 5g.
4. Put glutinous rice flour of 10g.
5. Put flour(wheat flour) of 200g.
6. Pour water of 130~150g.
7. Add salt of 2g.
8. Grind the ingredients finely in a blender.
9. Keep the color of the chives.
10. Pour the dough onto a pan.
11. Make the dough even in size.
12. Garnish the dish.

TIP 建议

5. 부침가루와 부추를 믹서기에 함께 갈면 혼합이 잘되어 색깔이 곱다.

5. 油炸粉和韭菜一起放在搅拌机里绞碎可以使其更好地混合，切颜色均匀美观。

Step 5. Grind the flour(wheat flour) and chives together in the blender to give a beautiful color to the jeon.

새우튀김

炸虾

Fried shrimp

재료 및 분량(2인분)
새우 13마리, 밀가루(박력분) 200g, 가지 10g, 무 10g, 깻잎 2장, 대파 5g, 실파 5g, 달걀 1개, 진간장 30g, 고춧가루 10g, 생강 3g, 레몬 3g
***응용요리재료** : 오징어튀김, 빙어튀김, 미꾸라지튀김 등에 활용할 수 있는 요리이다.

材料和份量(2人份)
虾 13条, 面粉(薄力粉) 200g, 茄子 10g, 萝卜 10g, 苏子叶 2张, 大葱 5g, 小葱 5g, 鸡蛋 1个, 浓酱油 30g, 辣椒面儿 10g, 生姜 3g, 柠檬 3g
***应用料理材料**：可用于制作炸鱿鱼、炸胡瓜鱼和炸泥鳅鱼等的料理。

Ingredients and portion(2 portions)
13 shrimps, 200g flour(weak flour), 10g eggplant, 10g white radish, 2 sesame leaves, 5g leek, 1 egg, 30g thick soy sauce, 10g chili powder, 3g ginger, 3g lemon
***Note** : You can fry squid, smelt and loach.

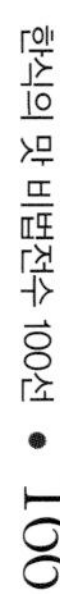

새우 손질 虾的处理方法 How to prepare shrimp

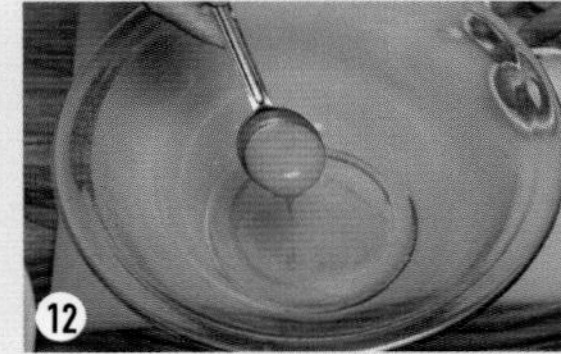

튀김옷 만들기 制作油炸外层 How to make frying cover

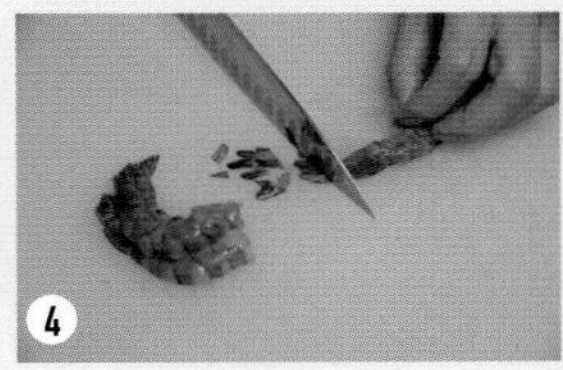

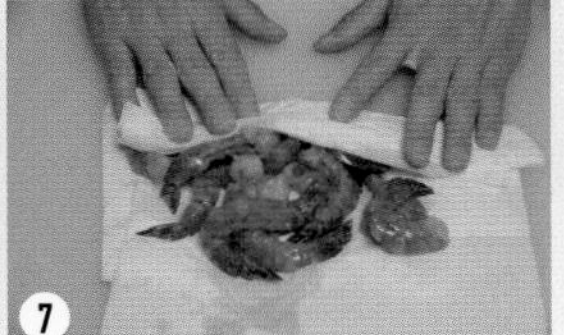

만드는 방법

1. 새우 머리를 손질한다.
2. 껍질을 제거한다.
3. 내장을 제거한다.
4. 꼬리를 다듬는다.
5. 소금으로 씻는다.
6. 찬물에 헹군다.
7. 물기를 제거한다.
8. 안쪽에 칼집을 넣는다.
9. 물기를 제거한다.
10. 새우 휘지 않게 손질
11. 밀가루를 입힌다.
12. 달걀 노른자를 2개 넣는다.

制作方法

1. 处理虾头。
2. 去除虾皮。
3. 去除内脏。
4. 处理虾尾。
5. 用盐水冲洗。
6. 用凉水冲洗。
7. 控除水份。
8. 在虾的内存开口。
9. 去除水分。
10. 处理虾，注意不要使虾弯曲。
11. 涂上面粉。
12. 放入2个鸡蛋黄。

How to cook

1. Cut the heads of shrimps.
2. Remove the shells.
3. Remove the internal organs.
4. Cut the tails.
5. Wash them with salt.
6. Wash them in cold water.
7. Drain well.
8. Make cuts inside the shrimp.
9. Drain well.
10. Keep the shrimp not to get curled up.
11. Cover the shrimps with flour.
12. Put two yolks.

퇴기기
油炸
How to fry

만드는 방법

13. 풀어준다.
14. 얼음을 넣는다.
15. 물 500L와 레몬을 넣는다.
16. 박력분을 넣는다.
17. 섞어준다.
18. 튀김옷을 완성한다.
19. 160℃의 온도에 튀긴다.
20. 튀김옷을 입혀 튀긴다.
21. 일정한 크기로 튀긴다.
22. 새우를 위, 아래로 움직이면서 튀긴다.
23. 기름 속에서 위, 아래로 움직여서 튀긴다.
24. 일정한 온도를 유지한다.

制作方法

13. 搅拌。
14. 放入冰块。
15. 倒入500L水和柠檬。
16. 放入薄力粉。
17. 混合搅拌。
18. 油炸外层制作完成。
19. 在160℃温度下油炸。
20. 涂上油炸外层后再炸。
21. 炸成一定的大小。
22. 将虾上下移动着炸。
23. 在油里上下移动着炸。
24. 保持一定的温度。

How to cook

13. Whip them.
14. Put ice cubes.
15. Put 500g of water and a lemon.
16. Add weak flour.
17. Mix them well.
18. You have the batter completed.
19. Fry the shrimps at 160℃.
20. Cover the shrimps in the batter.
21. Get the shrimps fried by even size.
22. Keep moving the shrimps up and down while frying them.
23. Move the shrimps up and down in oil.
24. Keep the temperature.

완성하기
完成
How to finish

만드는 방법

25. 기름이 튀지 않도록 주의한다.
26. 여유분 튀김옷을 넣어준다.
27. 튀김옷을 건져내고 튀김을 한다.
28. 여유분 튀김옷으로 꽃이 피게 한다.
29. 속까지 익힌다.
30. 노릇하게 튀긴 후 기름기를 제거한다.
31. 종이 위에 담는다.
32. 완성품을 담는다.

制作方法

25. 注意不要溅到油。
26. 使用剩余的油炸外层。
27. 炸2次。
28. 使剩余的油炸外层炸开花。
29. 使油炸料理的颜色呈现金黄色。
30. 炸熟后去掉油份。
31. 放在纸上。
32. 盛好完成的菜。

How to cook

25. Watch out the heated oil.
26. Fry the rest of the batter.
27. Take the pieces of batter out of the oil.
28. Fry the rest of the batter and make them look the shape of flowers.
29. Cook well.
30. Fry the shrimps until golden brown and drain the oil.
31. Put the shrimps on a sheet of paper.
32. Serve the dish well.

TIP
建议

8. 새우칼집을 안쪽으로 깊숙이 넣어야 새우등이 휘지 않는다.
14. 얼음을 이용하여 튀김의 바삭함을 살린다.
26. 여유분의 튀김옷을 떨어뜨려서 튀김옷의 두께와 바삭한 식감을 높인다.

8. 只有使虾的切口深入内侧，虾背才不会弯曲。
14. 利用冰块使油炸料理香脆。
26. 使用剩余的油炸外层使油炸料理的外层更厚和更香脆。

Step 8. Make a deep cut inside the shrimp so that the shrimp will not get curled up.
Step 14. Increase the crispiness of the fried shrimps with ice cubes.
Step 26. Drop the rest of the batter to increase the thickness and the crispiness of the fried shrimps.

韩食美"味"秘方传授 100选 实践篇

Transmission of a Hundred Secrets of Korean Dishes Flavor 碳烤类

한 식 의 맛 비 법 전 수 100 선

실기편

구이

고등어간장구이

醬汁烤鲅鱼

Mackerel roasted with soy sauce

재료 및 분량(2인분)
고등어 1마리, 레몬 2g, 진간장 15g, 청주 5g, 생강 5g, 후춧가루 0.5g
***육수** : 다시마 5g, 생강 3g, 대파 3g, 통후추 3g, 청주 10g
***응용요리재료** : 등푸른생선류와 흰살생선류 등에 다양하게 이용하여 구이를 할 수 있다.

材料和份量(2人份)
青花鱼 1条, 柠檬 2g, 浓酱油 15g, 白酒 5g, 生姜 5g, 胡椒粉 0.5g
***肉汤** : 海带 5g, 生姜 3g, 大葱 3g, 整个胡椒 3g, 白酒 10g
***应用料理材料** : 可用于烤背部蓝色的海鲜类和肉质白嫩的海鲜类等, 可多样的使用。

Ingredients and portion(2 portions)
1 mackerel, 2g lemon, 15g soy sauce, 5g clean rice-wine, 5g ginger, 0.5g ground black pepper
***Meat stock** : 5g dried sea tangle, 3g ginger, 3g leek, 3g whole black pepper, 10g clean rice-wine
***Note** : You can also use external blue colored fish and white fish.

고등어 손질 青花鱼的处理方法 How to prepare mackerel

간장소스 만들기 制作酱油调味料 How to make soy sauce

완성하기 完成 How to finish

만드는 방법

1. 고등어 뼈를 제거한다.
2. 내장 부위를 제거한다.
3. 길이 5cm로 토막낸다.
4. 껍질 쪽에 칼집을 넣는다.
5. 생강 5g, 청주 5g, 후춧가루 0.5g에 재운다.
6. 육수를 끓인다.
7. 육수 500g에 진간장 15g을 넣는다.
8. 청주 10g을 넣는다.
9. 레몬과 고등어를 넣는다.
10. 석쇠에서 직화구이한다.

11~12. 완성품을 담는다.

制作方法

1. 去除青花鱼骨。
2. 去除内脏部位。
3. 将青花鱼切成大小为5cm的块状。
4. 在鱼皮侧开口。
5. 放入5g生姜，5g白酒和0.5g胡椒粉调味腌制。
6. 煮肉汤。
7. 在肉汤500g中放入15g酱油。
8. 放入10g白酒。
9. 放入柠檬和青花鱼。
10. 用烤网直烤。

11~12. 盛菜。

How to cook

1. Remove bones of the mackerel.
2. Remove internal organs.
3. Cut the fish by 5cm.
4. Make cuts on the skin.
5. Season the fish with 5g of ginger, 5g of clean rice-wine and 0.5g of pepper powder.
6. Cook stock.
7. Put 15g of thick soy sauce to 500g of stock.
8. Add 10g of clean rice-wine.
9. Put a lemon and the mackerel.
10. Roast the fish right on a gridiron.

11~12. Serve the dish well.

TIP
建议

5. 고등어의 처음 비린내를 제거한다.
9. 간장양념에 담그면 고등어의 색감과 마지막 비린내까지 잡아준다.

5. 去除鲅鱼的鱼腥味。
9. 将青花鱼放在酱油调味料里可去除剩下的腥味，并且可以调节颜色。

Step 5. Reduce the smell of the mackerel.
Step 9. Keep putting the mackerel into the thick soy sauce several times. This will reduce the smell of the fish, preserving the original color of it.

꽁치소금구이

盐烤秋刀鱼

Pacific saury roasted with salt

재료 및 분량(2인분)
꽁치 2마리, 우엉 40g, 소금 10g, 청주 15g, 레몬 10g, 생강 5g, 후춧가루 0.5g
***우엉조림양념** : 진간장 50g, 설탕 5g, 올리고당 5g, 통깨 2g, 식용유 10g, 물 50g
***응용요리재료** : 고등어소금구이, 조기, 삼치, 갈치 등에 활용할 수 있는 요리이다.

材料和份量(2人份)
秋刀鱼 2条, 牛蒡 40g, 盐 10g, 白酒 15g, 柠檬 10g, 生姜 5g, 胡椒粉 0.5g
***煎牛蒡的调味料** : 50g浓酱油, 5g糖, 5g低聚糖, 2g芝麻, 10g食用豆油, 50g水。
***应用料理材料** : 可用于制作盐烤青花鱼、黄鱼、金枪鱼和刀鱼等的料理。

Ingredients and portion(2 portions)
2 pacific saury(mackerel pike), 40g Burdock root, 10g salt, 15g clean rice-wine, 10g lemon, 0.5g ginger, 0.5g ground black pepper
***Marinade for braising Burdock** : 50g thick soy sauce, 10g sugar, 10g oligosaccharide, 2g sesame, 10g salad oil, 50g water
***Note** : You can use the recipe to roast mackerel, croaker and hairtail.

꽁치 손질 秋刀鱼的处理方法 How to prepare a pacific saury

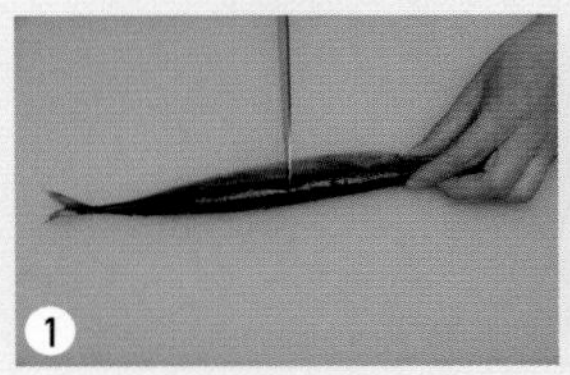

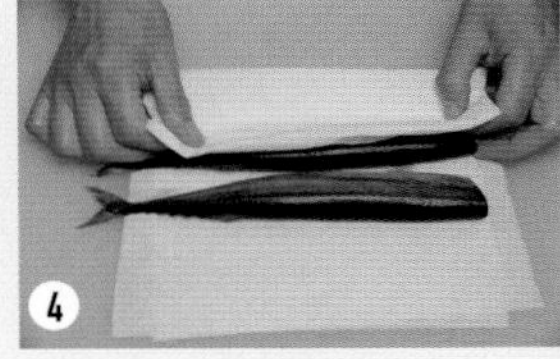

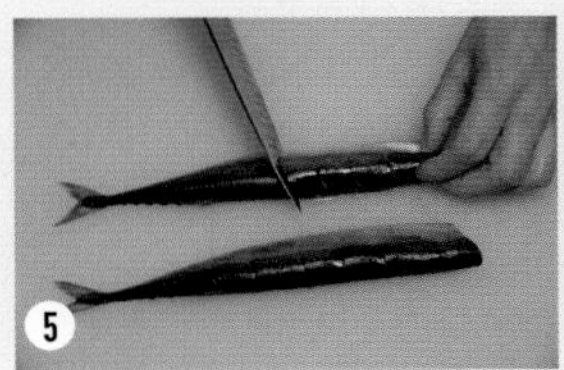

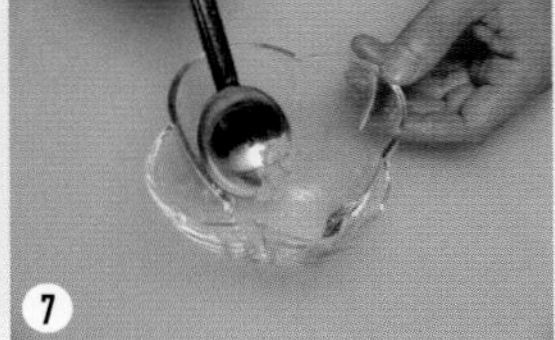

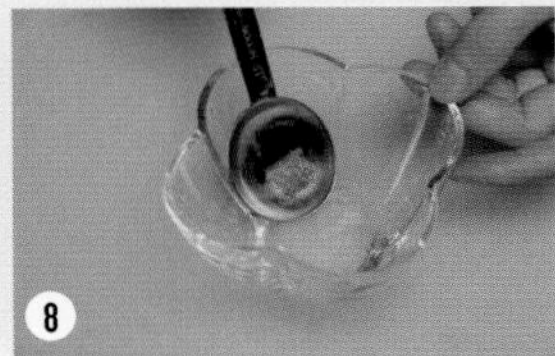

꽁치 굽기 烤秋刀鱼 How to roast a pacific saury

만드는 방법

1. 비늘을 제거한다.
2. 머리와 내장을 제거한다.
3. 소금을 뿌린다.
4. 물기를 제거한다.
5. 칼집을 넣는다.
6. 청주 10g을 넣는다.
7. 생강 3g을 넣는다.
8. 후춧가루 0.5g을 넣는다.
9. 꽁치 비린내를 제거한다.
10. 석쇠를 코팅처리한다.
11. 직화구이한다.
12. 소금을 뿌려 굽는다.

制作方法

1. 去除腥味。
2. 去除头部和内脏。
3. 撒上盐。
4. 控除水分。
5. 切口。
6. 放入10g白酒。
7. 放入3g生姜。
8. 放入0.5g胡椒。
9. 去除秋刀鱼的腥味。
10. 对烤网进行涂层处理。
11. 用直火烤。
12. 撒上盐后用火烤。

How to cook

1. Remove scales.
2. Remove the head and the internal organs.
3. Sprinkle salt.
4. Remove water.
5. Make cuts.
6. Put 10g of clean rice-wine.
7. Put 3g of ginger.
8. Add 0.5g of pepper.
9. Reduce the smell of a pacific saury.
10. Get a gridiron coated.
11. Roast the fish right on the gridiron.
12. Sprinkle salt while roasting the fish.

결들임채소 손질
装饰菜的处理方法
How to prepare the vegetables as garnish

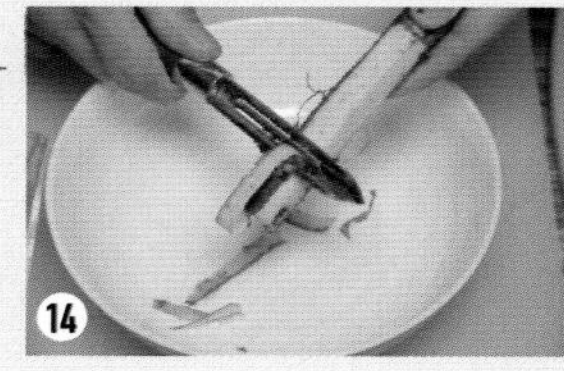

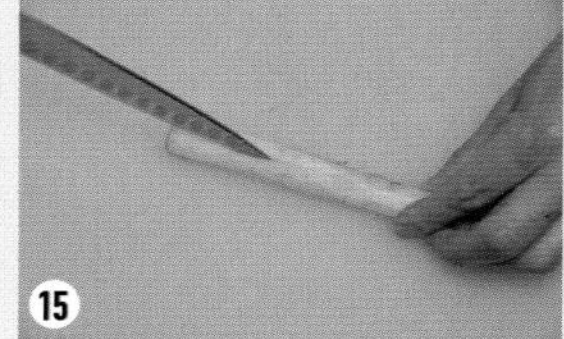

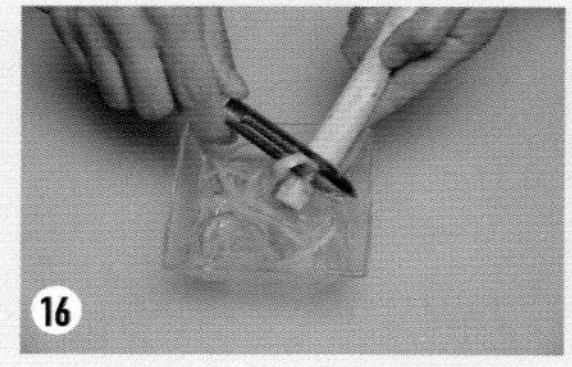

만드는 방법

13. 물에 식초를 넣는다.
14. 우엉의 껍질을 제거한다.
15. 가운데 칼집을 넣는다.
16. 얇게 편 썬다.
17. 갈변을 방지한다.
18. 식용유를 10g 넣는다.
19. 우엉을 볶는다.
20. 물 50g을 넣는다.
21. 진간장 10g을 넣는다.
22. 설탕 5g을 넣는다.
23. 올리고당 5g을 넣는다.
24. 통깨 2g을 넣는다.

制作方法

13. 在水中加入食醋。
14. 去除牛蒡皮。
15. 在中间切开。
16. 切成薄片。
17. 防止变成褐色。
18. 放10g食用油。
19. 炒牛蒡。
20. 倒入50g水。
21. 放入10g浓酱油。
22. 放入5g白糖。
23. 放入5g低聚糖。
24. 放入2g芝麻。

How to cook

13. Put vinegar into water.
14. Peel off burdock.
15. Make a cut in the middle.
16. Cut the burdock into thin slices.
17. Keep the burdock in the water not to turn brown.
18. Put 10g of cooking oil.
19. Roast the burdock.
20. Pour 50g of water.
21. Put 10g of thick soy sauce.
22. Put 5g of sugar.
23. Put 5g of oligosaccharide.
24. Add 2g of sesame.

25

28

26

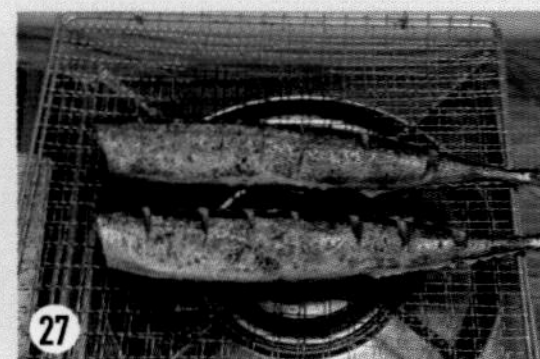
27

굽기
烤
How to roast

만드는 방법
25. 우엉볶음을 완성한다.
26. 레몬을 0.2cm로 슬라이스한다.
27. 꽁치를 노릇하게 굽는다.
28. 완성하여 담는다.

制作方法
25. 炒牛蒡完成。
26. 将柠檬切成大小0.2cm。
27. 将秋刀鱼烤成金黄色。
28. 完成后盛菜。

How to cook
25. You have your burdock done.
26. Cut the lemon into 0.2cm slices.
27. Get the pacific saury roasted golden brown.
28. Finish.

TIP
建议

9. 꽁치의 비린내를 제거한다.
12. 통째로 구워서 깨지지 않도록 주의한다.

9. 去除秋刀鱼的腥味。
12. 烤秋刀鱼时要整烤。

Step 9. Reduce the smell of the pacific saury.
Step 12. Have the fish roasted at a time as you pay extra attention not to break it.

더덕너비아니구이

沙参烤牛肉

Roasted Todok neobiany

재료 및 분량(2인분)
더덕 3뿌리, 소고기(채끝살) 150g, 양파 20g, 대파 5g, 실파 5g, 마늘 3g, 통잣 3g, 고추장 15g, 소갈비양념 20g, 통깨 3g, 참기름 5g, 설탕 3g

材料和份量(2人份)
沙参 3根, 牛肉(牛上腰) 150g, 洋葱 20g, 大葱 5g, 小葱 5g, 大蒜 3g, 松树籽 3g, 辣椒酱 15g, 牛排调味料 20g, 芝麻 3g, 香油 5g, 白糖 3g

Ingredients and portion(2 portions)
3 todok(Codonopsis lanceolata) roots, 150g beef(sirloin), 20g onion, 5g leek, 5g spring onion, 3g garlic, 3g pine nut, 15g red pepper paste(gochujang), 20g beef rib barbecue sauce, 3g sesame seed, 5g sesame seed oil, 3g sugar

더덕 손질 沙参的处理方法 How to prepare deodeok

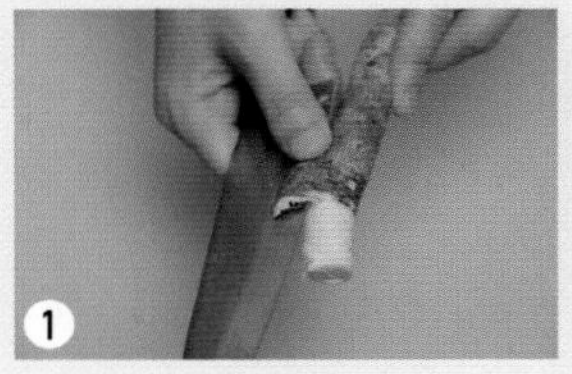

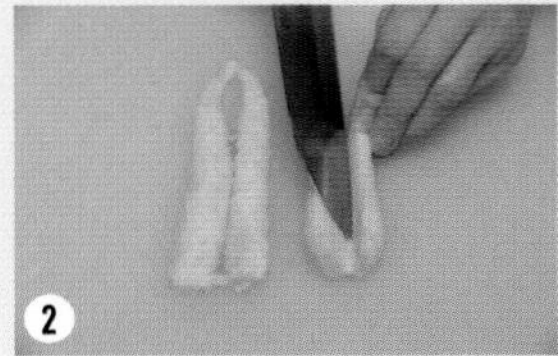

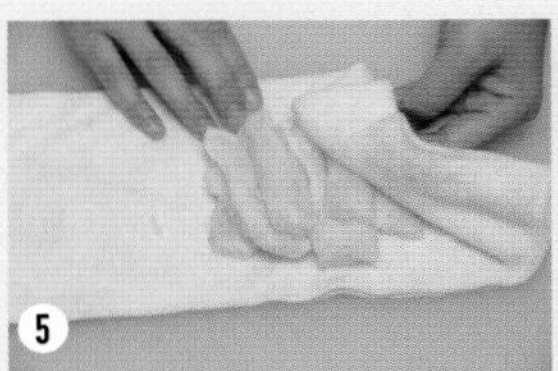

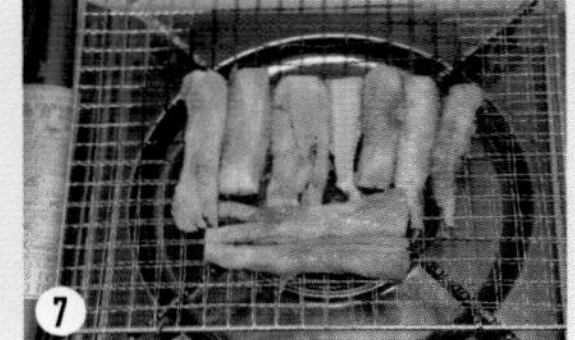

소고기 손질 牛肉的处理方法 How to prepare beef

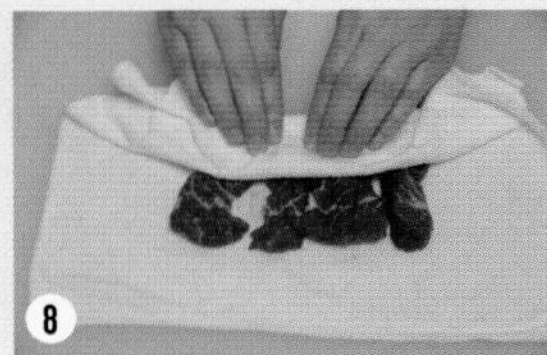

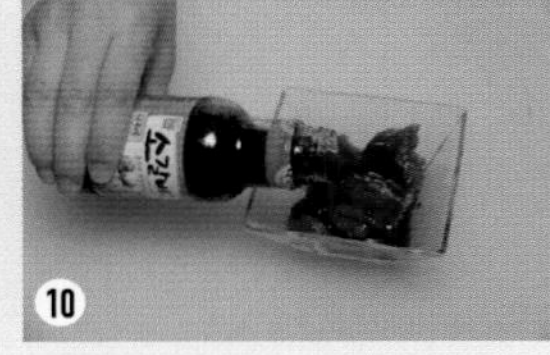

고추장구이 양념장 만들기 制作烧烤用辣椒酱调味料 How to make gochujang marinade

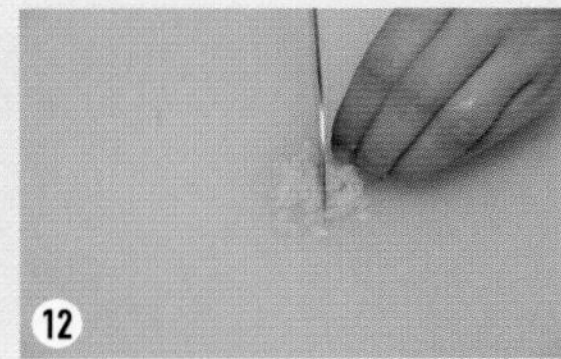

만드는 방법

1. 더덕 껍질을 제거한다.
2. 2등분하여 펼친다.
3. 방망이로 두들겨 두께를 조절한다.
4. 소금으로 쓴맛을 제거한다.
5. 물기를 제거한다.
6. 진간장과 참기름을 1:3 비율로 밑간한다.
7. 초벌구이한다.
8. 소고기 핏물을 제거한다.
9. 두들긴다.
10. 소불고기 소스로 양념한다.
11. 대파는 0.3cm로 다진다.
12. 마늘은 0.2cm로 다진다.

制作方法

1. 去除沙参皮。
2. 2等分。
3. 用棒子敲打。(调整厚度)
4. 用盐去除苦味。
5. 控除水分。
6. 用浓酱油和香油以1:3的比例混合调味。
7. 第一次烤。
8. 去除牛肉的血水。
9. 敲打牛肉。
10. 用烤牛肉调味料调味。
11. 将大葱切成大小0.3cm的葱沫。
12. 将大蒜切成大小0.2cm的蒜泥。

How to cook

1. Peel off todok.
2. Cut it in half and spread it.
3. Pound the todok to adjust the thickness.
4. Remove a bitter taste of the todok with salt.
5. Remove water.
6. Season the todok with thick soy sauce and sesame oil at a ratio of 1:3.
7. Get the todok slightly roasted at first.
8. Reduce the blood.
9. Pound the beef.
10. Season the beef with bulgogi sauce.
11. Chop up the spring onion 0.3cm.
12. Chop up the garlic 0.2cm.

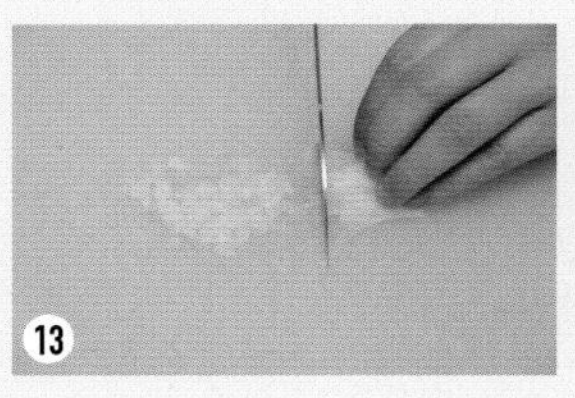
13

24

완성하기
完成
How to finish

14

23

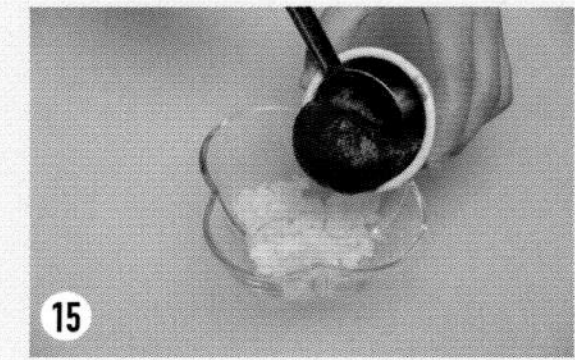
15

22

16

21

굽기
烤
How to roast

17

20

18

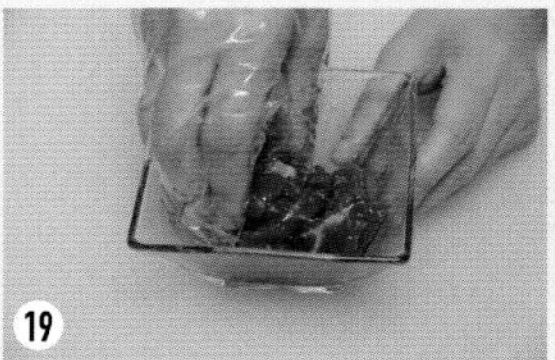
19

만드는 방법

13. 양파는 0.3cm로 다진다.
14. 함께 담는다.
15. 고추장을 15g 넣는다.
16. 통깨를 3g 넣는다.
17. 참기름을 5g 넣는다.
18. 설탕을 3g 넣는다.
19. 소고기양념에 재운다.
20. 더덕양념에 재운다.
21. 석쇠에서 양념을 바른다.
22. 직화구이한다.
23. 타지 않게 굽는다.
24. 통잣을 종이에서 다진다.

制作方法

13. 将洋葱切成大小0.3cm的丁。
14. 混合在一起。
15. 放入15g辣椒酱。
16. 放入3g芝麻。
17. 放入5g香油。
18. 放入3g白糖。
19. 将牛肉放入调味料中腌制。
20. 将沙参放入调味料中腌制。
21. 在烤网上涂抹调味料。
22. 用直火烤。
23. 注意不要烤焦。
24. 将松树籽放在纸上碾碎。

How to cook

13. Chop up the onion 0.3cm.
14. Put them all together in a small bowl.
15. Put 15g of gochujang.
16. Put 3g of sesame.
17. Put 5g of sesame oil.
18. Add 3g of sugar.
19. Marinate the beef.
20. Get the todok marinated.
21. Put the marinate on the beef and todok on an gridiron.
22. Grill them.
23. Be careful not to burn the ingredients.
24. Chop the pine nut covered in a sheet of paper.

완성하기
完成
How to finish

25

28

26

27

만드는 방법

25. 실파를 송송 다진다.
26. 너비아니구이에 잣가루를 뿌린다.
27. 더덕구이에 실파가루를 뿌린다.
28. 완성하여 담는다.

制作方法

25. 将小葱切成葱沫。
26. 在烤牛肉上撒上碾碎的松树籽。
27. 在沙参上撒上切好的葱沫。
28. 完成后盛菜。

How to cook

25. Chop up the small green onion.
26. Sprinkle the pine nut flour on the roasted neobiany.
27. Sprinkle the small green onion powder on the roasted todok.

TIP
建议

석쇠에 직화구이를 할 때는 양념장으로 인해 타지 않도록 불 조절을 해야 한다.

在烤网上用直火烤时，因为调味料的缘故，注意不要烤焦，应及时调整火候。

When you have your beef and todok roasted on the gridiron, be careful not to burn them because of the marinade.

돼지갈비구이

碳烤猪排

Roasted pork ribs

재료 및 분량(2인분)
돼지갈비(등갈비) 500g, 수삼 1뿌리, 배즙 20g, 양파즙 15g, 고추장 15g, 마늘 5g, 생강 3g, 황기 3g, 황설탕 15g, 후춧가루 0.5g, 참기름 10g, 청주 20g
***응용요리재료** : 육류요리에 활용된다.

材料和份量(2人份)
猪排(背部排骨) 500g, 水参 1根, 梨汁 20g, 洋葱汁 15g, 辣椒酱 15g, 大蒜 5g, 生姜 3g, 黄芪 3g, 黄糖 15g, 胡椒粉 0.5g, 香油 10g, 清酒 20g
***应用料理材料**：主要用于肉菜。

Ingredients and portion(2 portions)
500g pork rib(back ribs), 1 fresh ginseng root, 20g pear juice, 15g onion juice, 15g red pepper paste(gochujang), 5g garlic, 3g ginger, 3g milk vetch root, 15g brown sugar, 0.5g ground black pepper, 10g sesame seed oil, 20g rice-wine.
***Note** : You can use other kind of meat.

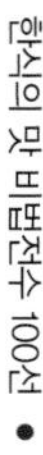

갈비 손질 排骨的处理方法 How to prepare the ribs

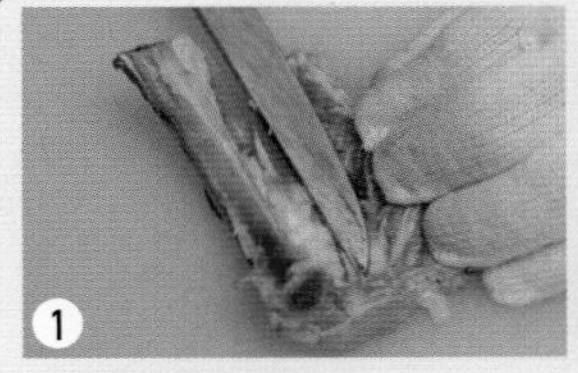

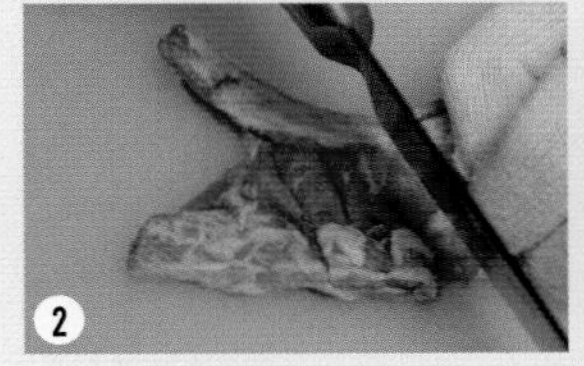

갈비구이 양념장 만들기
制作烤排骨用调味料
How to make roasted rib marinade

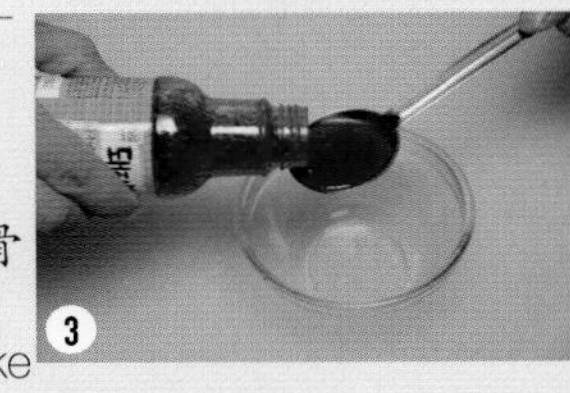

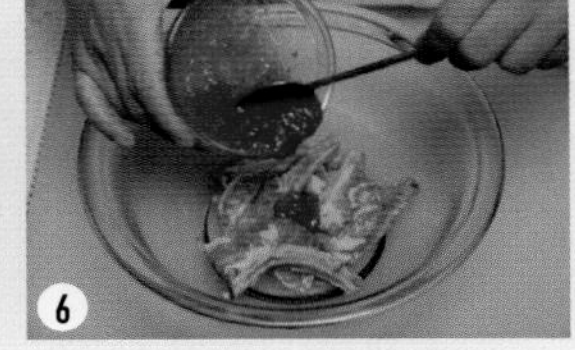

완성하기
完成
How to finish

만드는 방법

1. 갈비살을 다듬는다.
2. 살을 펼쳐 칼집을 넣는다.
3. 돼지불고기소스 20g을 넣는다.
4. 진간장을 5g 넣는다.
6. 양념에 재운다.
7. 초벌구이한다.
8. 양념을 발라주면서 굽는다.
9. 타지 않게 굽는다.

制作方法

1. 处理排骨肉。
2. 铺开排骨肉，在肉上切口。
3. 放入20g烤猪肉调味料。
4. 放入5g酱油。
6. 放入调味料中腌制。
7. 第一次烤。
8. 边烤边涂抹调味汁。
9. 注意不要烤焦。

How to cook

1. Trim the rib flesh.
2. Spread the flesh and make cuts.
3. Put 20g of pork bulgogi sauce.
4. Put 5g of thick soy sauce.
6. Marinate the rib.
7. Get the rib slightly roasted at first.
8. Roast the ribs as you keep adding the marinade.
9. Pay attention not to burn the ribs.

5-1. 황설탕을 15g 넣는다.

5-1. 加15g黄糖。

5-1. Put 15g of brown sugar.

5-2. 올리고당을 10g 넣는다.

5-2. 放入低聚糖10g。

5-2. Add 10g of oligosaccharide.

5-3. 다진 마늘 5g, 배즙 20g, 양파즙 15g을 넣는다.

5-3. 放入蒜泥5g、梨汁20g和洋葱汁15g。

5-3. Put 5g of crushed garlic, 20g of pear juice and 15g of onion juice.

5-4. 후춧가루 0.5g, 수삼즙 5g을 넣는다.

5-4. 放入胡椒粉0.5g和水参汁5g。

5-4. Put 0.5g of pepper and 5g of fresh ginseng grated.

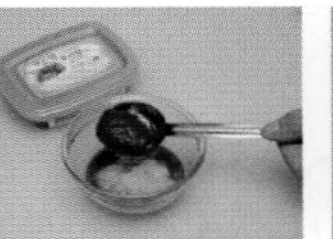

5-5. 고추장 15g을 넣는다.

5-5. 放入辣椒酱15g。

5-5. Add 15g of gochujang.

5-6. 참기름 10g과 청주 20g을 넣는다.

5-6. 加10g香油和20g清酒。

5-6. Put 10g of sesame oil and 20g of rice-wine.

TIP 建议

7. 양념을 약하게 발라서 초벌구이에 타지 않도록 주의하여 익힌다.

7. 调味料要涂抹得薄一些，烤的时候注意不要烤焦。

Step 7. Apply a small amount of the marinade in the beginning so that you will not burn the ribs.

돼지된장구이

酱烤猪肉

Roast pork with doenjang

재료 및 분량(2인분)
돼지(등심) 400g, 배 20g, 양파 10g, 황기 5g, 수삼 1뿌리, 달걀 1개, 마늘 5g, 생강 3g, 된장 20g, 고춧가루 15g, 청주 20g, 올리고당 30g
***응용요리재료** : 육류요리와 생선요리에 된장양념으로 사용하여 요리할 수 있다.

材料和份量(2人份)
猪肉(里脊) 400g, 梨 20g, 洋葱 10g, 黄芪 5g, 水参 1根, 鸡蛋 1个, 大蒜 5g, 生姜 3g, 大酱 20g, 辣椒面儿 15g, 清酒 20g, 低聚糖 30g
***应用料理材料** : 肉类的料理和海鲜料理可用大酱调味料。

Ingredients and portion(2 portions)
400g pork(loin), 20g pear, 10g onion, 5g milk vetch root, 1 fresh ginseng root, 1 egg, 5g garlic, 3g ginger, 20g soybean paste(doenjang), 15g chili powder. 30g oligosaccharide
***Note** : Doenjang can also be used in other dishes of meat and fish.

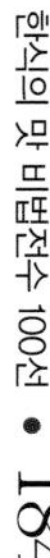

고기 손질 肉的处理方法 How to prepare the meat

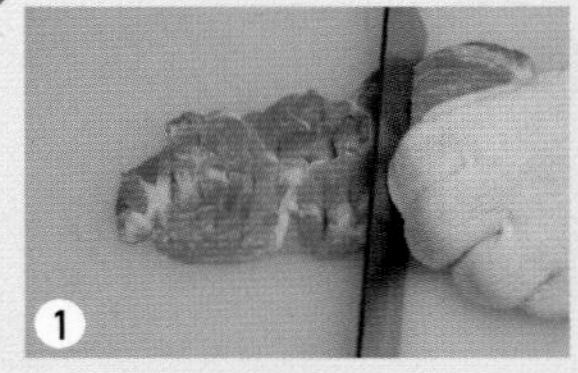

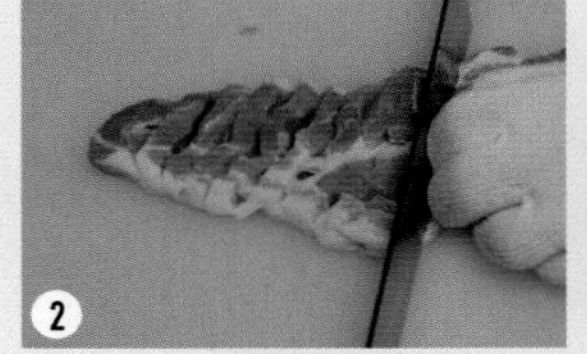

된장구이 양념장 만들기
制作酱烤用调味料
How to make doenjang marinade

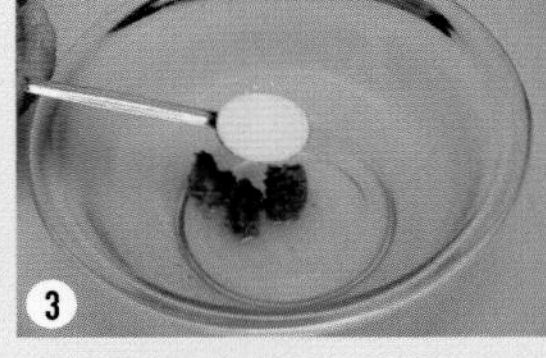

완성하기
完成
How to finish

만드는 방법

1. 돼지등심을 0.5cm 간격으로 칼집을 준다.
2. 앞, 뒤 모두 사선으로 칼집을 넣는다.
3. 된장 20g에 설탕 15g을 넣는다.
4. 배즙 20g을 넣는다.
6. 양념장을 바른다.
7. 재운다.
8. 직화구이한다.
9. 타지 않게 굽는다.

制作方法

1. 在猪里脊肉上切出间隔0.5cm的切口。
2. 前后切成射线切口。
3. 在20g大酱中放入15g白糖。
4. 放入20g梨汁。
6. 涂抹调味料。
7. 腌制。
8. 用直火烤。
9. 注意不要烤焦。

How to cook

1. Make cuts in the loin at each 0.5cm.
2. Make diagonal cuts on both sides.
3. Put 20g of doenjang and 15g of sugar.
4. Put 20g of pear juice.
6. Apply the marinade.
7. Marinate the ingredient.
8. Grill the ingredient right on a gridiron.
9. Be careful not to burn the ingredient.

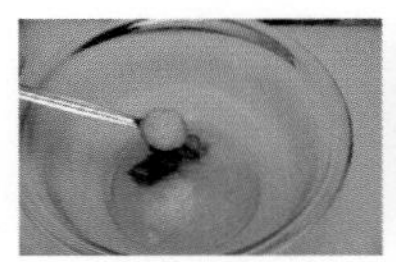

5-1. 양파즙 10g, 청주 20g, 올리고당 30g을 넣는다.

5-1. 放10g洋葱汁，20g清酒，30g低聚糖。

5-1. Put 10g of onion juice, 20g of rice-wine, amd 30g of oligosaccharide.

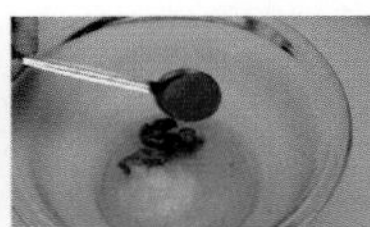

5-2. 고춧가루 15g을 넣는다.

5-2. 放入辣椒面儿15g。

5-2. Add 15g of chili powder.

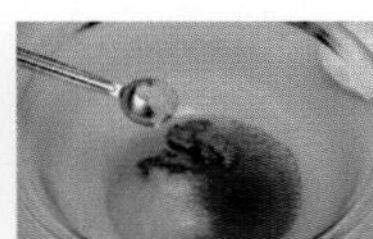

5-3. 다진 마늘 5g을 넣는다.

5-3. 放入蒜泥5g。

5-3. Put 5g of chopped garlic.

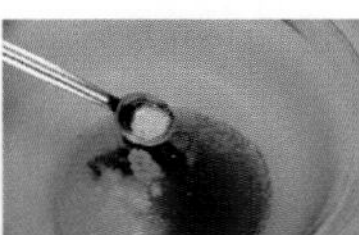

5-4. 황기가루 2g, 수삼즙 3g을 넣는다.

5-4. 放入黄芪粉末2g和水参汁3g。

5-4. Add 2g of milk vetch root powder and 3g of fresh ginseng juice.

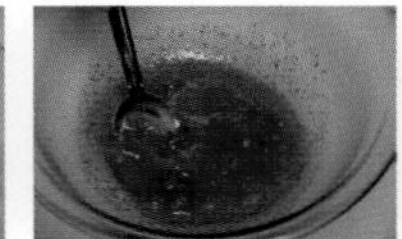

5-5. 골고루 저어준다.

5-5. 搅拌均匀。

5-5. Mix them all.

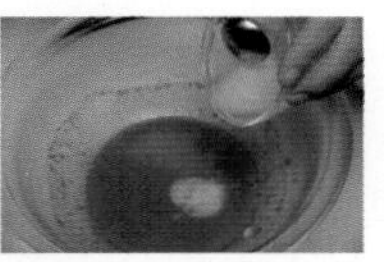

5-6. 양념장에 달걀 노른자 20g을 넣는다.

5-6. 在调味料中放入鸡蛋黄20g。

5-6. Put 20g of egg yolk into the marinade.

TIP
建议

1. 칼집을 일정하게 깊이 넣어야 양념장이 고루 밴다.
8. 타지 않게 불 조절하여 굽는다.

1. 只有使切口均匀和有深度才能使调味料进入其中。
8. 为了不会烤焦，烧烤时应注意调节火候。

Step 1. Make regular cuts to get the marinade absorbed even.
Step 8. Control the heat so that you will not get your pork burned.

병어된장구이

酱烤鲳鱼

Pomfret roasted with doenjang

재료 및 분량(2인분)
병어 1마리, 된장 40g, 다시마 5g, 가지 5g, 브로콜리 2g, 생강 2g, 레몬 3g, 청주 20g, 소금 3g, 설탕 5g
***응용요리재료** : 생선류에 요리된다.

材料和份量(2人份)
鲳鱼 1条, 大酱 40g, 海带 5g, 茄子 5g, 西兰花 2g, 生姜 2g, 柠檬 3g, 白酒 20g, 盐 3g, 白糖 5g
***应用料理材料** : 用于海鲜类料理。

Ingredients and portion(2 portions)
1 pomfret fish, 40g soybean paste(doenjang), 5g dried sea tangle, 5g eggplant, 2g broccoli, 2g ginger, 3g lemon, 20g clean rice-wine, 3g salt, 5g sugar
***Note** : You can apply on fishes.

병어 손질 鲳鱼的处理方法 How to prepare pomfret

된장양념장 만들기
制作大酱调味料
How to make doenjang

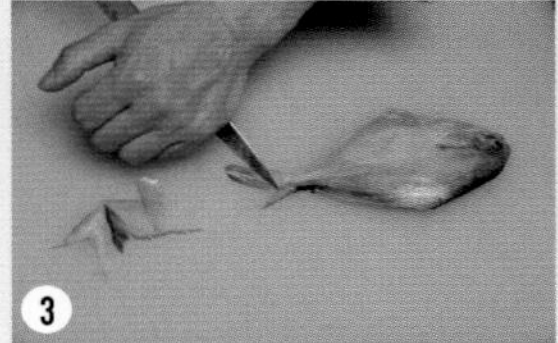

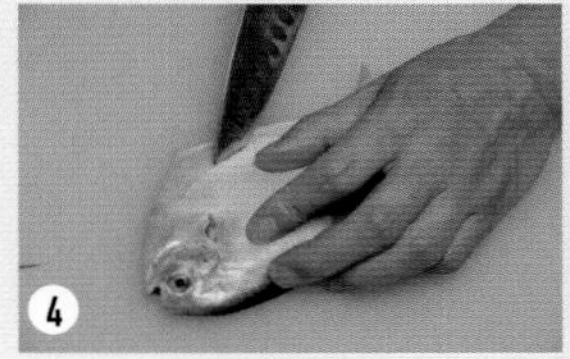

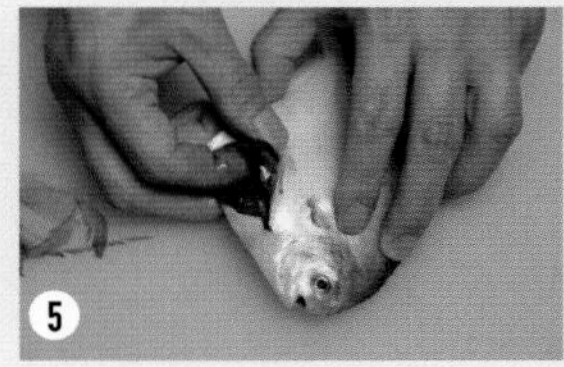

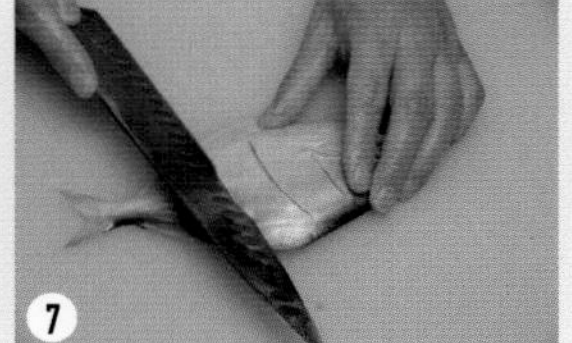

만드는 방법

1. 비늘을 제거한다.
2. 지느러미를 제거한다.
3. 꼬리를 손질한다.
4. 한 면만 칼집을 넣는다.
5. 내장을 제거한다.
6. 소금을 뿌려놓는다.
7. 칼집을 넣는다.
8. 다진 생강 5g, 청주 10g을 뿌린다.
9. 된장 20g, 청주 10g을 넣는다.
10. 설탕 5g을 넣는다.
11. 다진 마늘 5g을 넣는다.
12. 다시마와 된장양념을 준비한다.

制作方法

1. 去除鳞片。
2. 去除鱼鳍。
3. 处理鱼尾。
4. 只对一面切口。
5. 去除内脏。
6. 撒上盐。
7. 切口。
8. 撒上5g捣碎的生姜和10g白酒。
9. 放入20g大酱和10g白酒。
10. 放入5g白糖。
11. 放入5g蒜泥。
12. 准备好海带和大酱调味料。

How to cook

1. Remove the scales.
2. Remove the fin.
3. Trim the tail.
4. Make cuts on one side of the fish.
5. Remove the internal organs.
6. Sprinkle salt.
7. Make cuts.
8. Put 5g of chopped ginger and 10g of rice-wine.
9. Add 20g of doenjang and 10g of rice-wine.
10. Put 5g of sugar.
11. Put 5g of chopped garlic.
12. Get tangle and deonjang.

굽기
烤
How to roast

채소(가지) 손질
蔬菜(茄子)的处理方法
How to prepare the vegetables (eggplant)

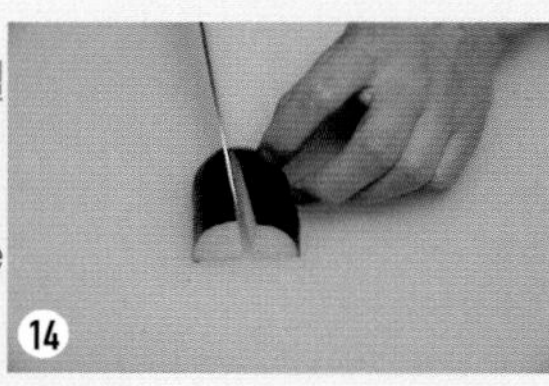

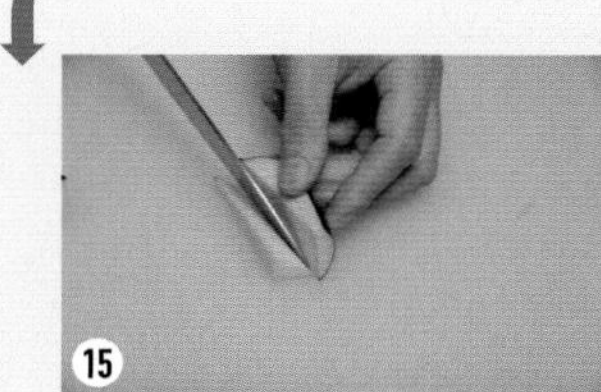

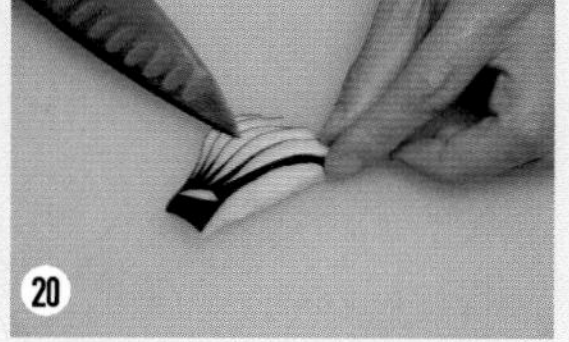

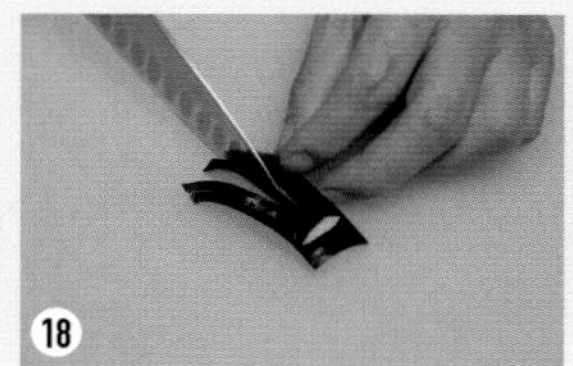

만드는 방법

13. 병어를 재운다.
14~20. 가지를 손질한다.
21. 가지 손질을 완성한다.
22. 된장양념을 발라 굽는다.
23. 된장양념 재운 모습
24. 석쇠에 통으로 굽는다.

制作方法

13. 腌制鲳鱼。
14~20. 处理茄子。
21. 处理完茄子。
22. 涂抹大酱调味料后用火烤。
23. 在大酱调味料中腌制的样子。
24. 在烤网上整烤。

How to cook

13. Marinate the pomfret.
14~20. Trim and cut eggplant.
21. You have the eggplant well trimmed.
22. Apply the doenjang and grill the fish.
23. A picture of the fish marinated with doenjang.
24. Get the fish grilled on a gridiron.

만드는 방법

25. 타지 않게 굽는다.
26. 깨지지 않게 담는다.
27. 완성품에 채소를 곁들인다.

制作方法

25. 注意不要烤焦。
26. 将烤好的整条鱼盛放在盘子上。
27. 最后在上面点缀上装饰菜。

How to cook

25. Be careful not to burn the fish.
26. Do not break the fish.
27. Garnish the dish with the vegetables.

TIP
建议

된장소스는 생선요리의 잡냄새를 제거하는 데 도움이 되며, 거즈주머니에 된장소스를 담아서 생선을 재우면, 구울 때 생선이 타는 것을 예방할 수 있다.

大酱调味料可去除海鲜的鱼腥味，并且烘烤时可避免烤焦鱼身。

Doenjang will reduce the smell of the fish. If you marinate the fish with doenjang in a cotton pocket, you can preserve the fish from being burned.

소갈비구이 (불고기양념)

烤牛排(调味烤肉)

Roasted beef rib (with bulgogi sauce)

재료 및 분량(1인분)

한우(소갈비) 250g, 배 30g, 양파 20g, 수삼 1뿌리, 대파 3g, (홍고추 5g), 구기자 3g, 황기 3g, 마늘 5g, 생강 2g, 올리고당 5g, 후춧가루 0.5g, 청주 5g, 참기름 3g, 맛술 5g, 통깨 2g, 소갈비 소스 10g

***응용요리재료** : 돼지고기(안심), 닭(가슴살), 소고기(채끝살) 등의 다양한 고기 부위의 양념하는 요리에 활용된다.

材料和份量(2人份)

韩牛(牛排) 250g, 梨 30g, 洋葱 20g, 水参 1根, 大葱 3g, (红辣椒 5g), 枸杞子 3g, 黄芪 3g, 大蒜 5g, 生姜 2g, 低聚糖 5g, 辣椒面儿 1g, 白酒 5g, 香油 3g, 味精 5g, 芝麻 2g, 牛排调味料 10g

***应用料理材料:** 可利用调味的猪肉(里脊)、鸡肉(鸡胸脯肉)和牛肉(牛上腰肉)等肉类。

Ingredients and portion(2 portions)

250g Korea beef(beef ribs), 30g pear, 20g onion, 1 fresh ginseng root, 3g leek(5g red chilli), 3g chinese wolfberries(goji), 3g milk vetch root, 5g garlic, 2g ginger, 5g oligosaccharide, 0.5g ground black pepper, 5g rice-wine, 3g sesame seed oil, 5g Cooking-wine, 2g sesame seed, 10g beef rib barbecue sauce

***Note** : You can use the recipe to cook various meats as pork(tenderloin), chicken(breast) and beef(sirloin).

소갈비 손질 牛肉的处理方法 How to prepare beef ribs

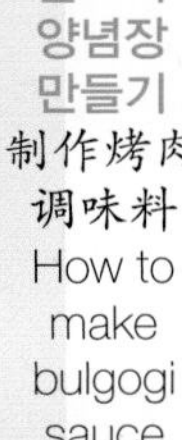

불고기 양념장 만들기 制作烤肉调味料 How to make bulgogi sauce

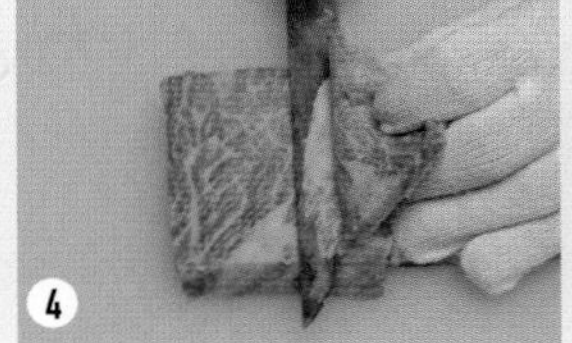

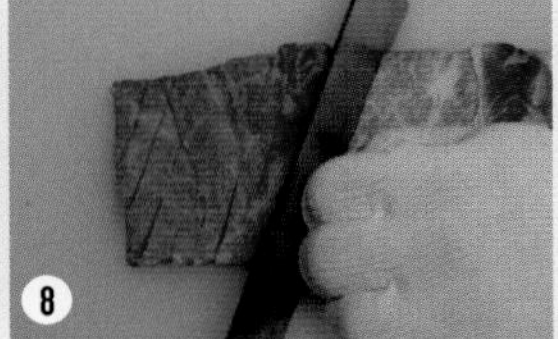

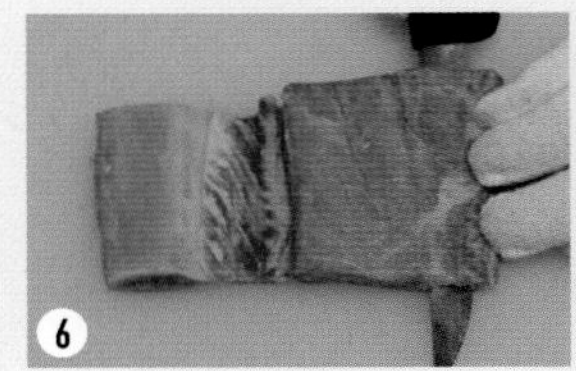

만드는 방법

1. 소갈비는 6cm로 토막내고 막을 제거한다.
2. 질긴 부분을 제거한다.
3. 뼈부터 포를 떠서 넘긴다.
4. 안쪽 살도 포를 떠서 넘긴다.
5. 뼈와 살을 포 떠서 연결시킨다.
6. 두꺼운 안쪽 살을 또 포 떠서 넘긴다.
7. 두께 0.7cm 정도로 포 떠서 손질한다.
8. 0.6cm 간격으로 어슷하게 칼집을 넣는다.
9. 뒤집어서 0.6cm 간격으로 칼집을 넣는다.
10. 살쪽만 앞, 뒤 칼집을 넣어 완성한다.
11. 배즙을 만든다.
12. 양파즙을 만든다.

制作方法

1. 将牛排骨切成大小为6cm的块并去膜。
2. 去除肉筋。
3. 将肉从骨头上剔下来切成片状。
4. 将内侧的肉也切成片状。
5. 将骨头和肉处理好后连接起来。
6. 将厚厚的内侧肉也切成片状。
7. 将肉切成0.7cm左右厚的肉片。
8. 在肉上切出间隔0.6cm的斜刀口。
9. 反过来切出间隔0.6cm的斜刀口。
10. 只在前后面有肉的部位切出刀口。
11. 制作梨汁。
12. 制作洋葱汁。

How to cook

1. Cut the beef ribs 6cm. Remove the film.
2. Cut the hard parts of the ribs.
3. Slice the beef and turn it over to the opposite side.
4. Slice the other side of the beef and turn it over.
5. Connect the rib and the flesh.
6. Slice the thick inner flesh and turn it over.
7. Make a slice of 0.7cm thick. Trim the meat.
8. Make diagonal cuts at each 0.6cm.
9. Turn the meat over and make diagonal cuts at each 0.6cm.
10. Make cuts on both side of the meat.
11. Make pear juice.
12. Make onion juice.

만드는 방법

13. 소갈비소스 10g을 넣는다.
14. 황기, 구기자가루를 각각 3g 넣는다.
15. 다진 마늘을 5g 넣는다.
16. 다진 생강을 3g 넣는다.
17. 대파는 흰 부분만 다진다.
18. 다진 대파를 3g 넣는다.
19. 올리고당을 5g 넣는다.
20. 청주를 5g 넣는다.
21. 후춧가루를 0.5g 넣는다.
22. 참기름을 3g 넣는다.
23. 맛술을 5g 넣는다.
24. 통깨를 2g 넣는다.

制作方法

13. 加入10g牛肉调味料。
14. 黄芪和枸杞子分别放入3g。
15. 放入5g蒜泥。
16. 放入3g捣碎的生姜。
17. 将大葱的白色部分切碎。
18. 放入3g切碎的大葱。
19. 放入5g低聚糖。
20. 加入5g白酒。
21. 放入0.5g胡椒粉。
22. 加入3g香油。
23. 放入5g料酒。
24. 放入2g芝麻。

How to cook

13. Put 10g of beef rib sauce.
14. Put 3g of each milk vetch root and Chinese matrimony vine powder.
15. Add 5g of chopped garlic.
16. Put 3g of chopped ginger.
17. Trim the spring onion only the white part.
18. Put 3g of the chopped spring onion.
19. Add 5g of oligosaccharide.
20. Pour 5g of clean rice-wine.
21. Spinkle 0.5g of pepper.
22. Put 3g of sesame oil.
23. Put 5g of cooking-wine.
24. Sprinkle 2g of sesame.

완성하기
完成
How to finish

만드는 방법

25. 양념장을 바른다.
26. 재운다.
27. 갈비는 직화구이한다.
28. 골고루 익힌다.
29. 타지 않게 굽는다.
30. 부드럽게 익힌다.
31. 굽기 완성
32. 완성하여 담는다.

制作方法

25. 制作调味酱。
26. 放置腌制。
27. 用火烤排骨。
28. 均匀地烤熟。
29. 注意不要烤焦。
30. 烤到肉变嫩的时候。
31. 烤完。
32. 完成后盛菜。

How to cook

25. Apply the sauce.
26. Marinate the ingredient.
27. Get the ribs right on a gridiron.
28. Grill the ribs even.
29. Be careful not to burn the ribs.
30. Grill the ribs soft.
31. Serve the dish well.
32. Finish.

TIP
建议

10. 칼집을 일정하게 사선으로 넣어 완성한다.
27. 숯불 또는 직화구이를 부드럽게 익혀 완성한다.

10. 切出的刀口要均匀和倾斜。
27. 用炭火或直火烤肉，直至肉变嫩。

Step 10. Make regular diagonal cuts in the ribs.
Step 27. Get the ribs grilled right on charcoal fire or a gridiron.

소갈비구이 (소금구이)

烤牛排(盐烤)

Grilled beef ribs(with salt)

재료 및 분량(1인분)
한우(소갈비) 250g, 배 30g, 양파 20g, 수삼 1뿌리, 구기자 3g, 황기 3g, 마늘 5g, 소금 5g, 양파즙 15g, 올리고당 10g, 마늘 5g, 생강 3g, 후춧가루 0.5g, 청주 5g, 참기름 5g, 대파 10g, 설탕 5g
***응용요리재료** : 육류요리에 사용가능하다.

材料和份量(2人份)
韩牛(牛排) 250g, 梨 30g, 洋葱 20g, 水参 1根, 大葱 3g, 枸杞子 3g, 黄芪 3g, 大蒜 5g, 生姜 2g, 盐 5g, 洋葱汁 15g, 低聚糖 10g, 大蒜 5g, 生姜 3g, 胡椒粉 1g, 白酒 5g, 香油 5g, 黄芪 2g, 大葱 10g, 白糖 5g
***应用料理材料**: 可用于肉类料理。

Ingredients and portion(2 portions)
250g Korea beef(beef ribs), 30g pear, 20g onion, 1 fresh ginseng root, 3g chinese wolfberries (goji), 3g milk vetch root, 5g garlic, 5g salt, 15g onion juice, 10g oligosaccharide, 5g garlic, 3g ginger, 0.5g ground black pepper, 5g clean rice-wine, 5g sesame seed oil, 10g leek, 5g of sugar
***Note** : You can apply the recipe to other meat dishes.

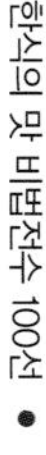

소갈비 손질 牛排的处理方法 How to prepare the ribs

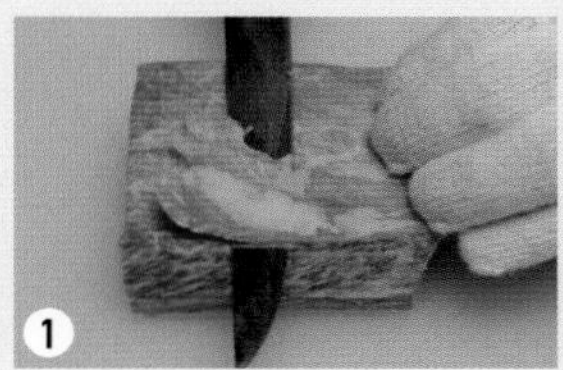

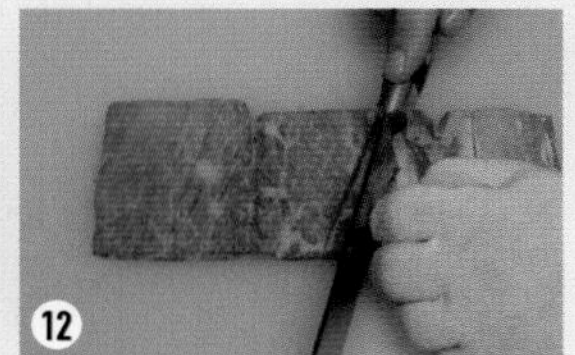

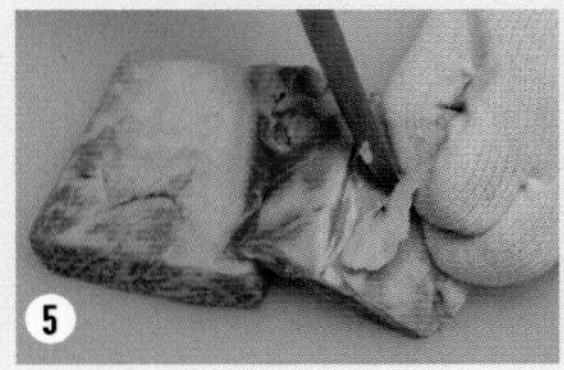

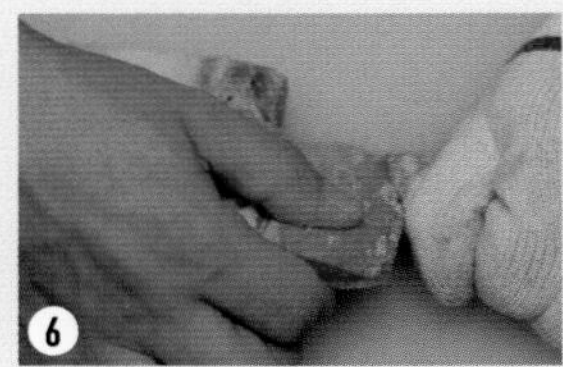

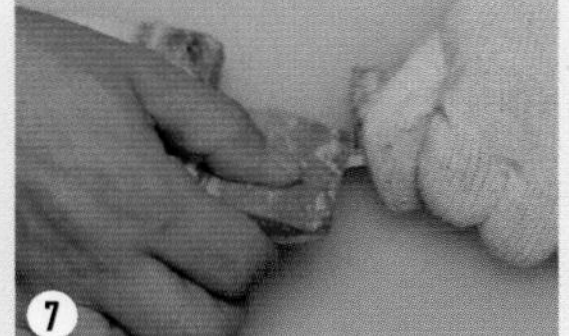

만드는 방법

1. 갈비는 6cm로 토막내고 겉의 기름기를 제거한다.
2. 뼈쪽 살을 포 뜬다.
3. 살을 포 떠서 연결시킨다.
4. 두께 1cm로 포 뜬다.
5. 질긴 부위를 손질한다.
6. 질긴 막을 제거한다.
7. 뼈에 붙은 막을 제거한다.
8. 두께 1cm로 포를 뜬다.
9. 포 떠서 펼친다.
10. 저민 과정 완성한다(앞).
11. 저민 과정 완성한다(뒤).
12. 어슷하게 0.7cm로 칼집을 넣는다.

制作方法

1. 将排骨切成大小为6cm的块状，去除表面的油脂。
2. 将骨头侧的肉切成片状。
3. 将肉切成片状后连接起来。
4. 将肉切成厚1cm的片状。
5. 处理肉筋。
6. 去除肉筋膜。
7. 去除粘在骨头上的膜。
8. 切成厚1cm的片状。
9. 切成肉片后铺平。
10. 切片过程完成。(前)
11. 切片过程完成。(后)
12. 切出长0.7cm的斜刀口。

How to cook

1. Cut the ribs 6cm and remove the fats.
2. Slice the meat inside.
3. Connect the slices.
4. Make the slices 1cm-thick.
5. Trim the hard parts of the slices.
6. Remove the hard film.
7. Remove the film from the ribs.
8. Slice the meat 1cm thick.
9. Slice the meat and spread it out.
10. How to cut the meat into slices. (the front)
11. How to cut the meat into slices. (the rear)
12. Make 0.7cm diagonal cuts.

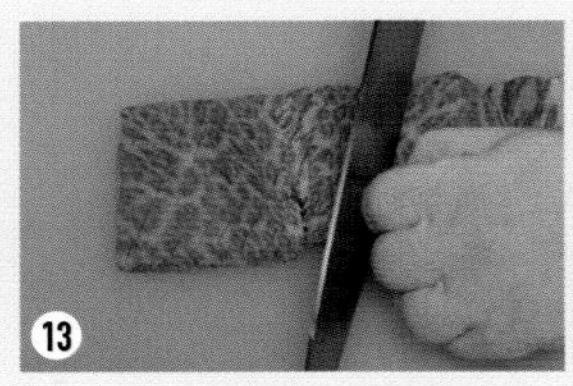

13

24

14

23

소금구이
양념장
만들기
制作盐烤用
调味料
How to make
sauce for the
dish

15

22

16

21

17

20

18

19

만드는 방법

13. 앞, 뒤 어슷하게 칼집을 넣는다.
14. 칼집을 완성한다.
15. 소금을 5g 넣는다.
16. 배즙을 20g 넣는다.
17. 양파즙을 15g 넣는다.
18. 올리고당을 10g 넣는다.
19. 다진 마늘을 5g 넣는다.
20. 다진 생강을 3g 넣는다.
21. 후춧가루를 0.5g 넣는다.
22. 청주를 5g 넣는다.
23. 참기름을 5g 넣는다.
24. 황기, 구기자가루를 각각 3g씩 넣는다.

制作方法

13. 在前、后切出斜刀口。
14. 切刀口过程完成。
15. 放入5g盐。
16. 加入20g梨汁。
17. 加入15g洋葱汁。
18. 加入10g低聚糖。
19. 放入5g蒜泥。
20. 放入3g捣碎的生姜。
21. 放入0.5g胡椒粉。
22. 加入5g白酒。
23. 加入5g香油。
24. 黄芪和枸杞子分别放入3g。

How to cook

13. Make diagonal cuts on the both sides.
14. You have done with making the cuts.
15. Put 5g of salt.
16. Put 20g of pear juice.
17. Add 15g of onion juice.
18. Put 10g of oligosaccharide.
19. Put 5g of crushed garlic.
20. Add 3g of chopped ginger.
21. Sprinkle 0.5g of pepper powder.
22. Pour 5g of clear rice-wine.
23. Put 5g of sesame oil.
24. Put 3g of each milk vetch root and Chinese matrimony vine powder.

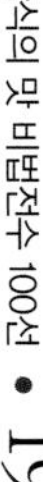

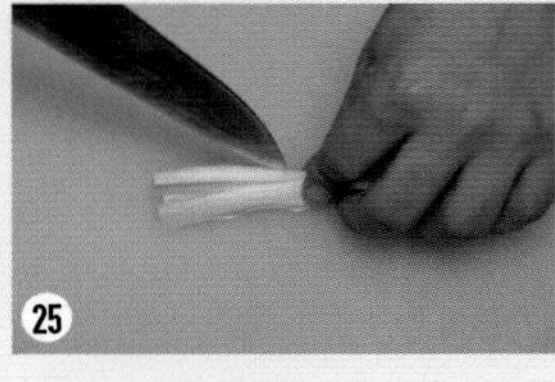

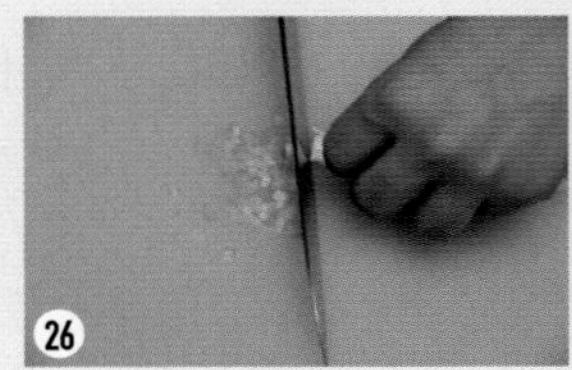

완성하기
完成
How to finish

만드는 방법

25. 대파를 채 썬다.
26. 대파는 흰 부분만 다진다.
27. 다진 대파 10g을 넣는다.
28. 설탕을 5g 넣는다.
29. 골고루 섞는다.
30. 양념장을 바른다.
31. 석쇠에 굽는다.
32. 타지 않게 굽는다.

制作方法

25. 将大葱切成丝细。
26. 将大葱的白色部分切碎。
27. 放入切碎的大葱10g。
28. 放入白糖5g。
29. 均匀混合搅拌。
30. 涂抹调味酱。
31. 放在烤网上烤。
32. 注意不要烤焦。

How to cook

25. Julienne the spring onion.
26. Chop the spring onion on the white part.
27. Put 10g of chopped spring onion.
28. Add 5g of sugar.
29. Mix them all together.
30. Apply the sauce.
31. Grill the ribs on a gridiron.
32. Pay attention not to burn the ribs.

TIP
建议

14. 갈비의 칼집 깊이를 일정한 간격으로 넣는다.
29. 갈비소스를 숙성하여 사용하면 더욱 깊은 맛을 낼 수 있다.
소금구이는 마늘, 생강, 대파 양념을 생략할 수도 있다.

14. 排骨的刀口要有一定的深度和间隔。
29. 若使排骨调味料熟成后再使用的话，味道会更可口。
盐烤时可省略大蒜、生姜、大葱等调料。

Step 14. Make cuts in the ribs at a regular interval.
Step 29. Get the sauce fermented even more for a richer flavor. With this salt-flavor recipe, you might pass the garlic, ginger, spring onion marinade part.

조기양념구이

烤黄花鱼

Grilled croaker

재료 및 분량(2인분)
조기 2마리, 실고추 2g, 대파 5g, 홍고추 3g, 풋고추 3g, 마늘 3g, 생강 3g, 진간장 30g, 고춧가루 15g
***육수** : 다시마 5g, 대파 3g, 마늘 2g, 생강 3g
***초벌양념** : 소금 2g, 생강즙 2g, 청주 2g
***양념장** : 진간장 30g, 고춧가루 15g, 설탕 5g, 마늘 5g, 청주 10g, 참기름 5g, 통깨 2g, 홍고추 10g, 풋고추 5g, 대파 5g
***응용요리재료** : 가자미, 이면수, 코다리, 고등어 등의 생선구이에 이용하여 요리한다.

材料和份量(2人份)
黄花鱼 2条, 辣椒丝 2g, 大葱 5g, 红辣椒 3g, 青辣椒 3g, 大蒜 3g, 生姜 3g, 浓酱油 30g, 辣椒面儿 15g
***肉汤** : 海带 5g, 大葱 3g, 大蒜 2g, 生姜 3g
***打底调味料** : 盐 2g, 生姜汁 2g, 白酒 2g
***调味酱**: 浓酱油 30g, 辣椒面儿 15g, 白糖 5g, 大蒜 5g, 白酒 10g, 香油 5g, 芝麻 2g, 红辣椒 10g, 青辣椒 5g, 大葱 5g
***应用料理材料**: 可利用鲽鱼肉、单鳍多线鱼、鳕鱼片和青花鱼等制作鱼类烧烤料理。

Ingredients and portion(2 portions)
2 white croakers fish, 2g shreded red pepper, 5g leek, 3g red chilli, 3g green chilli, 3g garlic, 3g ginger, 30g thick soy sauce, 15g chili powder
***Stock** : 5g dried sea tangle, 3g leek, 2g garlic, 3g ginger
***First grill sauce** : 2g salt, 2g ginger juice, 2g rice-wine
***Sauce** : 30g thick soy sauce, 15g chili powder, 5g sugar, 5g garlic, 10g rice-wine, 5g sesame oil, 2g sesame, 10g red chilli, 5g green chilli, 5g leek
***Note** : You can use the recipe to grill a halibut, atka-fish, kodari and mackerel.

육수 끓이기 煮肉汤 How to cook stock

양념장 만들기
制作调味酱
How to make the sauce

조기 손질
黄花鱼的处理方法
How to prepare croaker

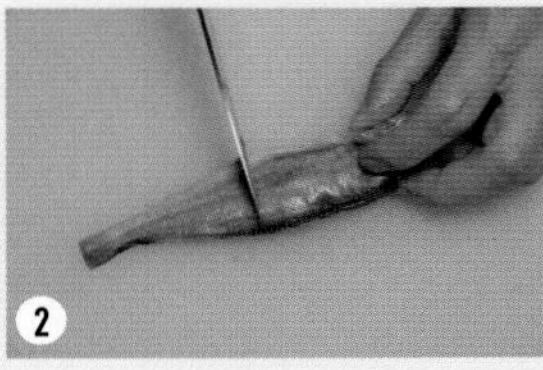

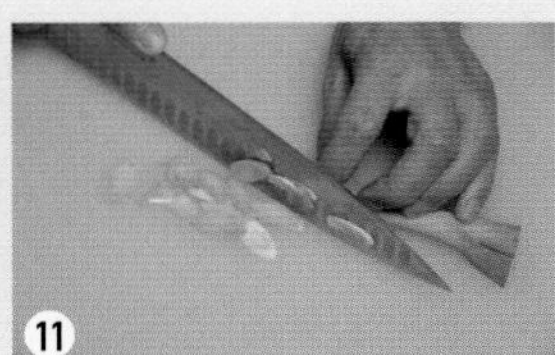

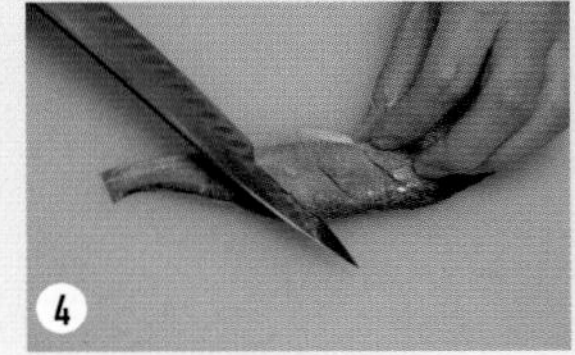

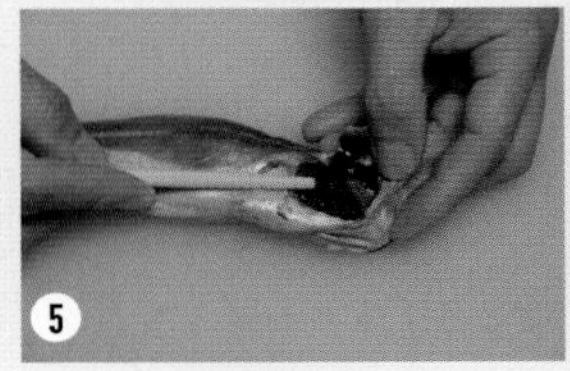

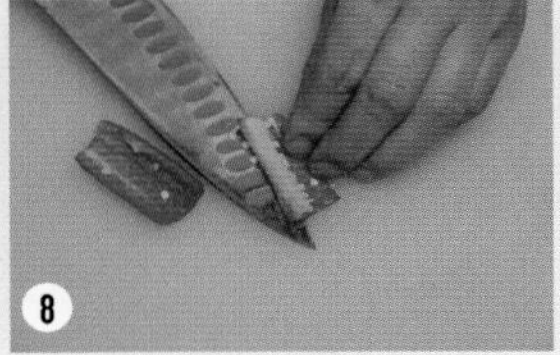

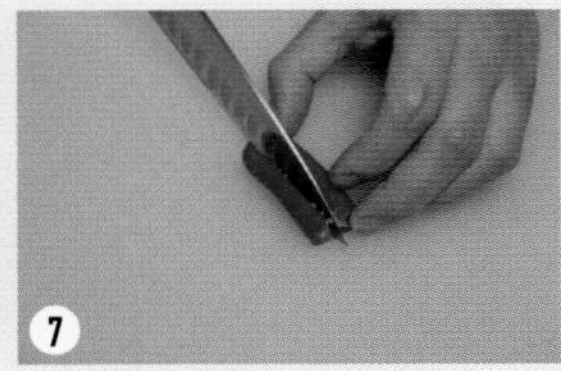

양념장 만들기
制作调味酱
How to make the sauce

만드는 방법

1. 육수를 만든다.
2. 조기의 비늘을 제거한다.
3. 조기에 칼집을 넣는다.
4. 반대편도 칼집을 넣는다.
5. 아가미를 꺼낸다.
6. 소금 2g, 청주 2g, 생강즙 2g에 밑간한다.
7. 고추를 반으로 자른다.
8. 고추씨를 제거한다.
9. 곱게 채를 썬다.
10. 곱게 다진다.
11. 파는 어슷썬다.
12. 육수에 진간장 30g을 넣는다.

制作方法

1. 煮好肉汤备用。
2. 去除黄花鱼的鱼鳞。
3. 在黄花鱼上切口。
4. 另一面也切口。
5. 取出鱼鳃。
6. 放入2g盐、2g白酒和2g生姜汁。
7. 将辣椒切成两半。
8. 去除辣椒籽。
9. 将辣椒切成丝细。
10. 将辣椒切成碎块。
11. 将大葱切成丝细。
12. 在肉汤中加入30g浓酱油。

How to cook

1. Cook stock.
2. Remove the scales.
3. Make cuts in the fish.
4. Make cuts on the other side of the fish.
5. Remove a gill.
6. Season the fish with 2g of salt, 2g of clean rice-wine and 2g of ginger juice.
7. Cut the chilli in half.
8. Remove the red-pepper seeds.
9. Julienne the chilli fine.
10. Chop them up.
11. Cut the spring onion diagonally.
12. Put 30g of thick soy sauce into the stock.

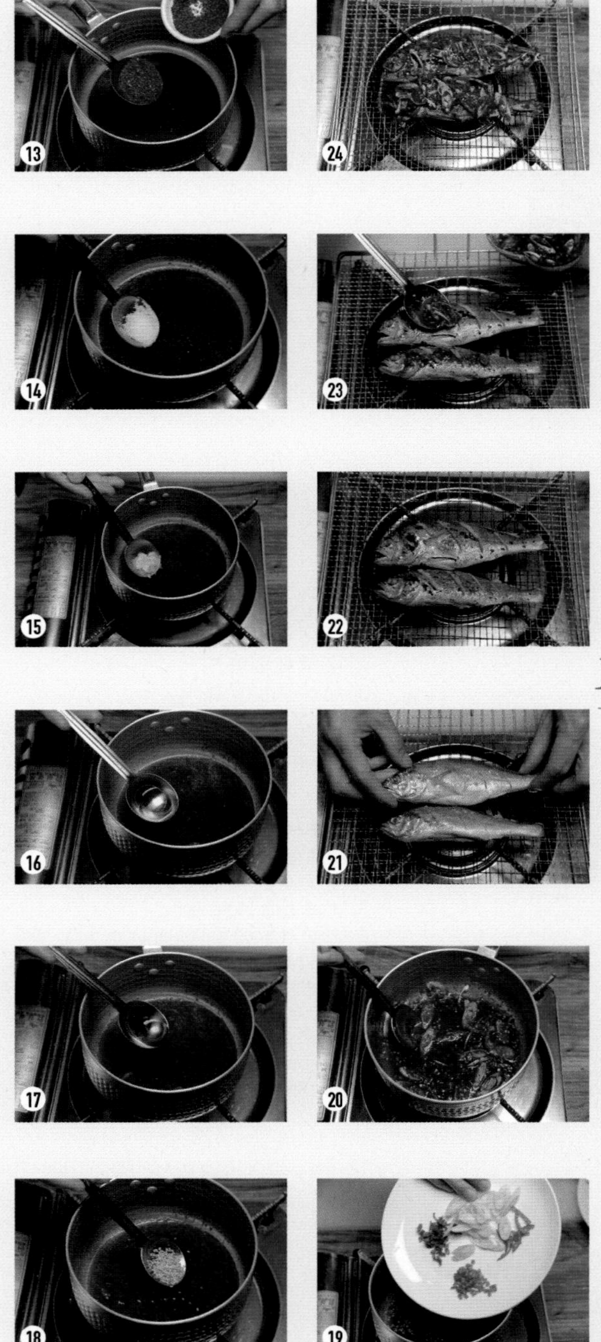

굽기
烤制
How to grill

만드는 방법

13. 고춧가루 15g을 넣는다.
14. 설탕 5g을 넣는다.
15. 다진 마늘 5g을 넣는다.
16. 청주 10g을 넣는다.
17. 참기름 5g을 넣는다.
18. 통깨 2g을 넣는다.
19. 채소를 넣는다.
20. 양념장을 한번 끓인다.
21. 달군 석쇠에 조기를 올린다.
22. 노릇하게 굽는다.
23. 초벌 양념장을 바른다.
24. 마지막 양념장을 바른다.

制作方法

13. 放入15g辣椒面儿。
14. 放入5g白糖。
15. 放入5g蒜泥。
16. 加入10g白酒。
17. 加入5g香油。
18. 放入2g芝麻。
19. 放入蔬菜。
20. 将调味酱煮一下。
21. 将黄花鱼放在烤网上。
22. 烤黄花鱼直至变成金黄色。
23. 涂抹打底调味酱。
24. 最后再涂抹一次调味酱。

How to cook

13. Put 15g of chili powder.
14. Add 5g of sugar.
15. Put 5g of chopped garlic.
16. Pour 10g of rice-wine.
17. Put 5g of sesame oil.
18. Put 2g of sesame.
19. Put the vegetables.
20. Boil the sauce once.
21. Place the croaker on the hot gridiron.
22. Grill the fish until it gets golden brown.
23. Apply the first fire sauce.
24. Apply the sauce for the last time.

만드는 방법
25. 조기를 통째로 담는다.
26. 완성한다.

制作方法
25. 完成后盛放在盘子里。
26. 完成。

How to cook
25.Serve the dish well.
26.Finish.

TIP
建议

23~24. 양념장을 여러 번 발라가면서 조기를 타지 않게 굽는다.

23~24. 多次涂抹调味料，注意不要烤焦。

Step 23~24. Apply the sauce several times and not to burn your croaker.

황태고추장양념구이

烤明太鱼(秘制辣酱)

Roasted alaska pollack with gochujang

재료 및 분량(2인분)
포북어 1마리, 당근 5g, 양파 3g, 실파 3g, 마늘 2g, 생강 2g
***양념장** : 고추장 40g, 고춧가루 5g, 올리고당 10g, 청주 10g, 통깨 3g, 후춧가루 0.5g
***응용요리재료** : 더덕, 고등어, 삼겹살 등으로 다양하게 요리에 활용된다.

材料和份量(2人份)
明太鱼 1条, 胡萝卜 5g, 洋葱 3g, 小葱 3g, 大蒜 2g, 生姜 2g
***调味酱** : 辣椒酱 40g, 辣椒面儿 5g, 低聚糖 10g, 白酒 10g, 芝麻 3g, 胡椒粉 0.5g
***应用料理材料** : 可利用沙参、青花鱼和五花肉等。

Ingredients and portion(2 portions)
1 dried pollack fillet, 5g carrot, 3g onion, 3g spring onion, 2g garlic, 2g ginger
***Sauce** : 40g red pepper paste (gochujang), 5g chili powder, 10g oligosaccharide, 10g clean rice-wine, 3g sesame seed, 0.5g ground black pepper
***Note** : You can cook with todok, mackerel and pork belly.

북어 손질 干明太鱼的处理方法 How to prepare the alaska pollack

양념장 만들기 制作调味酱 How to make sauce

만드는 방법

1. 마른 북어를 불린다.
2. 물기를 제거한다.
3. 5cm 크기로 자른다.
4. 뼈를 제거한다.
5. 유장처리 밑간한다.
6. 초벌구이한다.
7. 고추장 40g에 올리고당 10g을 넣는다.
8. 다진 당근, 양파, 마늘, 생강, 통깨를 넣는다.
9. 초벌구이한 북어를 양념장에 재운다.
10. 팬에 익힌다.
11. 윤기 나게 완성한다.
12. 담는다.

制作方法

1. 将干明太鱼泡在水里。
2. 控干水分。
3. 切成大小5cm。
4. 去掉骨头。
5. 用乳清打底调味。
6. 初烤。
7. 在辣椒酱40g中加入低聚糖10g。
8. 放入捣碎的胡萝卜、洋葱、大蒜和生姜。
9. 将初烤好的明太鱼放入调味酱中腌制。
10. 在平底锅中炸熟。
11. 完成。
12. 盛菜。

How to cook

1. Soak the dried pollack in water.
2. Drain water.
3. Cut the fish by 5cm.
4. Remove the bone.
5. Season with sesame oil and soy sauce at a ratio of 3:1.
6. Grill as the first time.
7. Put 10g of oligosaccharide and 40g of gochujang.
8. Put crushed carrot, onion, garlic, ginger and sesame.
9. Marinate the first grilled alaska pollack with sauce.
10. Cook the fish in a pan.
11. You have completed the dish.
12. Finish.

TIP
建议

6. 초벌구이할 때 완전히 익힌다.
10. 고추장양념만 익으면 완성된다.

6. 初烤时要完全烤熟。
10. 只烤熟辣椒酱调味料即可。

Step 6. Get the fish fully cooked when you do the first grilling.
Step 10. Enjoy the fish with the sauce made of gochujang.

韩食美“味”秘方传授 100选 实践篇

Transmission of a Hundred Secrets of Korean Dishes Flavor

煎菜，炒菜

한식의 맛 비법전수 100선

실기편

조림, 볶음, 편육

가지튀김조림

炖油炸茄子

Fried boiled eggplant

재료 및 분량(2인분)
가지 150g, 피망(3색) 10g, 양파 5g, 대파 5g, 마늘 3g, 달걀 1개, 전분가루 10g, 밀가루(박력분) 20g, 소금 2g, 참기름 5g, 검정깨 1g, 흰깨 1g, 후춧가루 0.5g
***양념장** : 고추장 20g, 토마토케첩 10g, 진간장 5g, 올리고당 20g, 청주 15g
***응용요리재료** : 수분 없는 채소요리에 가능하다.

材料和份量(2人份)
茄子 150g, 大辣椒(3色) 10g, 洋葱 5g, 大葱 5g, 大蒜 3g, 鸡蛋 1个, 淀粉 10g, 面粉(薄力粉) 20g, 盐 2g, 香油 5g, 黑芝麻 1g, 白芝麻 1g, 胡椒粉 1g
***调味酱**: 辣椒酱 20g, 番茄酱 10g, 浓酱油 5g, 低聚糖 20g, 白酒 15g
***应用料理材料**: 可利用没有水分的蔬菜。

Ingredients and portion(2 portions)
150g eggplant, 10g bell pepper(in 3 different colors), 5g onion, 5g leek, 3g garlic, 1 egg, 10g starch flour, 20g flour(weak flour), 2g salt, 5g sesame oil, 1g white sesame, 0.5g ground black pepper
***Sauce** : 20g red pepper paste, 10g tomato ketchup, 5g thick soy sauce, 20g oligosaccharide, 15g rice-wine
***Note** : You can apply the recipe to cook non-moist vegetable dishes.

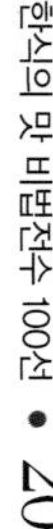

가지 손질 茄子的处理方法 How to prepare eggplant

튀기기
油炸
How to fry

만드는 방법

1. 가지를 한입 크기로 잘라 물기를 제거한다.
2. 소금, 후춧가루, 전분가루를 묻힌다.
3. 박력분과 달걀물을 묻힌다.
4. 180℃에 바삭하게 2번 튀겨낸다.
6. 조린 소스에 가지를 넣는다.
7. 참기름 5g을 넣는다.
8. 통깨를 넣는다.
9. 완성접시에 담는다.

制作方法

1. 将茄子切成适合一口吃下去的块状。
2. 撒上盐、胡椒粉和淀粉。
3. 撒上薄力粉和搅拌好的鸡蛋。
4. 在180℃的油里炸2次.
6. 在炖菜调味料中放入茄子。
7. 加入5g香油。
8. 放入芝麻。
9. 完成后盛放在容器里。

How to cook

1. Cut the eggplant into bite-sized pieces. Drain water.
2. Cover the eggplant with salt, pepper and starch flour.
3. Cover the eggplant with weak flour and egg water.
4. Fry at 180℃ twice.
6. Put the eggplant into the boiled sauce.
7. Put 5g of sesame oil.
8. Add sesame.
9. Serve the dish.

조림양념장 만들기
制作炖菜调味料
How to make marinade

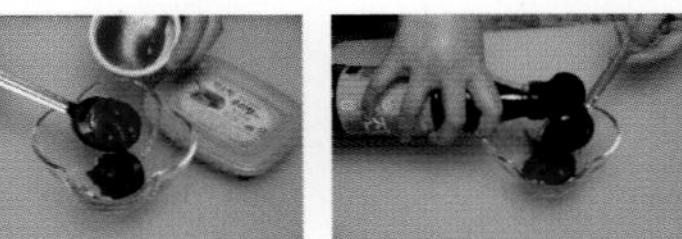

5-1. 고추장 20g, 케첩 10g을 넣는다.
5-1. 放入辣椒酱20g和番茄酱10g。
5-1. Put 20g of gochujang and 10g of ketchup.

5-2. 진간장을 5g 넣는다.
5-2. 加入酱油5g。
5-2. Put 5g of thick soy sauce.

5-3. 올리고당을 20g 넣는다.
5-3. 加入低聚糖20g。
5-3. Add 20g of oligosaccharide.

5-4. 냄비에 붓는다.
5-4. 倒入锅里。
5-4. Put the ingredients into a pot.

5-5. 청주를 15g 넣는다.
5-5. 加入白酒15g。
5-5. Pour 15g of clear rice-wine.

5-6. 양파, 피망, 대파를 넣는다.
5-6. 放入洋葱、大辣椒和大葱。
5-6. Put onion, bell pepper and green onion.

TIP
建议

2. 밑간한 가지는 전분가루부터 묻힌 뒤 밀가루(박력분)를 발라야 더욱 바삭거린다.

2. 只有将打底调味的茄子沾上淀粉后再涂抹上一层面粉(薄力粉)味道才会更香脆。

Step 2. For the seasoned eggplant, cover it with starch flour first and then, the flour(weak flour). This way, you will not lose the crispiness of the fried eggplant.

간장떡볶이

酱油炒年糕
Soy sauce tteokbbokki

재료 및 분량(2인분)
조랭이떡 200g, 소고기 60g, 당근 5g, 양파 3g, 피망 3g, 새송이버섯 2g, 대파 2g, 청경채 5g, 생강 2g, 마늘 2g, 아몬드슬라이스 2g, 불고기소스 10g, 찹쌀가루 6g, 들기름 3g

材料和份量(2人份)
料理用年糕(小葫芦年糕) 200g, 牛肉 60g, 胡萝卜 5g, 洋葱 3g, 大辣椒 3g, 杏鲍菇 2g, 大葱 2g, 青菜 5, 生姜 2g, 大蒜 2g, 杏仁片 2g, 烤肉调味料 10g, 粘米面儿 6g, 芝麻油 3g

Ingredients and portion(2 portions)
200g small rice cake (joraengyi tteok), 60g beef, 5g carrot, 3g onion, 3g bell pepper, 2g king oyster mushroom, 2g leek, 5g boy choy, 2g ginger, 2g garlic, 2g sliced almond, 10g bulgogi sauce, 6g Glutinous rice flour, 3g perilla seed oil , 3g sesame oil

재료 손질 材料处理方法 Clean ingredients

완성하기 完成 How to finish

만드는 방법
1. 조랭이떡을 데친다.
2. 들기름 3g에 소고기를 볶는다.
3. 불고기소스 10g을 붓는다.
4. 조린다.
5. 조랭이떡을 조린다.
6. 물 300g을 붓는다.
7. 다진 마늘 2g을 넣는다.
8. 채소를 넣는다.
9. 함께 조린다.
10. 청경채, 참기름을 넣는다.
11. 찹쌀가루와 1:1 비율로 물을 풀어 넣는다.
12. 완성한다.

制作方法
1. 料理用年糕(小葫芦年糕)糕焯一下。
2. 用3g芝麻油炒牛肉。
3. 加入10g烤肉调味料。
4. 炖。
5. 炖年糕。
6. 加入300g水。
7. 放入2g蒜泥。
8. 放入蔬菜。
9. 一起炖。
10. 放入青菜和香油。
11. 将粘米面儿和水以1:1的比例混合搅拌。
12. 完成。

How to cook
1. Blanch joraenggi tteok in water.
2. Pan-fry the beef with 3g of perilla oil.
3. Pour 10g of bulgogi sauce.
4. Boil the ingredient down.
5. Have the tteok boiled down.
6. Pour 300g of water.
7. Put 2g of crushed garlic.
8. Put the vegetables.
9. Get them boiled down together.
10. Put bok choy and sesame oil.
11. Put glutinous rice flour and water at a ratio of 1:1.
12. Finish.

TIP 建议

11. 찹쌀물로 농도를 맞춘다. 소량 첨가된다.

11. 用粘米面儿水调节浓度。添加少量。

Step 11. Adjust the consistence with the glutinous rice flour water. Put a small amount of it.

검정콩자반

韩式酱黑豆

Black beans cooked in soy sauce

재료 및 분량(2인분)
검정콩 100g, 진간장 30g, 설탕 15g, 올리고당 20g, 통깨 2g, 해바라기씨 2g
***응용요리재료** : 백태(흰콩), 콩 종류를 조려서 밑반찬으로 요리한다.

材料和份量(2人份)
黑豆 100g, 浓酱油 30g, 白糖 15g, 低聚糖 20g, 芝麻 2g, 葵花籽 2g
***应用料理材料:** 可利用眉豆(白豆)和豆类制成小菜。

Ingredients and portion(2 portions)
100g black soy bean, 30g thick soy sauce, 15g sugar, 20g oligosaccharide, 2g sesame seed, 2g sunflower seeds
***Note** : You can also cook with white beans and other beans for your side dish.

만드는 방법

1. 검정콩을 1시간 불린다.
2. 양념장에 진간장 30g을 넣는다.
3. 물 100g을 넣는다.
4. 설탕 15g을 넣는다.
5. 올리고당 20g을 넣는다.
6. 콩과 양념장을 넣는다.
7. 약불에서 조려준다.
8. 거품을 걷어낸다.
9. 통깨를 넣는다.
10. 해바라기씨를 2g 넣는다.
11. 윤기 나게 담는다.
12. 완성한다.

制作方法

1. 将黑豆泡在水里1个小时。
2. 放入30g调味酱–酱油。
3. 加入100g水。
4. 放入15g白糖。
5. 放入20g低聚糖。
6. 放入豆子和调味酱。
7. 用微火炖。
8. 去除产生的泡沫。
9. 放入芝麻。
10. 放入2g葵花籽。
11. 盛菜。
12. 完成。

How to cook

1. Soak the black bean in water for one hour.
2. Put the sauce and 30g of thick soy sauce.
3. Put 100g of water.
4. Add 15g of sugar.
5. Put 20g of oligosaccharide.
6. Put the bean and the marinade.
7. Boil it down over a low heat.
8. Skin foam.
9. Put sesame.
10. Put 2g of sunflower seeds.
11. Serve the dish in plate.
12. Finish.

TIP 建议

1. 콩을 충분히 불려서 요리해야 딱딱하지 않다.

1. 只有将都豆子充分泡开后豆子才不会太硬。

Step 1. Soak the beans as much as you can so that you can prevent the beans from being hard.

고등어무조림

萝卜炖鲅鱼

Mackerel cooked in soy sauce

재료 및 분량(2인분)
고등어 1마리, 무 150g, 당근 10g, 양파 5g, 풋고추 3g, 홍고추 3g, 양송이버섯 2개, 대파 4g, 고춧가루 5g, 청주 15g, 후춧가루 0.5g
***양념장** : 진간장 30g, 고춧가루 20g, 다진 마늘 7g, 설탕 7g, 참기름 5g, 통깨 3g
***육수** : 다시마 5g, 대파 2g, 생강 2g, 무 2g, 양파 2g
***응용요리재료** : 조기, 가자미, 삼치 등의 생선으로 조림요리가 된다.

材料和份量(2人份)
青花鱼 1条, 萝卜 150g, 胡萝卜 10g, 洋葱 5g, 青辣椒 3g, 红辣椒 3g, 洋香菇 2个, 大葱 4g, 辣椒面儿 5g, 白酒 15g, 胡椒粉 0.5g
***调味酱**: 浓酱油 30g, 辣椒面儿 20g, 蒜泥 7g, 白糖 7g, 香油 5g, 芝麻 3g
***肉汤**: 海带 5g, 大葱 2g, 生姜 2g, 萝卜 2g, 洋葱 2g
***应用料理材料**: 可利用黄花鱼、鲽鱼和鲅鱼等鱼类制成炖菜类料理。

Ingredients and portion(2 portions)
1 mackerel, 150g white radish, 10g carrot, 5g onion, 3g red chilli, 3g green chilli, 2 button mushroom, 4g leek, 5g chili powder, 15g clean rice-wine, 0.5g ground black pepper
***Sauce** : 30g thick soy sauce, 20g chili powder, 7g chopped garlic, 7g sugar, 5g sesame oil, 3g sesame
***Stock** : 5g dried sea tangle, 2g leek, 2g ginger, 2g white radish, 2g onion
***Note** : You can use the recipe to cook other fish dishes of croaker and halibut.

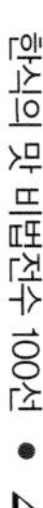

육수 끓이기 煮肉汤 How to cook stock

재료 손질 材料处理方法 Clean ingredients

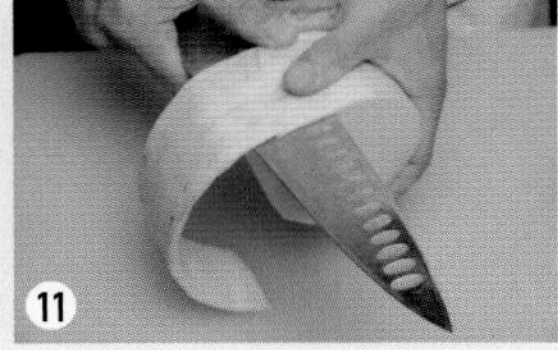

고등어 손질 青花鱼的处理方法 How to prepare mackerel

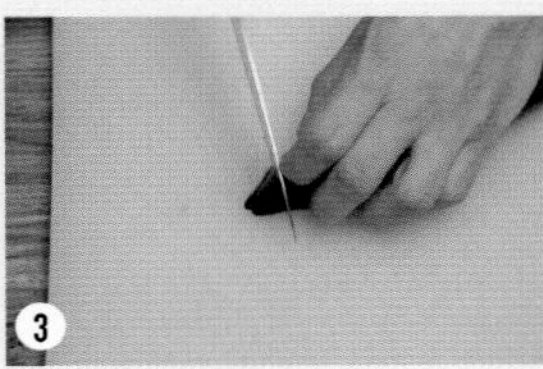

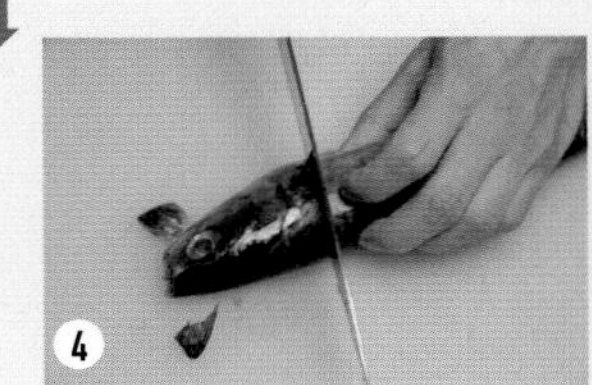

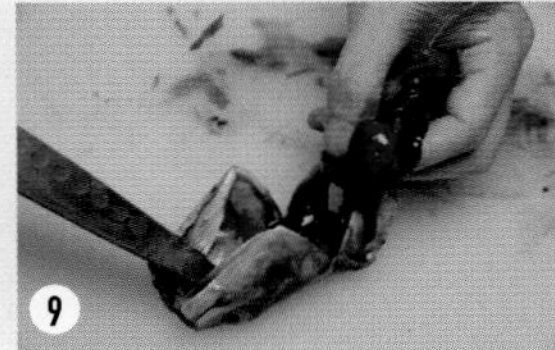

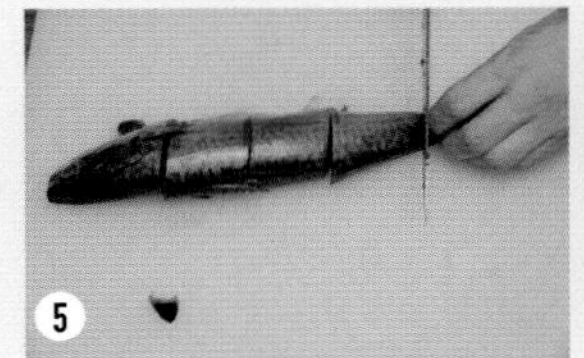

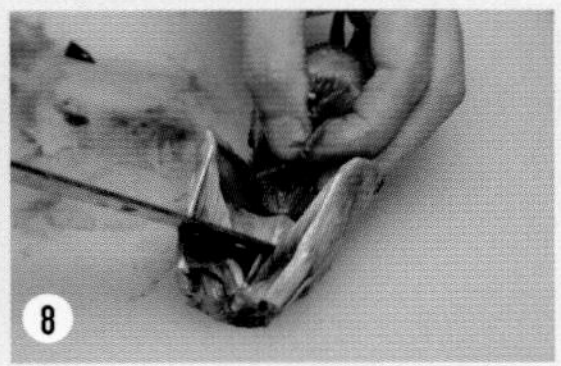

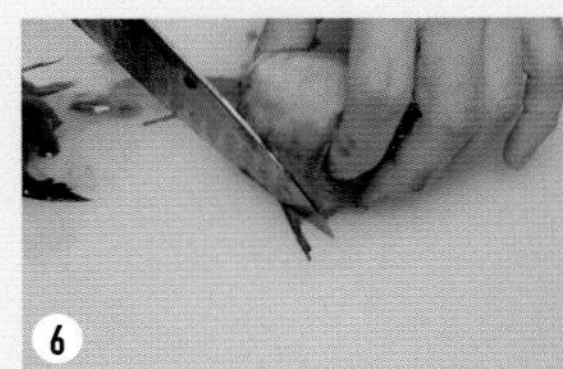

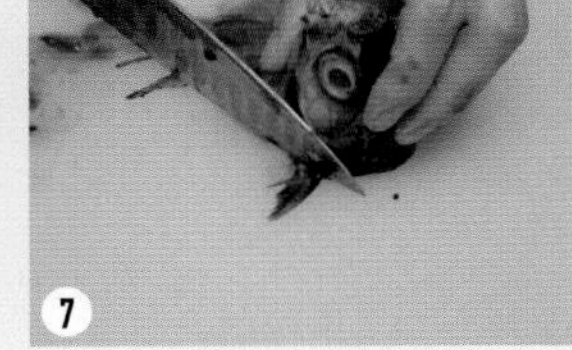

만드는 방법

1. 육수를 낸다.
2. 육수를 완성한다.
3. 고등어 입을 자른다.
4. 머리를 자른다.
5. 5cm로 토막낸다.
6. 지느러미를 제거한다.
7. 머리지느러미를 제거한다.
8. 아가미에 칼집을 넣는다.
9. 아가미를 제거한다.
10. 소금을 뿌린다.
11. 무 껍질을 벗긴다.
12. 굵직하게 편 썬다.

制作方法

1. 准备制作肉汤。
2. 煮好肉汤备用。
3. 切开青花鱼嘴。
4. 切掉鱼头。
5. 将青花鱼切成大小为5cm的块状。
6. 去除鱼鳍。
7. 去除头部鱼鳍。
8. 切开鱼鳃。
9. 去除鱼鳃。
10. 撒上盐。
11. 为萝卜去皮。
12. 将萝卜切成粗厚的片状。

How to cook

1. Cook stock.
2. Finish cooking stock.
3. Cut a mouth of the mackerel.
4. Cut a head.
5. Cut the fish 5cm.
6. Remove the fin.
7. Remove the head fin.
8. Make cuts in a gill.
9. Remove the gill.
10. Sprinkle salt.
11. Peel off the radish.
12. Cut the radish into thick slices.

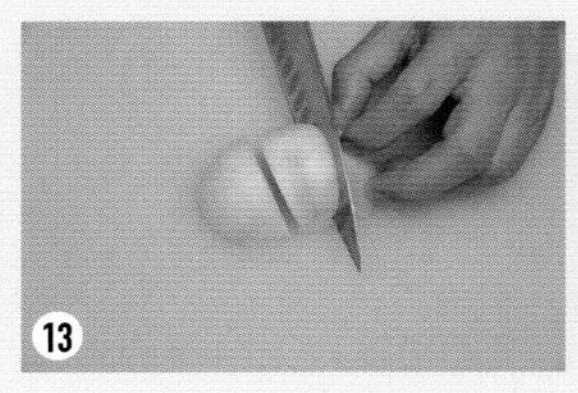

완성하기
完成
How to finish

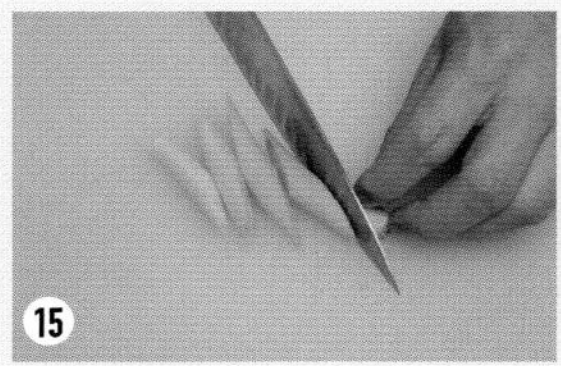

만드는 방법

13. 양파는 3x4cm로 썬다.
14. 당근은 1x4cm로 편 썬다.
15. 대파는 어슷썰기한다.
16. 고추는 어슷썰기한다.
17. 물+간장 30g을 넣는다.
18. 고춧가루를 20g 넣는다.
19. 다진 마늘을 7g 넣는다.
20. 설탕을 7g 넣는다.
21. 참기름을 5g 넣는다.
22. 통깨를 3g 넣는다.
23. 무를 바닥에 깐다.
24. 당근, 양파를 올린다.

制作方法

13. 将洋葱切成大小3x4cm。
14. 将胡萝卜切成大小为1x4cm的片状。
15. 斜切大葱。
16. 斜切辣椒。
17. 加入30g水+酱油。
18. 放入20g辣椒面儿。
19. 放入7g蒜泥。
20. 放入7g白糖。
21. 加入5g香油。
22. 放入3g芝麻。
23. 将萝卜垫在锅底。
24. 放上胡萝卜和洋葱。

How to cook

13. Cut the onion 3x4cm.
14. Cut the carrot 1x4cm.
15. Cut the spring onion diagonally.
16. Cut the chili diagonally.
17. Put water and 30g of thick soy sauce.
18. Put 20g of chili powder.
19. Put 7g of crushed garlic.
20. Add 7g of sugar.
21. Put 5g of sesame oil.
22. Add 3g of sesame.
23. Place the radish on the bottom.
24. Put the carrot and the onion on the top.

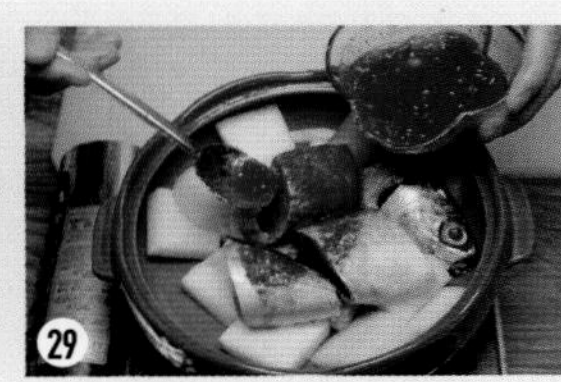

만드는 방법

25. 청주를 15g 넣는다.
26. 후춧가루를 0.5g 넣는다.
27. 고등어를 씻은 후 올린다.
28. 초벌 양념장을 올린다.
29. 양념장을 골고루 끼얹는다.
30. 골고루 붓는다.
31. 채소를 올려 끓인다.
32. 완성하여 담는다.

制作方法

25. 加入白酒15g。
26. 放入胡椒粉0.5g。
27. 将青花鱼洗干净后放在上面。
28. 首次加入调味酱。
29. 使调味酱均匀分布。
30. 均匀地倒入调味酱。
31. 放入蔬菜后一起煮。
32. 完成后盛放在容器里。

How to cook

25. Pour 15g of clean rice-wine.
26. Put 0.5g of pepper powder.
27. Wash mackerel and put it in a plate.
28. Do the first fire to the marinade.
29. Pour the marinade even to the fish.
30. Keep pouring.
31. Put the vegetables and boil the ingredients.
32. Serve the dish.

TIP
建议

31. 은근하게 오래 끓여서 깊은 맛을 낸다.

31. 多煮一些时间味道会更香浓。

Step 31. Boil the dish as long as you can add a richer flavor to the dish.

고추장떡볶이

辣椒酱炒年糕

Gochujang tteokbbokki

재료 및 분량(2인분)
떡볶이떡 200g, 양배추 20g, 당근 3g, 양파 3g, 대파 2g, 마늘 2g, 삶은 달걀 2개
***양념장** : 고추장 30g, 꿀 30g, 다진 마늘 10g

材料和份量(2人份)
炒年糕用年糕 200g, 大头菜 20g, 胡萝卜 3g, 洋葱 3g, 大葱 2g, 大蒜 2g, 煮好的鸡蛋 2个
***调味酱**: 辣椒酱 30g, 蜂蜜 30g, 蒜泥 10g

Ingredients and portion(2 portions)
200g Ddukbokki(thin rice cake), 20g cabbage, 3g carrot, 2g leek, 2g garlic, 2 boiled eggs
***Sauce** : 30g red pepper paste, 30g honey, 10g chopped garlic

재료 손질 材料处理方法 Clean ingredients

만드는 방법

1. 떡볶이떡을 데친다.
2. 어묵을 잘라 데친다.
3. 당근을 어슷썰기한다.
4. 양배추를 3x4cm로 자른다.
6. 양파를 넣는다.
7. 어묵을 넣는다.
8. 삶은 달걀을 넣는다.
9. 대파를 넣어 완성한다.

制作方法

1. 将炒年糕用年糕用水焯一下。
2. 将魔芋切好后用水焯一下。
3. 斜切胡萝卜。
4. 将大头菜切成大小3x4cm。
6. 放入洋葱。
7. 放入魔芋。
8. 放入煮好的鸡蛋。
9. 放入大葱后完成。

How to cook

1. Blanch tteok in water.
2. Cut fish cake and blanch them in water.
3. Cut the carrot diagonally.
4. Cut the cabbage 3x4cm.
6. Add the onion.
7. Put the fish cake.
8. Put the boiled egg.
9. Add the spring onion for the last touch.

끓이기
煮粥
How to boil

5-1. 물 600g에 고추장을 30g 넣는다.
5-1. 在水600g中放入辣椒酱30g。
5-1. Put 30g of gochujang into 600g of water.

5-2. 꿀을 30g 넣는다.
5-2. 加入蜂蜜30g。
5-2. Add 30ml of honey.

5-3. 고춧가루를 30g 넣는다.
5-3. 放入辣椒面儿30g。
5-3. Put 30g of chili powder.

5-4. 다진 마늘을 10g 넣는다.
5-4. 放入蒜泥10g。
5-4. Put 10g of chopped garlic.

5-5. 떡볶이를 넣는다.
5-5. 放入年糕。
5-5. Put tteobbokki tteok.

5-6. 채소를 넣는다.
5-6. 放入蔬菜。
5-6. Add the vegetables.

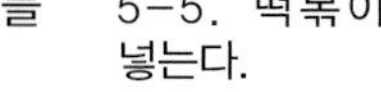

TIP
建议

5-5. 떡볶이를 완전히 넉넉히 조려야 색깔과 간이 맞는다.

5-5. 年糕要多煮一段时间颜色才会看起来有食欲，味道才会更可口。

Step 5-5. Use tteokbbokki as much as you can to cook the best tteokbbokki ever in your life.

굵은멸치풋고추 볶음

青椒炒鳀鱼

Anchovy fried with green chili

재료 및 분량(2인분)
굵은멸치 200g, 풋고추 10g, 참기름 3g
***양념장** : 고추장 15g, 올리고당 15g, 진간장 10g, 설탕 5g, 청주 15g
***응용요리재료** : 마른 오징어, 꼴뚜기 등의 요리에 활용된다.

材料和份量(2人份)
鳀鱼 200g, 青辣椒 10g, 香油 3g
***调味酱**: 辣椒酱 15g, 低聚糖 15g, 浓酱油 10g, 白糖 5g, 白酒 15g
***应用料理材料**: 可利用干鱿鱼和墨斗鱼等。

Ingredients and portion(2 portions)
200g dried big anchovy, 10g green chilli, 3g sesame seed oil
***Sauce** : 15g red pepper paste(gochujang), 15g oligosaccharide, 10g thick soy sauce, 5g sugar, 15g clean rice-wine
***Note** : You can use the recipe to cook with dried squid and baby octopus.

만드는 방법

1. 멸치의 내장을 제거한다.
2. 팬에 멸치를 볶는다.
3. 고추장 15g을 넣는다.
4. 올리고당을 15g 넣는다.
5. 진간장을 10g 넣는다.
6. 설탕을 5g 넣는다.
7. 두꺼운 팬에 양념장을 넣는다.
8. 청주를 15g 넣는다.
9. 멸치를 넣고 통깨를 넣는다.
10. 참기름을 3g 넣는다.
11. 풋고추를 넣어 완성한다.
12. 완성하여 담는다.

制作方法

1. 去除鳀鱼的内脏。
2. 用平底锅炒鳀鱼。
3. 放入15g辣椒酱。
4. 加15g低聚糖。
5. 加10g浓酱油。
6. 放入5g白糖。
7. 将调味酱放在厚平底锅上。
8. 加15g白酒。
9. 放入鳀鱼，然后放入芝麻。
10. 加入3g香油。
11. 放入青辣椒后完成。
12. 完成后盛菜。

How to cook

1. Remove the internal organs of the anchovy.
2. Fry the anchovy in a pan.
3. Put 15g of gochujang.
4. Add 15g of oligosaccharide.
5. Pour 10g of thick soy sauce.
6. Put 5g of sugar.
7. Pour the marinade into a big pan.
8. Pour 15g of clean rice-wine.
9. Put the anchovy and sesame.
10. Pour 3g of sesame oil.
11. Put green chili and finish cooking.
12. Finish.

TIP 建议

2. 멸치의 비린내를 제거한다.

2. 去除鳀鱼的腥味。

Step 2. The fishy smell of the anchovy is removed.

꽁치신김치조림

泡菜炖秋刀鱼

Pacific saury cooked with over-fermented Kimchi

재료 및 분량(2인분)
꽁치 2마리, 신김치 100g
***양념장** : 육수 50g, 설탕 5g, 다진 마늘 5g, 다진 생강 3g, 청주 15g, 통깨 2g, 후춧가루 0.5g
***육수** : 다시마 5g, 국멸치 3마리, 대파 2g, 생강 5g
***응용요리재료** : 고등어, 삼치, 갈치 등과 돼지고기류에 요리된다.

材料和份量(2人份)
秋刀鱼 2条, 新腌制的泡菜 100g
***调味酱**: 肉汤 50g, 白糖 5g, 蒜泥 5g, 捣碎的生姜 3g, 白酒 15g, 芝麻 2g, 胡椒粉 0.5g
***肉汤**: 海带 5g, 汤用鳀鱼 3条, 大葱 2g, 生姜 5g
***应用料理材料**: 可利用青花鱼、鲅鱼和刀鱼等和猪肉类。

Ingredients and portion(2 portions)
2 pacific saury(mackerel pike), 100g over-fermented Kimchi
***Sauce** : 50g stock, 5g sugar, 5g chopped garlic, 3g chopped ginger, 15g rice-wine, 2g sesame seed, 0.5g ground black pepper
***Stock** : 5g dried sea tangle, 3 anchovies, 2g leek, 5g ginger
***Note** : You can also cook with mackerel and hairtail.

육수 끓이기 煮肉汤 How to cook stock

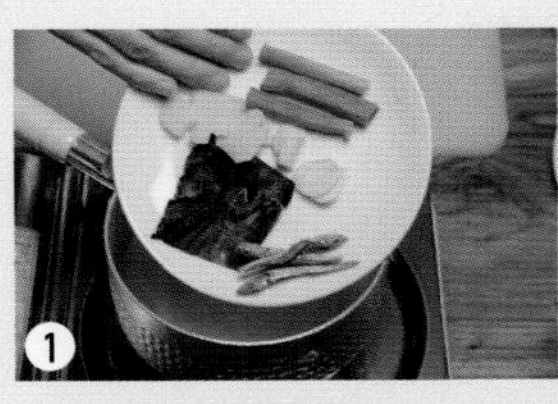

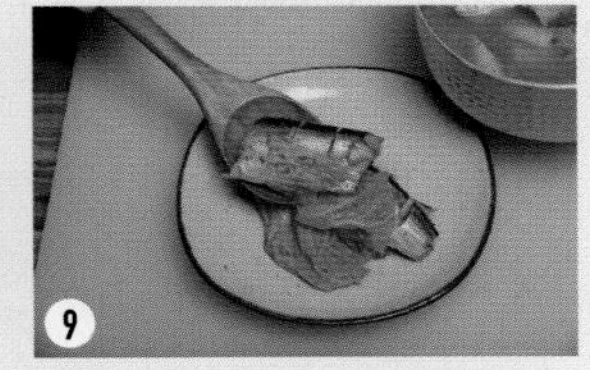

꽁치 손질 秋刀鱼的处理方法 How to prepare pacific sauries

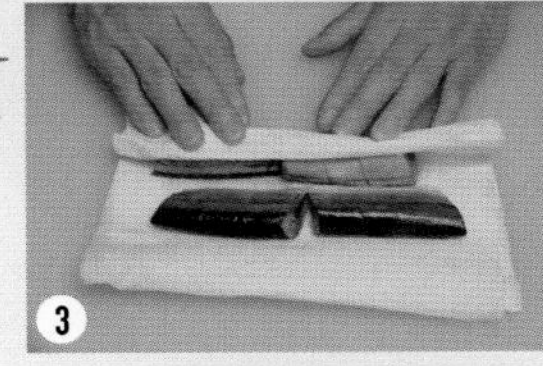

만드는 방법

1. 육수를 낸다.
2. 꽁치를 손질한다.
3. 물기를 제거한다.
4. 냄비에 꽁치, 신김치를 담는다.
6. 육수를 붓는다.
7. 대파를 넣는다.
8. 약불에서 조린다.
9. 완성접시에 담는다.

制作方法

1. 煮好肉汤备用。
2. 处理秋刀鱼。
3. 控干水分。
4. 将秋刀鱼和新腌制的白菜放入锅里。
6. 倒入肉汤。
7. 放入大葱。
8. 用微火炖。
9. 完成后盛放在容器里。

How to cook

1. Cook stock.
2. Wash pacific sauries.
3. Remove water.
4. Put the fish and over-feremented Kimchi into a pot.
6. Pour the stock.
7. Put the spring onion.
8. Cook at a low heat.
9. Serve the dish.

양념장 만들기 煮粥 How to boil

5-1. 육수 50g과 설탕 5g을 넣는다.
5-1. 加入肉汤50g和白糖5g。
5-1. Put 50g of stock and 5g of sugar.

5-2. 다진 마늘을 5g 넣는다.
5-2. 放入蒜泥5g。
5-2. Put 5g of chopped garlic.

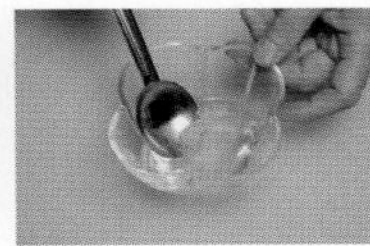

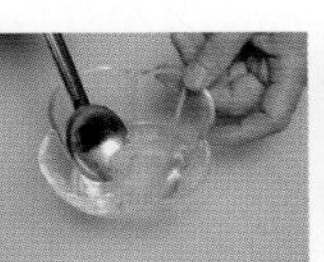

5-3. 다진 생강을 3g 넣는다.
5-3. 放入捣碎的生姜3g。
5-3. Add 3g of chopped ginger.

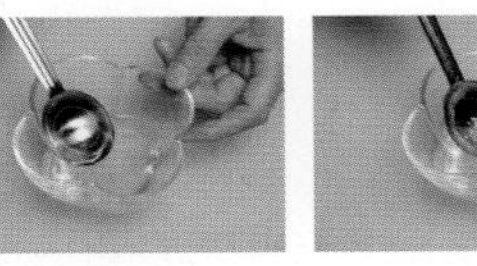

5-4. 청주를 15g 넣는다.
5-4. 加入白酒15g。
5-4. Pour 15g of rice-wine.

5-5. 통깨를 2g 넣는다.
5-5. 放入芝麻2g。
5-5. Put 2g of sesame.

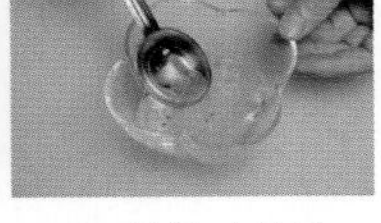

5-6. 후춧가루를 0.5g 넣는다.
5-6. 放入胡椒粉1g。
5-6. Sprinkle 0.5g of pepper powder.

TIP 建议

8. 30분 정도 끓여 김치의 맛이 생선의 맛과 어울리게 완성한다.

8. 煮30分钟左右，使泡菜的味道和鱼的味道充分地融合后即完成。

Step 8. Boil the dish 30 minutes to make Kimchi harmonized with the fish.

낙지채소볶음

蔬菜炒八爪鱼

Small octopus roasted with vegetables

재료 및 분량(2인분)
낙지 3마리, 애호박 30g, 양파 10g, 당근 5g, 느타리버섯 5g, 표고버섯 1장, 홍고추 3g, 풋고추 3g, 실파 3g, 마늘 2g, 생강 1g, 청주 15g, 참기름 5g
***양념장** : 진간장 10g, 고추장 15g, 고춧가루 20g, 통깨 1g, 까나리액젓 5g
***응용요리재료** : 주꾸미, 오징어 등으로 요리한다.

材料和份量(2人份)
八爪鱼 3条, 西葫芦 30g, 洋葱 10g, 胡萝卜 5g, 平菇 5g, 香菇 1个, 红辣椒 3g, 青辣椒 3g, 小葱 3g, 大蒜 2g, 生姜 1g, 白酒 15g, 香油 5g
***调味酱**: 浓酱油 10g, 辣椒酱 15g, 辣椒面儿 20g, 芝麻 1g, 玉筋鱼酱 5g
***应用料理材料**: 可利用小章鱼和鱿鱼等。

Ingredients and portion(2 portions)
3 small octopus, 30g long squash, 10g onion, 5g carrot, 5g oyster mushroom, 1 shiitake mushroom(mushroom pyogo), 3g red chilli, 3g green chilli, 3g spring onion, 2g garlic, 1g ginger, 15g rice-wine, 5g sesame oil
***Sauce** : 10g thick soy sauce, 15g red pepper paste(gochujang), 20g chili powder, 1g sesame, 5g sand eel sauce
***Note** : You can use the recipe to cook baby octopus and squid.

낙지 손질 八爪鱼的处理方法 How to prepare the small octopus

채소 손질 蔬菜的处理方法 How to prepare the vegetables

만드는 방법

1. 낙지를 소금으로 닦는다.
2. 낙지머리를 뒤집는다.
3. 내장을 제거한다.
4. 낙지는 5~7cm로 자른다.
5. 양파는 3x5cm로 자른다.
6. 단호박을 1/2로 자른다.
7. 편으로 썬다.
8. 당근을 편으로 썬다.
9. 실파를 5cm로 자른다.
10. 풋고추는 어슷썰기
11. 표고버섯은 편 썬다.
12. 느타리버섯은 찢는다.

制作方法

1. 用盐清洗八爪鱼。
2. 翻出八爪鱼头。
3. 去除内脏。
4. 将八爪鱼切成大小5~7cm。
5. 将洋葱切成大小3x5cm。
6. 将西葫芦切成两半。
7. 切成片状。
8. 将胡萝卜切成片状。
9. 将小葱切成大小5cm。
10. 斜切青辣椒。
11. 将香菇切成片状。
12. 用手撕好平菇。

How to cook

1. Rub the small octopus with salt.
2. Have the head turn inside out.
3. Remove internal organs.
4. Cut the small octopus 5~7cm.
5. Cut the onion 3x5cm.
6. Cut the squash in half.
7. Cut it into slices.
8. Cut the carrot into thin slices.
9. Cut the small green onion 5cm.
10. Have the green chili cut diagonally.
11. Cut the shiitake into thin slices.
12. Tear oyster mushroom.

양념장 만들기
制作调味酱
How to make marinade

만드는 방법

13. 채소를 준비한다.
14. 다진 마늘, 다진 생강을 준비한다.
15. 팬에 식용유를 두른다.
16. 호박을 볶는다.
17. 당근을 넣는다.
18. 양파, 버섯을 볶는다.
19. 풋고추를 넣는다.
20. 낙지를 볶는다.
21. 진간장 10g, 고추장 15g, 통깨 1g를 넣는다.
22. 액젓을 5g, 고춧가루 20g을 넣는다.
23. 양념장을 넣는다.
24. 청주 15g, 마늘 2g, 생강 1g을 넣는다.

制作方法

13. 准备好蔬菜。
14. 准备好蒜泥和捣碎的生姜。
15. 在平底锅上倒入食用油。
16. 炒西葫芦。
17. 放入胡萝卜。
18. 放入洋葱和蘑菇一起炒。
19. 放入青辣椒。
20. 炒八爪鱼。
21. 加10g浓酱油、15g辣椒酱和1g芝麻。
22. 放入5g鱼酱和20g辣椒面儿。
23. 放入调味酱。
24. 加入15g白酒和2g大蒜及1g生姜。

How to cook

13. Prepare the vegetables.
14. Have the choped garlic and chopped ginger with you.
15. Put cooking oil in a pan.
16. Fry squash.
17. Put the carrot.
18. Roast the onion and the mushroom.
19. Put the green chilli.
20. Stir-fry the small octopus.
21. Put 10g of thick soy sauce, 15g of gochujang and 1g of sesame.
22. Put 5g of sand eel sauce and 20g of chili powder.
23. Put the marinade.
24. Put 15g of clear rice-wine, garlic and ginger.

만드는 방법

25. 참기름 5g을 넣는다.
26. 익으면 담는다.
27. 완성하여 담는다.

制作方法

25. 加入香油5g。
26. 待熟后盛放在容器里。
27. 完成后盛菜。

How to cook

25. Pour 5g of sesame oil.
26. When the content is fully cooked, put it in a plate.
27. Enjoy the dish.

TIP
建议

낙지볶음을 할 때 수분이 나오지 않도록 신속하게 볶는다.
낙지의 양이 많을 때는 살짝 데쳐서 사용한다.

炒八爪鱼时要快速地炒，以防止出现水分。
当八爪鱼的量较多时，稍微焯一下再使用。

Stir-fry the small octopus quickly so that the dish will not get moist because of water from the small octopus.
When you have too many octopuses, start cooking after you have your small octopus blanched in water.

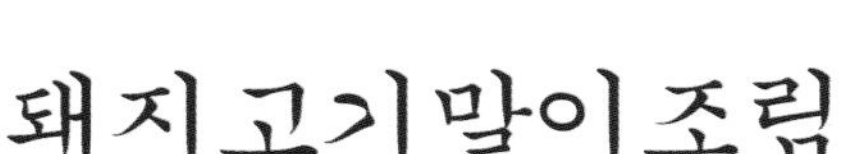

돼지고기말이조림

炖猪肉卷

Rolled pork cooked in soy sauce

재료 및 분량(2인분)
돼지고기(앞다리) 200g, 피망(3색) 6g, 매실즙 3g, 소금 2g, 참기름 5g, 후춧가루 0.5g
***간장양념** : 진간장 30g, 설탕 10g, 올리고당 10g, 채소류 10g
***육수** : 다시마 5g
***응용요리재료** : 소고기, 돼지고기, 닭(가슴살, 안심살) 등으로 요리한다.

材料和份量(2人份)
猪肉(前腿) 200g, 大辣椒(3色) 6g, 梅子汁 3g, 盐 2g, 香油 5g, 胡椒粉 0.5g
***酱油调味料** : 浓酱油 30g, 白糖 10g, 低聚糖 10g, 蔬菜 10g
***肉汤**: 海带 5g
***应用料理材料**: 可利用牛肉、猪肉和鸡肉(鸡胸脯肉和里脊肉)等。

Ingredients and portion(2 portions)
200g pork(arm shoulder), 6g bell pepper(in 3 different colors), 3g plum juice, 2g salt, 5g sesame seed oil, 0.5g ground black pepper
***Soy sauce** : 30g thick soy sauce, 10g sugar, 10g oligosaccharide, 10g vegetables
***Stock** : 5g dried sea tangle
***Note** : You can cook also with beef, other parts of pork and chicken(breast, tenderloin).

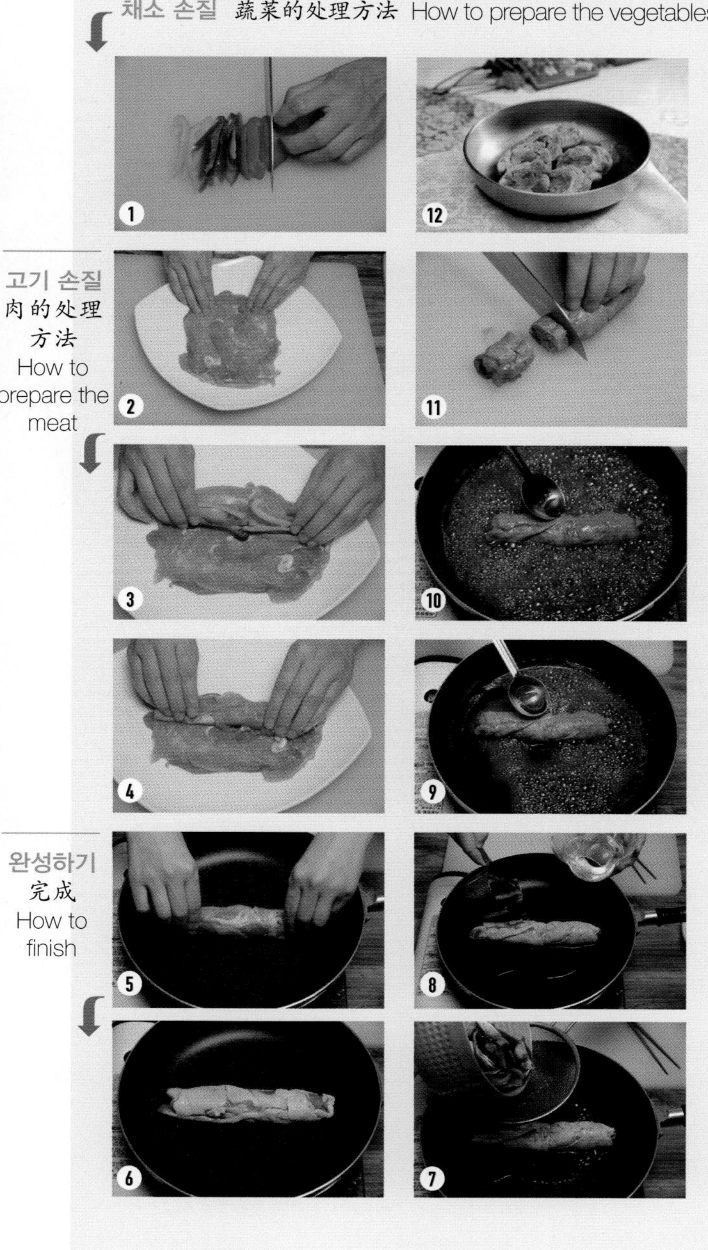

만드는 방법

1. 피망은 색색으로 채 썬다.
2. 돼지고기에 소금, 후춧가루를 넣는다.
3. 피망을 넣는다.
4. 둥그렇게 말이를 한다.
5. 프라이팬에 붙는 부분을 지진다.
6. 노릇하게 익혀 준비한다.
7. 간장소스를 붓는다.
8. 다시마육수를 100g 넣는다.
9. 참기름을 5g 넣는다.
10. 매실즙을 3g 넣는다.
11. 2cm 정도로 잘라서 준비한다.
12. 완성하여 담는다.

制作方法

1. 将各色大辣椒分别切好后备用。
2. 在猪肉上撒上盐和胡椒粉。
3. 放入大辣椒。
4. 卷成圆卷状。
5. 将粘在平底锅上的部分煎一下。
6. 煎成金黄色后准备好备用。
7. 倒入酱油调味料。
8. 加入100g海带汤。
9. 加入5g香油。
10. 加入3g梅子汁。
11. 切成大小2cm左右后准备好备用。
12. 完成后盛菜。

How to cook

1. Julienne the bell peppers.
2. Season the pork with salt and pepper.
3. Put the bell pepper.
4. Roll the meat over.
5. Fry a surface which would be contacted in a frying-pan.
6. Cook the meat until it gets golden brown.
7. Put thick soy sauce.
8. Put 100g of kelp stock.
9. Add 5g of sesame oil.
10. Pour 3g of plum juice.
11. Cut the beef about 2cm.
12. Finish.

TIP
建议

6. 팬에서 겉만 완전히 익혀 단단하게 고정시킨다.

6. 在平底锅上只须将表面部分完全煎熟后即可固定住。

Step 6. Get the pork fully cooked only outside in a pan.

돼지불고기볶음

炒猪肉

Fried pork bulgogi

재료 및 분량(2인분)
돼지고기(등심) 200g, 돼지갈비소스 50g, 느타리버섯 3g, 양파 2g, 쥬키니호박 2g, 사과 5g, 대파 2, 생강 1g, 마늘 3g, 찹쌀가루 5g, 청주 15g, 참기름 3g
***응용요리재료** : 소고기, 돼지고기, 닭고기류 등에 다양하게 요리한다.

材料和份量(2人份)
猪肉(里脊) 200g, 猪肉调味料 50g, 平菇 3g, 洋葱 2g, 菜瓜 2g, 苹果 5g, 大葱 2, 生姜 1g, 大蒜 3g, 粘米面儿 5g, 白酒 15g, 香油 3g
***应用料理材料**: 可利用牛肉、猪肉和鸡肉类等制作各种料理。

Ingredients and portion(2 portions)
200g pork(loin), 50g pork rib sauce, 3g oyster mushroom, 2g onion, 2g zucchini, 5g apple, 2g leek, 1g ginger, 3g garlic, 5g Glutinous rice flour, 15g rice-wine, 3g sesame oil
***Note** : You can use the recipe to cook with beef, pork and chicken.

고기 손질 肉的处理方法 How to prepare the pork

양념장 만들기
制作调味酱
How to make the marinade

만드는 방법

1. 돼지고기를 부드럽게 손질한다.
2. 돼지갈비소스를 50g 넣는다.
3. 생강즙을 넣는다.
4. 청주를 15g 넣어 볶는다.
5. 채소, 버섯류를 넣는다.
6. 사과즙을 5g 넣는다.
7. 다진 마늘을 3g 넣는다.
8. 참기름을 3g 넣는다.
9. 찹쌀가루를 5g 넣는다.
10. 농도를 부드럽게 완성한다.
11. 따뜻하게 담는다.
12. 완성하여 담는다.

制作方法

1. 处理猪肉使其软化。
2. 加入50g猪排调味料。
3. 加入生姜汁。
4. 加入15g白酒后炒一下。
5. 放入蔬菜和蘑菇类。
6. 加5g苹果汁。
7. 放入3g蒜泥。
8. 加入3g香油。
9. 加5g糯米粉。
10. 调整浓度并使味道柔嫩。
11. 盛放在容器里。
12. 完成后盛菜。

How to cook

1. Make the pork soft before cooking.
2. Put 50g of pork ribs sauce.
3. Put ginger juice.
4. Pour 15g of rice-wine and stir-fry the ingredients.
5. Put the vegetables and mushroom.
6. Add 5g of apple juice.
7. Put 3g of chopped garlic.
8. Put 3g of sesame oil.
9. Add 5g of glutinous rice flour.
10. Make the consistency soft.
11. Serve the dish warm.
12. Finish.

TIP 建议

9. 찹쌀가루를 소량 넣으면 양념장국물이 부드럽게 완성된다.

9. 放入少量的糯米粉会使调料酱汤口味更加柔和。

Step 9. You can make the consistency soft by putting a small amount of glutinous rice flour.

두부돼지고기조림

猪肉炖豆腐

Tofu-pork cooked in soy sauce

재재료 및 분량(2인분)
두부 100g, 돼지고기(등심) 20g, 양파 5g, 홍고추 3g, 풋고추 3g, 표고버섯 2g, 애호박 2g, 느타리버섯 2g, 들기름 3g
***양념장** : 고추장 20g, 국간장 3g, 설탕 5g, 올리고당 5g, 다진 마늘 2g, 청주 10g, 통깨 2g
***육수** : 다시마 5g, 국멸치 5마리, 마늘 1g, 홍고추 1g, 풋고추 1g

材料和份量(2人份)
豆腐 100g, 猪肉(里脊) 20g, 洋葱 5g, 红辣椒 3g, 青辣椒 3g, 香菇 2g, 西葫芦 2g, 平菇 2g, 芝麻油 3g
***调味酱** : 辣椒酱 20g, 汤用酱油 3g, 白糖 5g, 低聚糖 5g, 蒜泥 2g, 白酒 10g, 芝麻 2g
***肉汤**: 海带 5g, 汤用鳀鱼 5条, 大蒜 1g, 红辣椒, 青辣椒 1g

Ingredients and portion(2 portions)
100g bean curd(tofu), 20g pork(loin), 5g onion, 3g red chilli, 3g green chilli, 2g shiitake mushroom(mushroom shiitake), 2g long squash, 2g oyster mushroom, 3g perilla oil
***Sauce** : 20g red pepper paste(gochujang), 3g soy sauce, 5g sugar, 5g oligosaccharide, 2g chopped garlic, 10g rice-wine, 2g sesame
***Meat stock** : 5g dried sea tangle, 5 anchovies, 1g garlic, 1g red chilli, 1g green chilli

육수 끓이기 煮肉汤 How to cook stock

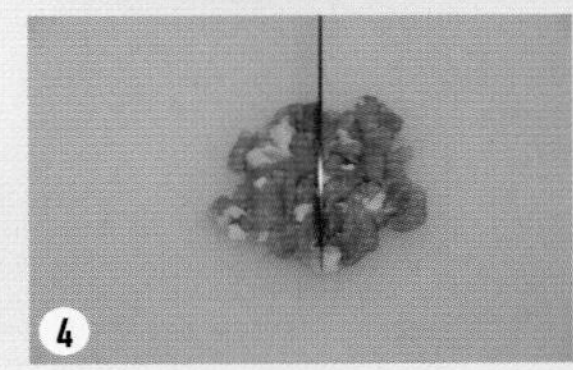

만드는 방법

1. 육수를 끓인다.
2. 두부를 편 썬다.
3. 팬에 노릇하게 굽는다.
4. 돼지고기를 부드럽게 두들긴다.
6. 냄비에 돼지고기, 양념장을 볶는다.
7. 모든 재료를 올린다.
8. 육수 150g을 넣고 끓인다.
9. 들기름 3g을 넣고 완성한다.

制作方法

1. 煮肉汤。
2. 将豆腐切成片状。
3. 在平底锅上煎成金黄色。
4. 敲打猪肉使其软化。
6. 将猪肉和调味酱放入锅里一起炒。
7. 放入所有材料。
8. 加入肉汤150g后一起煮。
9. 加3g芝麻油后完成。

How to cook

1. Cook stock.
2. Cut tofu into thin slices.
3. Cook the ingredients in a pan until it gets golden brown.
4. Pound the pork soft.
6. Stir-fry the pork in a pot with the marinade.
7. Put all the ingredients together.
8. Pour 150g of stock and have them boiled.
9. Pour 3g of sesame oil and have the dish completed.

양념장 만들기 制作调味酱 How to cook marinade

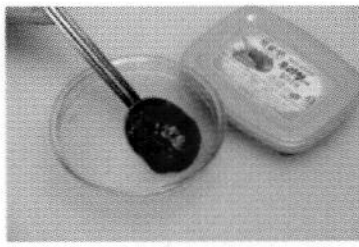

5-1. 고추장을 20g 넣는다.

5-1. 放入辣椒酱20g。

5-1. Put 20g of gochujang.

5-2. 국간장을 3g 넣는다.

5-2. 加入汤用酱油3g。

5-2. Put 3g of thick soy sauce.

5-3. 설탕을 5g 넣는다.

5-3. 放入白糖5g。

5-3. Add 5g of sugar.

5-4. 올리고당을 5g 넣는다.

5-4. 加入低聚糖5g。

5-4. Put 5g of oligosaccharide.

5-5. 다진 마늘을 2g 넣는다.

5-5. 放入蒜泥2g。

5-5. Add 2g of chopped garlic.

5-6. 청주 10g, 통깨를 2g 섞는다.

5-6. 加入白酒10g和芝麻2g后混合。

5-6. Mix them with 10g of rice-wine and 2g of sesame.

TIP 建议

3. 두부를 구워서 요리하면 모양이 깨지지 않는다.

3. 如果将豆腐煎好后再使用的话，豆腐的模样就不会变形。

Step 3. You can keep the tofu as it is supposed to be if you roast it beforehand.

두부소고기조림

牛肉炖豆腐

Tofu-beef cooked in soy sauce

재료 및 분량(2인분)
두부 150g, 소고기(안심) 20g, 양파 5g, 실파 5g, 불고기양념장 15g, 국간장 5g, 마늘 2g, 흰깨 1g, 검정깨 1g, 참기름 2g, 후춧가루 0.5g
***육수** : 다시마 5g, 국멸치 4마리, 마늘 1g, 마른 새우 1g, 감초 1g

材料和份量(2人份)
豆腐 150g, 牛肉(里脊) 20g, 洋葱 5g, 小葱 5g, 烤肉调味酱 15g, 汤用酱油 5g, 大蒜 2g, 白芝麻 1g, 黑芝麻 1g, 香油 2g, 胡椒粉 0.5g
***肉汤**: 海带 5g, 汤用鳀鱼 4条, 大蒜 1g, 干虾 1g, 甘草 1g

Ingredients and portion(2 portions)
150g bean curd(tofu), 20g beef(tenderloin), 5g onion, 5g spring onion, 15g bulgogi sauce, 5g soy sauce, 2g garlic, 1g white sesame, 1g black sesame, 2g sesame oil, 0.5g ground black pepper
***Stock** : 5g dried sea tangle, 4 anchovies, 1g garlic, 1g dried shrimp, 0.5g licorice

육수 끓이기 煮肉汤 How to cook stock

소고기 손질 牛肉的处理方法 How to prepare the beef

만드는 방법

1. 육수를 끓인다.
2. 소고기를 두들긴다.
3. 불고기양념장 15g을 넣는다.
4. 소고기를 볶아준다.
6. 실파를 송송 썰어 5g 넣는다.
7. 흰깨, 검정깨를 1g씩 넣는다.
8. 따뜻하게 담는다.
9. 완성하여 담는다.

制作方法

1. 煮肉汤。
2. 敲打牛肉。
3. 放入15g烤肉调味酱。
4. 炒牛肉。
6. 放入5g小葱花。
7. 白芝麻和黑芝麻分别放入1g。
8. 盛放在容器里。
9. 完成后盛菜。

How to cook

1. Cook stock.
2. Pound the beef.
3. Put 15g of bulgogi sauce.
4. Stir-fry the beef.
6. Add 5g of chopped small green onion.
7. Put 1g of white sesame and black sesame respectively.
8. Serve the dish warm.
9. Enjoy.

끓이기 煮粥 How to boil

5-1. 두부와 다진 양파를 넣는다.
5-1. 放入豆腐和切碎的洋葱。
5-1. Put tofu and chopped onion.

5-2. 육수를 200g 넣는다.
5-2. 加入肉汤200g。
5-2. Pour 200g of stock.

5-3. 국간장 5g으로 간 맞춘다.
5-3. 加入汤用酱油5g调味。
5-3. Season the content with 5g of soy sauce.

5-4. 다진 마늘을 2g 넣는다.
5-4. 放入蒜泥2g。
5-4. Put 2g of chopped garlic.

5-5. 후춧가루를 1g 넣는다.
5-5. 放入胡椒粉1g。
5-5. Put 0.5g of pepper powder.

5-6. 참기름을 2g 넣는다.
5-6. 加入香油2g。
5-6. Pour 2g of sesame oil.

TIP 建议

6. 두부조림이 완전히 완성되었을 때 실파를 넣어야 색을 살릴 수 있다.

6. 只有在完全炖熟后放入葱花颜色才会更新鲜。

Step 6. Put the small green onion after the tofu is fully cooked with soy sauce. You can preserve the beautiful green color of the small green onion.

뚝배기불고기

石锅烤肉

Ttukbaegi bulgogi

재료 및 분량(2인분)
소고기(안심) 300g, 당면 5g, 느타리버섯 5g, 표고버섯 2g, 실파 2g, 당근 2g, 양파 2g, 대파 2g, 참기름 5g
***양념장** : 소갈비양념 80g, 다시마 3g, 국멸치 2마리, 양파 3g, 마늘 2g
***응용요리재료** : 뚝배기궁중떡볶이 요리 등으로 활용된다.

材料和份量(2人份)
牛肉(里脊) 300g, 粉丝 5g, 平菇 5g, 香菇 2g, 小葱 2g, 胡萝卜 2g, 洋葱 2g, 大葱 2g, 香油 5g
***调味酱** : 调味牛排 80g, 海带 3g, 汤用鳀鱼 2条, 洋葱 3g, 大蒜 2g
***应用料理材料**: 可用于制作石锅宫中炒年糕等料理。

Ingredients and portion(2 portions)
300g beef(tenderloin), 5g starch noodle, 5g oyster mushroom, 2g shiitake mushroom(mushroom pyogo), 2g spring onion, 2g carrot, 2g onion, 2g leek, 5g sesame oil
***Sauce** : 80g beef rib barbecue sauce, 3g dried sea tangle, 2 anchovies, 3g onion, 2g garlic
***Note** : You can use the recipe to cook the royal ttukbbaegi tteokbbokki.

재료 손질 材料处理方法 Clean ingredients

만드는 방법

1. 소갈비양념소스를 80g 넣는다.
2. 다시마, 국멸치, 양파, 마늘을 넣는다.
3. 체에 내려 소스를 완성한다.
4. 소고기에 양념장을 넣는다.
5. 볶아준다.
6. 육수를 100g 넣는다.
7. 버섯류를 넣는다.
8. 채소류를 넣는다.
9. 당면을 불린다.
10. 당면은 마지막에 올린다.
11. 참기름을 5g 넣는다.
12. 완성하여 담는다.

制作方法

1. 放入80g牛排调味料。
2. 放入海带、青鳞鱼、洋葱和大蒜。
3. 用过滤网过滤出调味汁。
4. 在牛肉中放入调味酱。
5. 一起炒。
6. 加入100g肉汤。
7. 放入蘑菇类。
8. 放入蔬菜。
9. 将粉丝泡在水里。
10. 最后放入粉丝。
11. 加入5g香油。
12. 完成后盛菜。

How to cook

1. Put 80g of beef ribs sauce.
2. Put kelp, anchovies, onion and garlic.
3. Strain the sauce through a sieve.
4. Put marinade to the beef.
5. Stir-fry the ingredients.
6. Pour 100g of stock.
7. Put mushroom.
8. Add the vegetables.
9. Soak starch noodle in water.
10. Put the noodle at the last.
11. Add 5g of sesame oil.
12. Finish.

TIP
建议

9. 당면은 불려서 사용해야 부드럽게 익는다.

9. 粉丝只有用水泡开后口感才会柔嫩。

Step 9. Have the noodle soaked to cook it soft.

멸치고추장볶음

辣炒小银鱼

Anchovy fried with gochujang

재료 및 분량(2인분)
잔멸치 100g, 고추장 30g, 호박씨 5g, 마늘 2g, 실파 3g, 검정깨 1g, 흰깨 1g, 참기름 3g
***고추장양념장** : 고추장 30g, 올리고당 15g, 청주 15g

材料和份量(2人份)
小鳀鱼 100g, 辣椒酱 30g, 南瓜籽 5g, 大蒜 2g, 小葱 3g, 黑芝麻 1g, 白芝麻 1g, 香油 3g
***调味辣椒酱** : 辣椒酱 30g, 低聚糖 15g, 白酒 15g

Ingredients and portion(2 portions)
100g dried small anchovies, 30g red pepper paste (gochujang), 5g pumpkin seeds, 2g garlic, 3g spring onion, 1g black sesame, 1g white sesame, 3g sesame oil
***Red pepper paste sauce(gochujang sauce)** : 30g red pepper paste(gochujang), 15g oligosaccharide, 15g rice-wine

재료 손질 材料处理方法 Clean ingredients

만드는 방법

1. 고추장을 30g 넣는다.
2. 올리고당을 15g 넣는다.
3. 청주를 15g 넣는다.
4. 골고루 섞는다.
5. 실파는 살짝 데친다.
6. 팬에 고추장양념장을 붓는다.
7. 잔멸치를 볶는다.
8. 검정깨, 흰깨를 각각 1g씩 넣는다.
9. 참기름 3g, 마늘 2g을 넣는다.
10. 데친 실파를 넣는다.
11. 볶음멸치를 올려 완성한다.
12. 완성하여 담는다.

制作方法

1. 放入30g辣椒酱。
2. 放入15g低聚糖。
3. 加入15g白酒。
4. 均匀地混合。
5. 将小葱焯一下。
6. 在锅里加入调味辣椒酱。
7. 炒小鳀鱼。
8. 黑芝麻和白芝麻分别放入1g。
9. 加入3g香油和2g大蒜。
10. 放入焯好的小葱。
11. 将炒好的小银鱼盛放在容器里。
12. 完成后盛菜。

How to cook

1. Put 30g of gochujang.
2. Put 15g of oligosaccharide.
3. Add 15g of rice-wine.
4. Mix them all.
5. Slightly blanch the small green onion in water.
6. Pour gochujang marinade into a pan.
7. Stir-fry the little anchovies.
8. Put 1g of black sesame and white sesame.
9. Add 3g of sesame oil and 2g of garlic.
10. Put the blanched small green onion.
11. Put the fried anchovy and you have finished cooking.
12. Finish.

TIP 建议

6. 고추장양념장의 농도가 잡히면 잔멸치가 들어간다. 오래 볶으면 비린내가 난다.

6. 等到调味辣椒酱的浓度变浓后，把小鳀鱼放进去。如果炒得时间太长的话会有腥味。

Step 6. When you have right consistency for gochujang marinade, put the anchovies. Do not stir-fry the fish long. It will smell fishy.

병어무조림

萝卜炖鲳鱼

Pomfret-radish cooked with soy sauce

재료 및 분량(2인분)
병어(1kg) 1마리, 무 40g, 양파 5g, 대파 5g, 홍고추 3g, 풋고추 3g
***양념장** : 진간장 30g, 멸치육수 200g, 고춧가루 10g, 올리고당 10g, 다진 마늘 3g, 생강 2g, 청주 5g, 통깨 3g, 참기름 3g, 후춧가루 0.5g

材料和份量(2人份)
鲳鱼(1条) 1kg, 萝卜 40g, 洋葱 5g, 大葱 5g, 红辣椒 3g, 青辣椒 3g
***调味酱** : 浓酱油 30g, 鳀鱼汤 200g, 辣椒面儿 10g, 低聚糖 10g, 蒜泥 3g, 生姜 2g, 白酒 5g, 芝麻 3g, 香油 3g, 胡椒粉 0.5g

Ingredients and portion(2 portions)
1kg pomfret fish, 40g white radish, 5g onion, 5g leek, 3g red chilli, 3g green chilli
***Sauce** : 30g thick soy sauce, 200g anchovy stock, 10g chili powder, 10g oligosaccharide, 3g chopped garlic, 2g ginger, 5g rice-wine, 3g sesame, 3g sesame oil, 0.5g ground black pepper

병어 손질 鲳鱼的处理方法 How to prepare pomfret

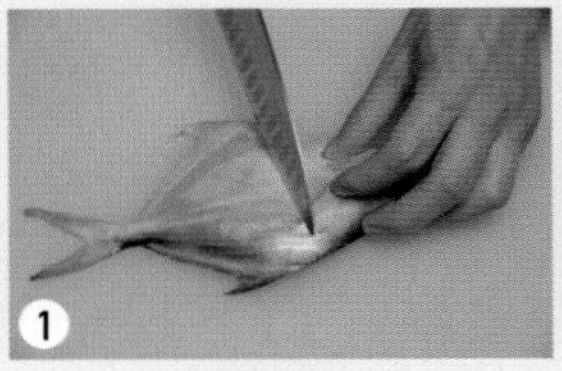

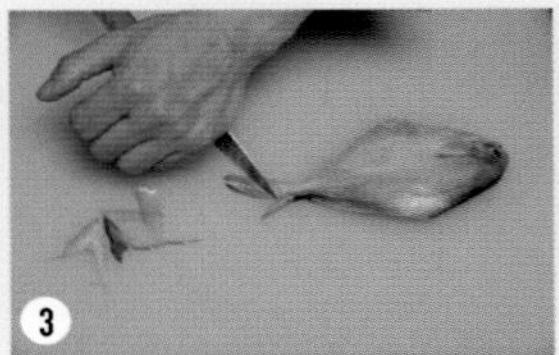

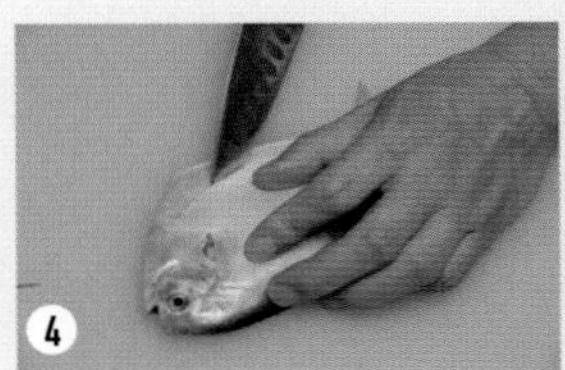

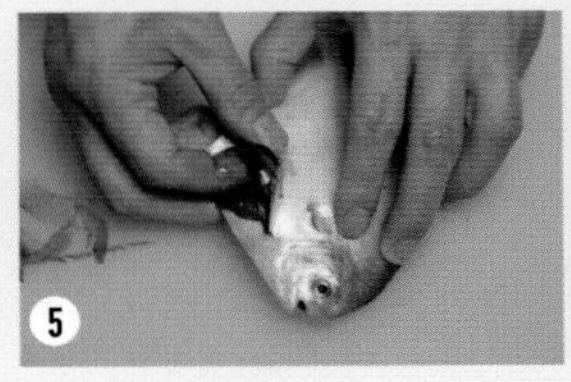

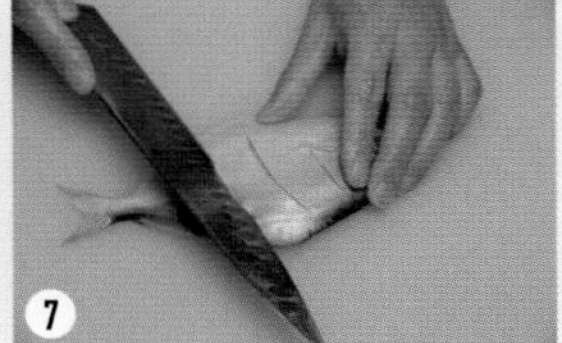

만드는 방법

1. 비늘을 제거한다.
2. 지느러미를 제거한다.
3. 꼬리를 손질한다.
4. 한쪽만 칼집을 넣는다.
5. 내장을 제거한다.
6. 소금을 뿌린다.
7. 칼집을 넣는다.
8. 생선 손질을 완성한다.
9. 진간장 30g과 육수 200g을 붓는다.
10. 고춧가루를 10g 넣는다.
11. 설탕을 5g 넣는다.
12. 다진 마늘 3g, 생강 2g을 넣는다.

制作方法

1. 去除鱼鳞。
2. 去除鱼鳍。
3. 处理鱼尾。
4. 只在一侧切出刀口。
5. 去除内脏。
6. 撒上盐。
7. 切出刀口。
8. 处理鲳鱼完成。
9. 加入30g浓酱油和200g肉汤。
10. 放入10g辣椒面儿。
11. 放入5g白糖。
12. 放入3g蒜泥和2g生姜。

How to cook

1. Remove scales.
2. Remove fin.
3. Trim the tail.
4. Make cuts on one side only.
5. Remove the internal organs.
6. Sprinkle salt.
7. Make cuts.
8. Have the fish well trimmed.
9. Pour 30g of thick soy sauce and 200g of stock.
10. Add 10g of chili powder.
11. Put 5g of sugar.
12. Add 3g of chopped garlic and 2g of ginger.

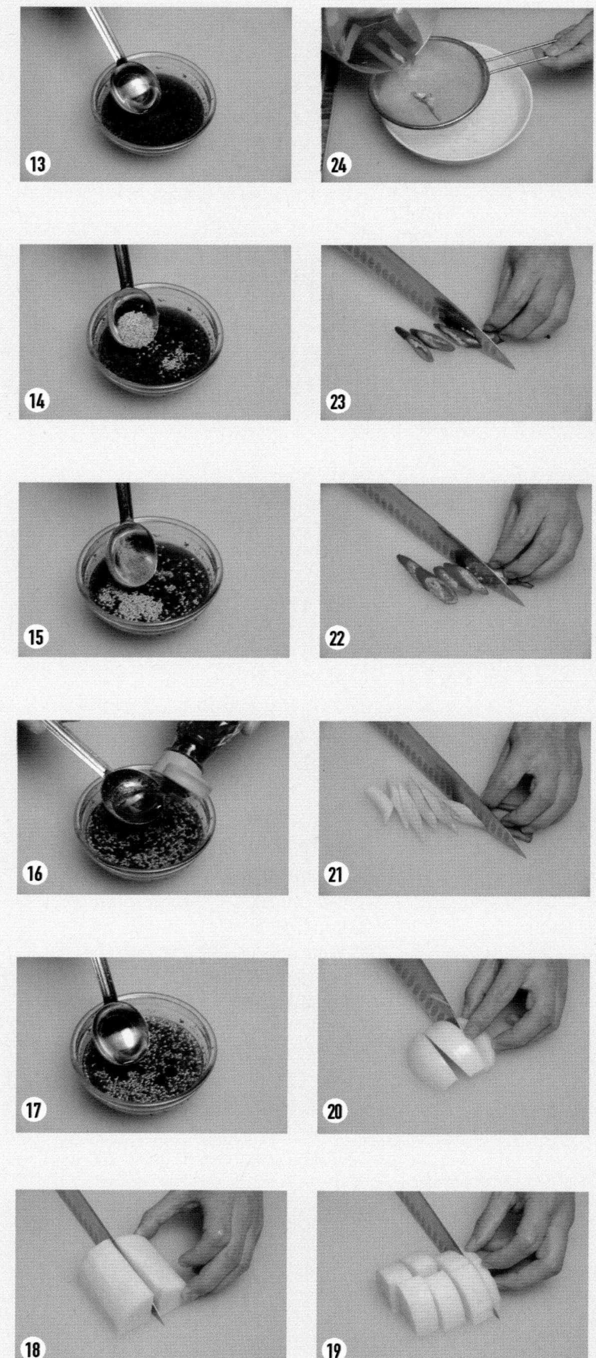

만드는 방법

13. 청주를 5g 넣는다.
14. 통깨를 3g 넣는다.
15. 후춧가루를 0.5g 넣는다.
16. 올리고당을 10g 넣는다.
17. 참기름을 3g 넣는다.
18. 무를 2등분한다.
19. 굵직하게 편 썬다.
20. 양파를 편 썬다.
21. 대파는 0.5cm로 어슷썬다.
22. 홍고추는 0.5cm로 어슷썬다.
23. 풋고추는 0.5cm로 어슷썬다.
24. 육수를 끓여 준비한다.

制作方法

13. 加入5g白酒。
14. 放入3g芝麻。
15. 放入0.5g胡椒粉。
16. 放入10g低聚糖。
17. 加入3g香油。
18. 将萝卜2等分。
19. 将萝卜切成粗厚的片状。
20. 将洋葱切成片状。
21. 将大葱斜切成大小0.5cm。
22. 将红辣椒斜切成大小0.5cm。
23. 将青辣椒斜切成大小0.5cm。
24. 煮好肉汤后备用。

How to cook

13. Put 5g of clean rice-wine.
14. Put 3g of sesame.
15. Add 0.5g of pepper powder.
16. Put 10g of oligosaccharide.
17. Put 3g of sesame oil.
18. Cut the radish in half.
19. Cut the radish into thick slices.
20. Cut the onion into slices.
21. Cut the spring onion 0.5cm diagonally.
22. Cut the red chilli 0.5cm diagonally.
23. Cut the green chilli 0.5cm diagonally.
24. Boil stock.

끓이기
煮制
How to boil

만드는 방법
25. 무를 넣는다.
26. 무 위에 채소를 올린다.
27. 병어를 넣는다.
28. 병어 위에 채소류를 올린다.
29. 양념장을 붓는다.
30. 무가 익을 때까지 끓인다.
31. 병어를 통째로 담는다.
32. 완성하여 담는다.

制作方法
25. 放入萝卜。
26. 放上蔬菜。
27. 放入鲳鱼。
28. 放上蔬菜。
29. 倒入调味酱。
30. 一起煮直至萝卜熟透。
31. 盛放在容器里。
32. 完成后盛菜。

How to cook
25. Put the radish.
26. Put the vegetables.
27. Put the pomfret.
28. Put the vegetables on it.
29. Pour the marinade.
30. Boil the content until you get the radish fully cooked.
31. Serve the dish warm.
32. Enjoy.

TIP
建议

병어는 내장만 제거하고 통째로 요리한다.

只去除鲳鱼的内脏，然后将整条鱼放入锅里炖。

Cook the whole pomfret at a time without internal organs only.

삼합

鱼肉鲜

Samhap

재료 및 분량(2인분)
홍어 300g, 묵은 김치 200g, 삼겹살 편육 70g, 홍고추 3g, 풋고추 3g
***양념장** : 소금, 고춧가루, 맛장 소량

材料和份量(2人份)
魟鱼 300g, 陈泡菜 200g, 五花肉片 70g, 红辣椒 3g, 青辣椒 3g
***调味酱**：盐，辣椒粉，少量调味酱

Ingredients and portion(2 portions)
300g skate(fish), 200g over-fermented Kimchi, 70g boiled pork belly slices, 3g red chilli, 3g green chilli
***Sauce** : A pinch of salt, chili powder and makjang.

홍어 손질 魟鱼的处理方法 How to prepare skate

묵은 김치 손질 陈泡菜的处理方法 How to prepare the over-fermented Kimchi

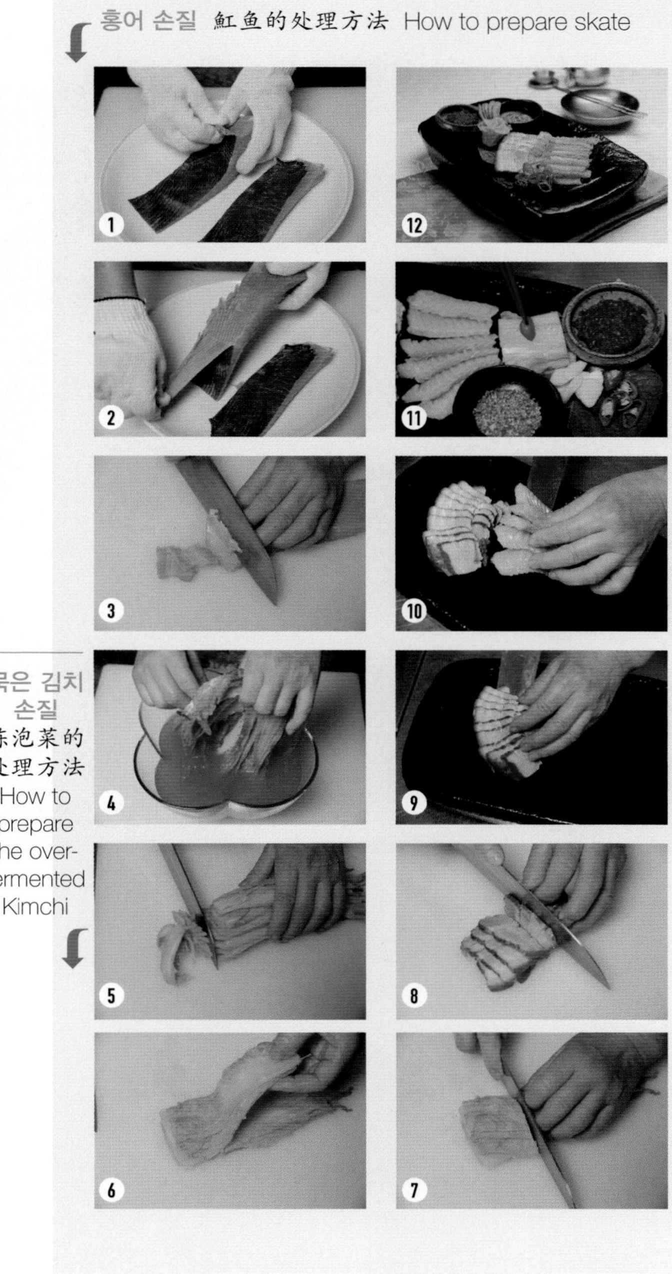

만드는 방법

1. 숙성된 홍어의 껍질을 제거한다.
2. 장갑을 끼고 미끄러지지 않게 제거한다.
3. 두께 0.3cm로 편 썬다.
4. 묵은 김치는 물에 헹궈 물기를 제거한다.
5. 기둥을 제거한다.
6. 한 장씩 겹친다.
7. 길이 5cm로 썬다.
8. 삼겹살 편육을 0.3cm로 썬다.
9. 완성접시에 담는다.
10. 홍어회를 올린다.
11. 묵은 김치를 담아 완성한다.
12. 완성하여 담는다.

制作方法

1. 将熟成的魟鱼去皮。
2. 为防滑最好戴着手套处理。
3. 切成厚0.3cm的片状。
4. 用水冲洗陈泡菜，然后控干水分。
5. 切掉泡菜根。
6. 将泡菜一张一张地铺开。
7. 切成长5cm。
8. 将五花肉片切成大小0.3cm。
9. 完成后盛放在盘子里。
10. 将魟鱼片放上去。
11. 放上陈泡菜后完成。
12. 完成后盛菜。

How to cook

1. Remove the peel of fermented skate.
2. Wear a pair of gloves so that it will not get slipped out of your hands.
3. Cut the fish into slices as thick as 0.3cm.
4. Wash the Kimchi in water and drain off the water.
5. Remove the Kimchi stalk.
6. Put a sheet of Kimchi on one another.
7. Cut Kimchi 5cm long.
8. Cut boiled pork belly by 0.3cm.
9. Serve the dish in a plate.
10. Put the skate raw in the plate.
11. Put Kimchi and serve the dish well.
12. Finish.

TIP
建议

1. 숙성된 홍어를 사용한다.

1. 使用熟成的魟鱼。

Step 1. Use fermented skate.

연근볶음

炒莲藕
Roasted lotus root

재료 및 분량(2인분)
연근 200g, 소고기(안심) 50g, 파프리카 4g, 브로콜리 2g, 찹쌀가루 5g, 마늘 2g

材料和份量(2人份)
莲藕 200g, 牛肉(里脊) 50g, 红灯笼辣椒 4g, 西兰花 2g, 粘米面儿 5g, 大蒜 2g

Ingredients and portion(2 portions)
200g lotus root, 50g beef(tenderloin), 4g paprika, 2g broccoli, 5g Glutinous rice flour, 2g garlic

재료 손질 材料处理方法 Clean ingredients

소고기 손질 牛肉的处理方法 How to prepare the beef

만드는 방법

1. 연근 껍질을 제거한다.
2. 연근 0.5cm로 편 썬다.
3. 끓는 물에 소금을 넣고 데친다.
4. 소고기는 2x3cm로 썬다.
5. 소갈비양념소스 15g으로 볶는다.
6. 청주를 15g 넣고 소고기를 볶는다.
7. 다진 마늘을 3g 넣는다.
8. 파프리카와 브로콜리를 넣는다.
9. 통깨를 2g 넣는다.
10. 참기름을 3g 넣는다.
11. 찹쌀가루 5g으로 농도를 조절한다.
12. 완성하여 담는다.

制作方法

1. 去掉莲藕的皮。
2. 将莲藕切成厚0.5cm的片。
3. 在烧开的水中放入盐焯一下。
4. 将牛肉切成大小2x3cm。
5. 放入15g牛肉调味料炒一下。
6. 加入15g白酒后再炒一下。
7. 放入3g蒜泥。
8. 放入红灯笼辣椒和西兰花。
9. 放入2g芝麻。
10. 加入3g香油。
11. 放入5g糯米粉调整浓度。
12. 完成后盛菜。

How to cook

1. Peel off the lotus root.
2. Cut the lotus root into 0.5cm slices.
3. Put salt into boiling water and blanch the lotus root.
4. Cut beef 2x3cm.
5. Stir-fry the beef with 15g of beef ribs sauce.
6. Pour 15g of rice-wine and the beef.
7. Put 3g of chopped garlic.
8. Put paprika and broccoli.
9. Add 2g of sesame.
10. Pour 3g of sesame oil.
11. Adjust the consistency with 5g of glutinous rice flour.
12. Finish.

TIP 建议

7. 연근을 충분히 익힌 다음 색깔 채소를 넣는다.

7. 待莲藕熟透后再放入有色蔬菜。

Step 7. Put the vegetables after the lotus root is fully cooked.

주꾸미볶음

辣炒小章鱼

Roasted baby octopus

재료 및 분량(2인분)
주꾸미 200g, 새우 50g, 피망 30g, 대파 2g, 홍고추 3g, 풋고추 3g, 양파 30g, 삶은 소면 50g, 다진 마늘 2g, 불고기양념소스 40g, 올리고당 5g, 검정깨 2g, 참기름 3g
***응용요리재료** : 낙지, 오징어, 해물류, 육류에 이용 가능한 요리이다.

材料和份量(2人份)
小章鱼 200g, 虾 50g, 大辣椒 30g, 大葱 2g, 红辣椒 3g, 青辣椒 3g, 洋葱 3g, 煮好的阳春面 50g, 蒜泥 2g, 烤肉调味料 40g, 低聚糖 5g, 黑芝麻 2g, 香油 3g
***应用料理材料** ：可用于章鱼，鱿鱼，海鲜类，肉类等料理。

Ingredients and portion(2 portions)
200g baby octopus, 50g shrimp, 30g bell peppers, 2g leek, 3g red chilli, 3g green chilli, 50g boiled starch noodle, 2g chopped garlic, 40g bulgogi sauce, 5g oligosaccharide, 2g black sesame, 3g sesame oil
***Note** : You can use the ingredients to cook other dishes of small octopus, squid, seafood and meat.

재료 손질 材料处理方法 Clean ingredients

만드는 방법

1. 주꾸미의 내장을 제거하고 데친다.
2. 팬에 주꾸미를 넣는다.
3. 불고기양념장을 40g 붓는다.
4. 다진 마늘을 5g 넣는다.
5. 피망을 손질한다.
6. 홍고추, 풋고추를 어슷썰기한다.
7. 대파를 어슷썰기한다.
8. 새우살과 채소류를 넣는다.
9. 올리고당을 5g 넣는다.
10. 참기름을 3g 넣는다.
11. 검정깨를 2g 넣는다.
12. 완성하여 담는다.

制作方法

1. 去除小章鱼的内脏，然后将小章鱼焯一下。
2. 将小章鱼放在平底锅上。
3. 加入40g烤肉调味料。
4. 放入5g蒜泥。
5. 处理大辣椒。
6. 斜切红辣椒和青辣椒。
7. 斜切大葱。
8. 放入虾肉和蔬菜类。
9. 放入5g低聚糖。
10. 加入3g香油。
11. 放入2g黑芝麻。
12. 完成后盛菜。

How to cook

1. Remove internal organs of the baby octopus and blanch it in water.
2. Put the baby octopus in pan.
3. Pour 40g of bulgogi sauce.
4. Put 5g of chopped garlic.
5. Trim paprika.
6. Cut the red chilli and the green chilli diagonally.
7. Cut the leek diagonally.
8. Put shrimp and vegetables.
9. Add 5g of oligosaccharide.
10. Pour 3g of sesame oil.
11. Spinkle 2g of black sesame.
12. Finish.

TIP
建议

1. 주꾸미는 데치면 완성요리에서 물을 줄일 수 있다.

1. 将小章鱼焯一下可以减少最终完成的料理中的水分。

Step 1. You can reduce water in a completed dish with the baby octopus blanched.

韩食美“味”秘方传授 100选 实践篇

Transmission of a Hundred Secrets of Korean Dishes Flavor 凉拌菜

한 식 의 맛 비 법 전 수 1 0 0 선

실기편

생채, 숙채

고사리들깨나물

芝麻蕨菜

Bracken-perilla herbs

재료 및 분량(2인분)
불린 고사리 300g, 들깨가루 30g, 들깨기름 30g, 실파 5g, 대파 2g, 마늘 3g, 통깨 2g, 국간장 3g
***육수** : 다시마 5g, 국멸치 2마리, 양파 2g

材料和份量(2人份)
泡好的蕨菜 300g, 野芝麻粉 30g, 芝麻油 30g, 小葱 5g, 大葱 2g, 大蒜 3g, 芝麻 2g, 汤用酱油 3g
***肉汤**: 海带 5g, 汤用鳀鱼 2条, 洋葱 2g

Ingredients and portion(2 portions)
300g soaked bracken(gosari), 30g perilla flour, 30g perilla oil, 5g of spring onion, 2g leek, 3g garlic, 2g sesame, 3g soy sauce
***Meat stock** : 5g dried sea tangle, 2 anchovies, 2g onion

고사리 손질 蕨菜的处理方法 How to prepare bracken

완성하기
完成
How to finish

만드는 방법

1. 고사리는 1시간 불린다.
2. 고사리를 끓는 물에 데친다.
3. 물기를 제거한다.
4. 길이 4cm로 자른다.
6. 육수를 80g 넣는다.
7. 약불에서 뚜껑 덮고 뜸 들인다.
8. 들깨가루 30g을 넣는다.
9. 송송 썬 대파, 실파를 넣는다.

制作方法

1. 将蕨菜放在水里泡1个小时。
2. 在烧开的水中将蕨菜焯一下。
3. 控干水分。
4. 切成长4cm。
6. 加入80g肉汤。
7. 盖上盖子在微火上炖。
8. 放入30g野芝麻粉。
9. 放入切好的大葱和小葱。

How to cook

1. Have the bracken blanched in water for one hour.
2. Blanch the bracken in boiling water.
3. Drain well.
4. Cut the bracken 4cm long.
6. Pour 80g of stock.
7. Put a lid on and leave the content get steamed over a low heat.
8. Put 30g of perilla flour.
9. Put the chopped spring onion and small green onion.

볶기
炒制
How to roast

5-1. 국간장을 3g 넣는다.
5-1. 加入汤用酱油3g。
5-1. Put 3g of soy sauce.

5-2. 다진 마늘을 3g 넣는다.
5-2. 放入蒜泥3g。
5-2. Put 3g of chopped garlic.

5-3. 통깨 3g을 넣는다.
5-3. 放入芝麻3g。
5-3. Add 3g of sesame.

5-4. 무쳐서 준비한다.
5-4. 混合好后备用。
5-4. Season the bracken with the ingredients.

5-5. 들기름 30g을 넣는다.
5-5. 加入芝麻油30g。
5-5. Put 30g of perilla oil.

5-6. 두꺼운 냄비에 볶아준다.
5-6. 在锅里炒。
5-6. Roast the content in a big pot.

TIP
建议

1. 마른 고사리는 하루 불리며, 불린 고사리는 데쳐서 바로 사용한다.

1. 将干蕨菜用水泡一天，将泡好的蕨菜用热水焯一下后马上使用。

Step 1. Keep the dried bracken in water overnight. Use the soaked bracken and blanch it for immediate use.

골뱅이무침

海螺拌菜

Whelk with vegetables and vinegar

재료 및 분량(2인분)
골뱅이 100g, 대파채 30g, 청오이 50g, 홍피망 20g, 깻잎 2g, 홍고추 3g, 풋고추 3g, 대파 2g, 국간장 3g, 설탕 5g, 들기름 5g
***양념장** : 배 50g, 마늘 5g, 양파 20g, 고추장 30g, 홍초 45g, 설탕 8g, 매실식초 10g, 올리고당 10g

材料和份量(2人份)
海螺 100g, 大葱丝 30g, 青黄瓜 50g, 红色大辣椒 20g, 苏子叶 2g, 红辣椒 3g, 青辣椒 3g, 大葱 2g, 汤用酱油 3g, 白糖 5g, 芝麻油 5g
***调味酱** : 梨 50g, 大蒜 5g, 洋葱 20g, 辣椒酱 30g, 红醋 45g, 白糖 8g, 梅子醋 10g, 低聚糖 10g

Ingredients and portion(2 portions)
100g whelk(snail), 30g julienne leek, 50g cucumber, 20g red bell pepper, 2g sesame leaf, 3g red chilli, 3g green chilli, 2g leek, 3g soy sauce, 5g sugar, 5g perilla oil
***Sauce** : 50g pear, 5g garlic, 20g onion, 30g red pepper paste(gochujang), 45g red vinegar, 8g sugar, 10g plum fruit vinegar, 10g oligosaccharide

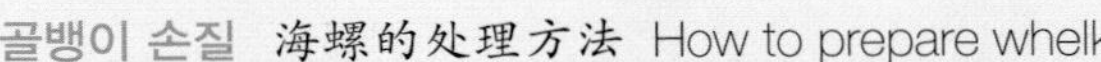

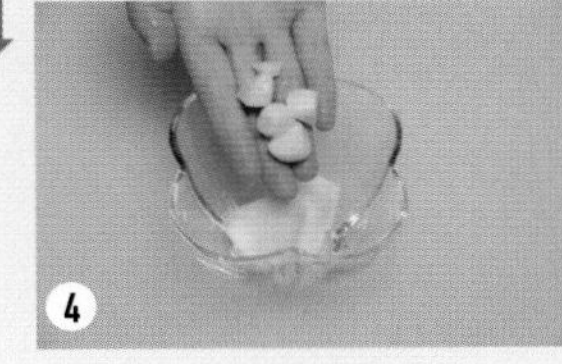

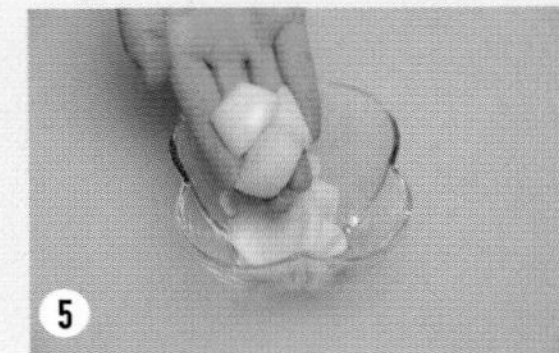

만드는 방법

1. 골뱅이를 3등분한다.
2. 매실식초 15g으로 헹군다.
3. 배는 편 썰어 20g 넣는다.
4. 마늘을 5g 넣는다.
5. 양파를 20g 넣는다.
6. 고추장을 30g 넣는다.
7. 홍초를 45g 넣는다.
8. 다진 마늘을 5g 넣는다.
9. 설탕을 8g 넣는다.
10. 매실식초를 10g 넣는다.
11. 올리고당을 10g 넣는다.
12. 양념재료를 섞는다.

制作方法

1. 将海螺3等分。
2. 用15g梅子醋冲洗一下。
3. 放入20g梨片。
4. 放入5g大蒜。
5. 放入20g洋葱。
6. 放入30g辣椒酱。
7. 加入45g红醋。
8. 放入5g蒜泥。
9. 放入8g白糖。
10. 加入10g梅子醋。
11. 放入10g低聚糖。
12. 将调味材料混合。

How to cook

1. Cut whelk into three pieces.
2. Wash the whelk in 15g of plum vinegar.
3. Cut the pear into thin slices and use only 20g of them.
4. Put 5g of garlic.
5. Put 20g of onion.
6. Add 30g of gochujang.
7. Add 45g of red vinegar.
8. Put 5g of chopped garlic.
9. Put 8g of sugar.
10. Pour 10g of plum vinegar.
11. Add 10g of oligosaccharide.
12. Mix the content all together.

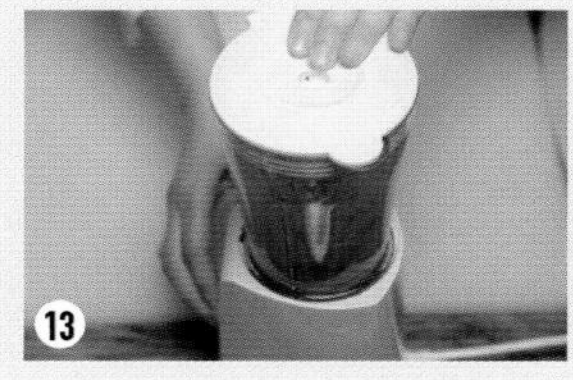

채소 손질
蔬菜的处理方法
How to prepare the vegetables

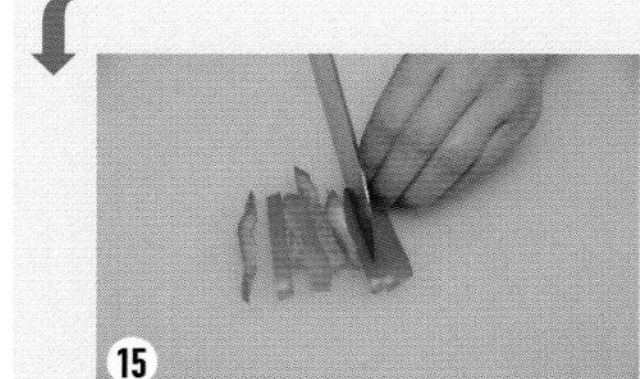

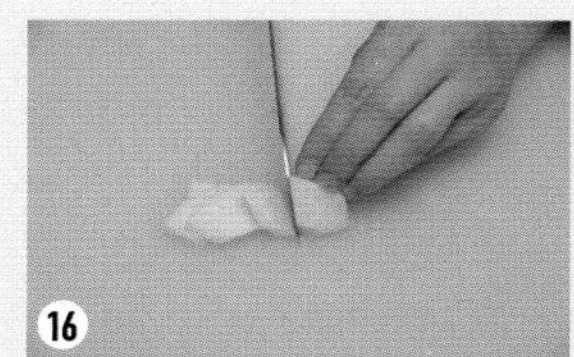

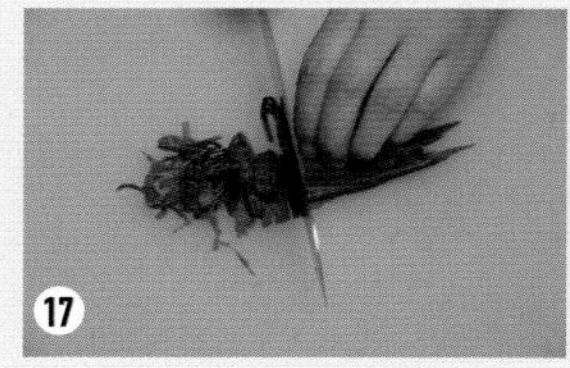

만드는 방법

13. 믹서기에 갈아 양념장을 만든다.
14. 오이를 자른다.
15. 피망을 0.5cm 채로 자른다.
16. 배를 2x3cm로 자른다.
17. 깻잎을 0.5cm 채로 자른다.
18. 양념장으로 무친다.
19. 무쳐 통깨를 넣는다.
20. 파채를 홍초 15g으로 무친다.
21. 국간장을 3g 넣는다.
22. 설탕을 5g 넣는다.
23. 들기름을 5g 넣는다.
24. 파채무침을 담는다.

制作方法

13. 用搅拌机搅碎后制成调味酱。
14. 将黄瓜切好备用。
15. 将大辣椒切成大小0.5cm。
16. 将梨切成大小2x3cm。
17. 将苏子叶切成大小0.5cm。
18. 用调味酱混合搅拌。
19. 放入芝麻。
20. 在葱丝中加入15g红醋。
21. 加入3g汤用酱油。
22. 放入5g白糖。
23. 加入5g芝麻油。
24. 将拌好的葱丝盛出来。

How to cook

13. Grind the content all in a blender.
14. Cut the cucumber.
15. Cut bell pepper into 0.5cm slices.
16. Cut the pear 2x3cm.
17. Julienne the sesame leaf 0.5cm.
18. Mix the ingredients with the marinade.
19. Mix the ingredients and put sesame.
20. Pour 15g of red vinegar to the spring onion slices.
21. Pour 3g of soy sauce.
22. Add 5g of sugar.
23. Pour 5g of perilla oil.
24. Put spring onion slices cooked with vegetables and vinegar.

만드는 방법

25. 골뱅이무침을 올린다.
26. 완성접시에 담는다.

制作方法

25. 放上拌好的海螺。
26. 完成后盛放在盘子里。

How to cook

25. Put the whelk cooked with vegetables and vinegar.
26. Place the food in a plate.

TIP
建议

20. 파 냄새를 제거하기 위해 홍초를 뿌려 놓으면 향이 살아난다.

20. 为了去除葱的味道，若撒上红醋的话，香味会更浓。

Step 20. Pour red vinegar to remove the smell of the spring onion. It will add sweet scent to the dish.

느타리버섯나물

平菇拌菜

Oyster mushroom with herbs

재료 및 분량(2인분)
느타리버섯 100g, 홍고추 3g, 풋고추 3g, 마늘 2g, 통깨 1g, 식용유 10g, 참기름 3g

材料和份量(2人份)
平菇 100g, 红辣椒 3g, 青辣椒 3g, 大蒜 2g, 芝麻 1g, 食用油 10g, 香油 3g

Ingredients and portion(2 portions)
100g oyster mushroom, 3g red chilli, 3g green chilli, 2g garlic, 1g sesame, 10g salad oil, 3g sesame oil

재료 손질 材料处理方法 Clean ingredients

만드는 방법

1. 느타리버섯을 찢어 준비한다.
2. 끓는 소금물에 데친다.
3. 물기를 짜서 준비한다.
4. 팬에 식용유를 두른다.
5. 데친 버섯을 볶는다.
6. 다진 마늘 2g을 넣는다.
7. 홍고추를 잘라 씨를 제거한다.
8. 풋고추, 대파를 어슷썬다.
9. 고추, 대파를 넣는다.
10. 참기름을 3g 넣는다.
11. 볶아서 완성한다.
12. 완성하여 담는다.

制作方法

1. 用手撕好平菇备用。
2. 在烧开的盐水中焯一下。
3. 控干水分后备用。
4. 在锅中倒入食用油。
5. 将焯好的蘑菇炒一下。
6. 放入蒜泥2g。
7. 切开红辣椒，去除辣椒籽。
8. 斜切青辣椒和大葱。
9. 放入辣椒和大葱。
10. 加入香油3g。
11. 炒好后完成。
12. 完成后盛菜。

How to cook

1. Tear oyster mushroom.
2. Blanch the mushroom in boiling water.
3. Drain well.
4. Put cooking oil into a pan.
5. Stir-fry the blanched mushroom.
6. Put 2g of chopped garlic.
7. Cut red chilli and remove seeds.
8. Cut the green chilli and spring onion diagonally.
9. Put chili and spring onion.
10. Pour 3g of sesame oil.
11. Stir-fry the dish and you are done cooking.
12. Enjoy.

TIP 建议

5. 느타리버섯은 가볍게 볶아야 부드러운 나물이 된다.

5. 平菇只有须稍微炒一下口味才会柔嫩。

Step 5. Try to stir-fry the oyster mushroom as slightly as possible. The mushroom will feel soft.

무나물

萝卜拌菜

Radish with herbs

재료 및 분량(2인분)
무 200g, 통마늘 5g, 통생강 5g, 대파 10g, 통깨 3g, 소금 5g, 참기름 5g

材料和份量(2人份)
萝卜 200g, 大蒜 5g, 生姜 5g, 大葱 10g, 芝麻 3g, 盐 5g, 香油 5g

Ingredients and portion(2 portions)
200g white radish, 5g garlic, 5g ginger, 10g leek, 3g sesame, 5g salt, 5g sesame oil

재료 손질 材料处理方法 Clean ingredients

만드는 방법

1. 무를 0.2cm로 채 썬다.
2. 냄비에 물을 넣고 삶는다.
3. 다진 마늘 3g을 넣는다.
4. 생강물을 넣는다.
5. 소금 간을 한다.
6. 10분 정도 삶는다.
7. 참기름을 5g 넣는다.
8. 대파를 0.1cm로 채 썬다.
9. 통깨를 넣는다.
10. 송송 자른 대파를 넣는다.
11. 무와 국물을 함께 담는다.
12. 완성하여 담는다.

制作方法

1. 将萝卜切成大小0.2cm的丝状。
2. 在锅里放入水后煮一下。
3. 放入3g蒜泥。
4. 放入生姜水。
5. 用盐调味。
6. 煮10分钟左右。
7. 加入5g香油。
8. 将大葱切成大小0.1cm。
9. 放入芝麻。
10. 放入切好的大葱。
11. 完成后盛放在容器里。
12. 完成后盛菜。

How to cook

1. Julienne the radish 0.2cm.
2. Boil the radish in a pot over a heat.
3. Put 3g of chopped garlic.
4. Pour ginger juice.
5. Season the content with salt.
6. Have the content boiled for about 10 minutes.
7. Pour 5g of sesame oil.
8. Cut the spring onion 0.1cm slices.
9. Put sesame.
10. Put chopped spring onion.
11. Serve the dish in a plate.
12. Finish.

TIP
建议

무나물은 약불에서 푹 익어야 향과 맛이 깊다.

萝卜只有在微火上熟透后香味和口味才会更浓→把萝卜用微火煮熟之后味道会变的更浓郁。

Cook the radish over a low heat so that you can add richer scent and flavor to the dish.

시래기나물

拌干菜

Dried radish greens

재료 및 분량(2인분)
불린 시래기나물 200g, 대파 5g, 홍고추 5g, 풋고추 5g, 마늘 3g, 된장 10g, 국간장 3g, 고춧가루 3g
***육수** : 다시마 5g, 국멸치 3마리, 양파 3g

材料和份量(2人份)
泡好的干菜 200g, 大葱 5g, 红辣椒 5g, 青辣椒 5g, 大蒜 3g, 大酱 10g, 汤用酱油 3g, 辣椒面儿 3g.
***肉汤**: 海带 5g, 汤用鳀鱼 3条, 洋葱 3g

Ingredients and portion
200g soaked radish leaves, 5g leek, 5g red chilli, 5g green chilli, 3g garlic, 10g soybean paste (doenjang), 3g soy sauce, 3g chili powder
***Meat stock** : 5g dried sea tangle, 3 anchovies, 3g onion

재료 손질 材料处理方法 Clean ingredients

만드는 방법
1. 뜨거운 물에 불린다.
2. 끓는 물에 데친다.
3. 물기를 제거한다.
4. 4cm로 자른다.
6. 육수를 넣는다.
7. 채로 자른 양파를 넣는다.
8. 고추를 넣는다.
9. 들깨가루를 넣어 완성한다.

制作方法
1. 泡在热水中。
2. 在烧开的水中焯一下。
3. 用过滤网控干水分。
4. 切成大小4cm。
6. 加入肉汤。
7. 放入切成大块的洋葱。
8. 放入辣椒。
9. 放入野芝麻粉。

How to cook
1. Soak the dried radish greens in hot water.
2. Blanch the greens in boiling water.
3. Strain water through a sieve.
4. Cut the greens by 4cm.
6. Pour stock.
7. Julienne the onion and put them.
8. Add the chili.
9. Put perilla flour.

5-1. 된장을 10g 넣는다.

5-1. 放入大酱10g。

5-1. Put 10g of doenjang.

5-2. 다진 마늘을 3g 넣는다.

5-2. 放入蒜泥3g。

5-2. Put 3g of chopped garlic.

5-3. 국간장을 3g 넣는다.

5-3. 加入汤用酱油3g。

5-3. Add 3g of soy sauce.

5-4. 무친다.

5-4. 混合搅拌。

5-4. Mix the ingredients.

5-5. 두꺼운 냄비에 볶는다.

5-5. 放在锅里炒。

5-5. Roast the ingredients in a pot.

5-6. 고춧가루를 3g 넣는다.

5-6. 放入辣椒面儿3g。

5-6. Sprinkle 3g of chili powder.

TIP 建议

시래기나물은 부드럽게 많이 불려서 완전히 삶아 익힌다.

干菜应在水里充分泡开后完全煮熟。

For the dried radish greens, have them fully soaked and cooked.

잡채

杂菜

Japchae

재료 및 분량(2인분)
불린 당면 150g, 소고기(안심) 60g, 느타리버섯 3g, 새송이버섯 2g, 목이버섯 2g, 당근 3g, 양파 2g, 홍피망 2g, 통마늘 2g, 대파 2g, 실파 2g, 달걀 1개, 식용유 10g, 불고기소스 20g, 소금 5g, 통깨 2g, 참기름 5g

材料和份量(2人份)
泡好的粉丝 150g, 牛肉(里脊) 60g, 平菇 3g, 杏鲍菇 2g, 木耳 2g, 胡萝卜 3g, 洋葱 2g, 红色大辣椒 2g, 大蒜 2g, 大葱 2g, 小葱 2g, 鸡蛋 1个, 食用油 10g, 烤肉调味料 20g, 盐 5g, 芝麻 2g, 香油 5g

Ingredients and portion(2 portions)
150g soaked starch noodle, 60g beef(tenderloin), 3g oyster mushroom, 2g oyster mushroom, 2g black mushroom, 3g carrot, 2g onion, 2g red bell pepper, 2g garlic, 2g spring onion, 1 egg, 10g cooking oil, 20g bulgogi sauce, 5g salt, 2g sesame seed, 5g sesame seed oil

재료 손질 材料处理方法 How to prepare

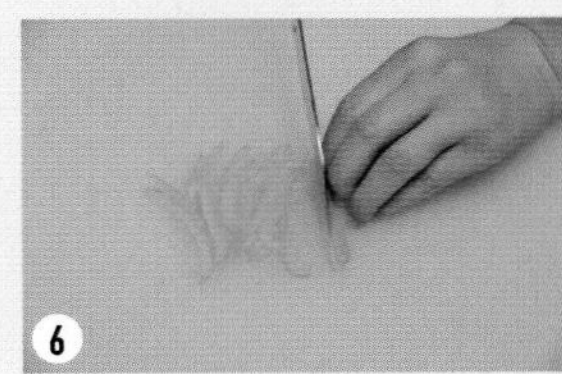

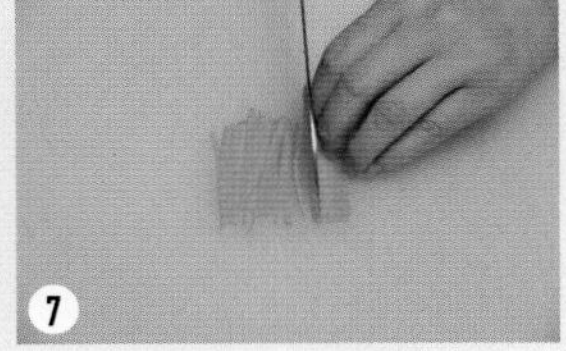

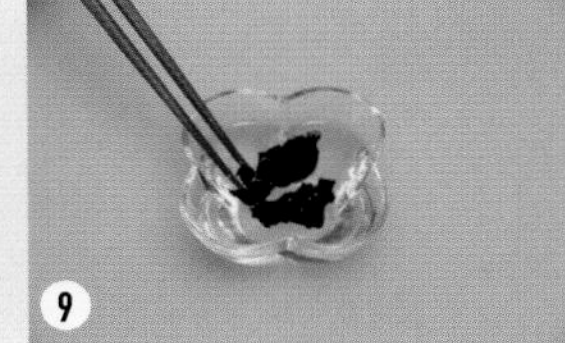

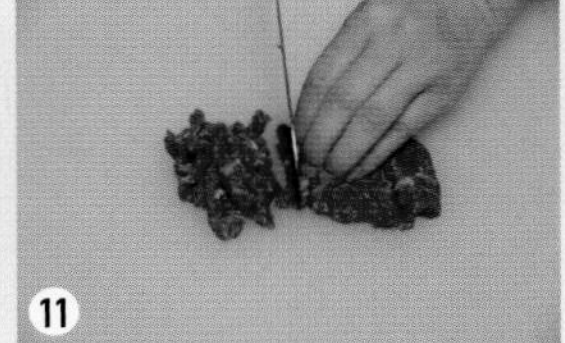

만드는 방법

1. 새송이버섯을 0.2cm로 자른다.
2. 실파를 4cm로 자른다.
3. 버섯류, 실파를 데친다.
4. 버섯, 실파를 소금 간한다.
5. 양파를 0.2cm로 자른다.
6. 홍피망을 0.2cm로 자른다.
7. 당근은 0.2cm 채로 자른다.
8. 팬에 노릇하게 볶는다.
9. 목이버섯을 불린다.
10. 불고기소스를 양념으로 볶아낸다.
11. 소고기를 0.2cm 채로 자른다.
12. 불고기소스 20g을 양념한다.

制作方法

1. 将杏鲍菇切成大小0.2cm。
2. 将小葱切成大小4cm。
3. 将蘑菇类和小葱焯一下。
4. 在蘑菇和小葱中放入盐调味。
5. 将洋葱切成大小0.2cm。
6. 将大辣椒切成大小0.2cm。
7. 将胡萝卜切成大小0.2cm的丝状。
8. 放在锅里炒。
9. 将木耳泡在水里。
10. 放入烤肉调味料炒一下。
11. 将牛肉切成大小0.2cm。
12. 放入20g烤肉调味料。

How to cook

1. Cutting oyster mushroom into 0.2cm.
2. Cut the small green onion into 4cm.
3. Blanch mushroom and small green onion.
4. Season mushroom and small green onion with salt.
5. Cut the onion into 0.2cm.
6. Cut the bell pepper into 0.2cm.
7. Cut the carrot into 0.2cm slices.
8. Fry the ingredients in a pan until they turn golden brown.
9. Get the black mushroom soaked.
10. Fry the ingredients with bulgogi sauce.
11. Cut the beef into 0.2cm slices.
12. Season the ingredients with 20g of bulgogi sauce.

당면 손질
粉丝的处理方法
How to prepare starch noodle

당면양념장 만들기
制作粉丝调味酱
How to make starch marinade

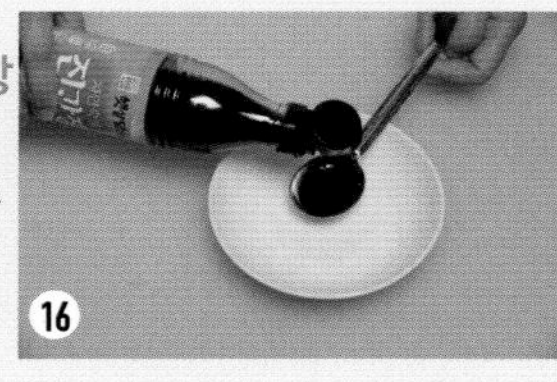

완성하기
完成
How to finish

만드는 방법

13. 팬에 볶는다.
14. 불린 당면을 준비한다.
15. 끓는 물에 데친다.
16. 진간장을 15g 넣는다.
17. 설탕을 10g 넣는다.
18. 참기름을 3g 넣는다.
19. 데친 당면을 섞는다.
20. 당면에 맛을 들인다.
21. 무친 당면을 섞는다.
22. 양념한 채소류를 섞는다.
23. 참기름을 무친다.
24. 버무린다.

制作方法

13. 放在锅里炒。
14. 准备泡好的粉丝备用。
15. 放在烧开的水中焯一下。
16. 加入15g浓酱油。
17. 放入10g白糖。
18. 加入3g香油。
19. 与焯好的粉丝混合在一起。
20. 使味道渗入粉丝。
21. 混合搅拌粉丝。
22. 将调好味的蔬菜类混合在一起。
23. 加入香油混合搅拌。
24. 混合搅拌。

How to cook

13. Cook the ingredients in a pan.
14. Prepare soaked starch noodle with you.
15. Blanch the noodle in boiling water.
16. Put 15g of thick soy sauce.
17. Put 10g of sugar.
18. Put 3g of sesame oil.
19. Mix the ingredients with starch noodle.
20. Add some flavors to the starch noodle.
21. Mix the starch noodle all together.
22. Put the seasoned vegetables.
23. Pour some sesame oil.
24. Mix the ingredients all together.

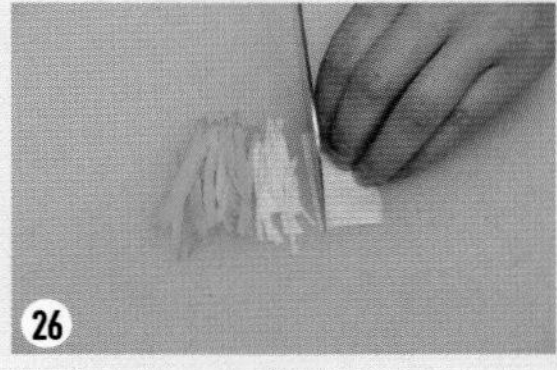

만드는 방법
25. 황, 백 지단을 부친다.
26. 0.2cm의 채로 자른다.
27. 담아 완성한다.

制作方法
25. 分别煎鸡蛋黄和鸡蛋清。
26. 切成大小0.2cm的丝状。
27. 放入杂菜中完成。

How to cook
25. Make an egg into garnish in yellow and white.
26. Julienne it 0.2cm.
27. Serve the dish.

TIP
建议

잡채에 섞는 채소는 색스럽게 다양하게 준비하고, 오래 볶으면 아삭거리지 않으므로 살짝만 볶아 준비한다.

为杂菜用蔬菜最好准备各种颜色的蔬菜，如果炒得太久，味道就不会香脆，稍微炒一下即可。

Have vegetables in many colors prepared with you. Get the vegetables slightly cooked because if you roast them too long, they will not feel crispy.

죽순나물

竹笋拌菜
Bamboo shootnamul

재료 및 분량(2인분)
불린 죽순 150g, 적양파 3g, 대파 2g, 실파 1g, 들깨가루 5g, 통깨 1g, 국간장 3g, 참기름 3g
***육수** : 다시마 5g, 국멸치 3마리, 양파 3g

材料和份量(2人份)
泡好的竹笋 150g, 红色洋葱 3g, 大葱 2g, 小葱 1g, 野芝麻粉 5g, 芝麻 1g, 汤用酱油 3g, 香油 3g
***肉汤**: 海带 5g, 汤用鳀鱼 3条, 洋葱 3g

Ingredients and portion(2 portions)
150g soaked bamboo shoots, 3g red onion, 2g leek, 1g spring onion, 5g perilla flour, 1g sesame, 3g soy sauce, 3g sesame oil
***Stock** : 5g dried sea tangle, 3 anchovies, 3g onion

재료 손질 材料处理方法 Clean ingredients

만드는 방법

1. 죽순을 불린다.
2. 끓는 물에 삶아낸다.
3. 체에 거른다.
4. 4cm로 자른다.
6. 참기름을 3g 넣는다.
7. 통깨를 1g 넣는다.
8. 고소하게 볶아낸다.
9. 완성접시에 담는다.

制作方法

1. 将竹笋泡在水里。
2. 用烧开的水焯一下。
3. 捞出来。
4. 切成大小4cm。
6. 加入3g香油。
7. 放入1g芝麻。
8. 炒一下。
9. 完成后盛放在盘子里。

How to cook

1. Soak the bamboo shoots in water.
2. Boil the bamboo shoots in boiling water.
3. Strain the water through a sieve.
4. Cut the bamboo shoots into 4cm.
6. Pour 3g of sesame oil.
7. Put 1g of sesame.
8. Roast the sesame until it smells sweet.
9. Put the dish in a plate.

끓이기
煮粥
How to boil

5-1. 국간장을 3g 넣는다.

5-1. 加入汤用酱油3g。

5-1. Put 3g of soy sauce.

5-2. 멸치육수를 100g 넣는다.

5-2. 加入鳀鱼汤100g。

5-2. Pour 100g of anchovy stock.

5-3. 어슷썬 대파를 넣는다.

5-3. 放入斜切好的大葱。

5-3. Cut the spring onion diagonally and put them into the stock.

5-4. 채 썬 적양파를 넣는다.

5-4. 放入切好的红色洋葱。

5-4. Julienne the red onion and put them too.

5-5. 들깨가루를 5g 넣는다.

5-5. 放入野芝麻粉5g。

5-5. Add 5g of perilla flour.

5-6. 냄비에 20분 정도 열을 가한다.

5-6. 加热锅20分钟左右。

5-6. Apply heat to a pot for about 20 minutes.

TIP
建议

1. 말린 죽순은 충분히 불려서 사용한다.

1. 将干竹笋充分泡开后再使用。

Step 1. You need to have dried bamboo shoots fully soaked in water.

참나물무침

山芹菜拌菜

Chamnamul mix

재료 및 분량(2인분)
참나물 100g, 마늘 3g, 대파 3g, 통깨 1g, 소금 2g, 국간장 2g, 참기름 3g

材料和份量(2人份)
山芹菜 100g, 大蒜 3g, 大葱 3g, 芝麻 1g, 盐 2g, 汤用酱油 2g, 香油 3g

Ingredients and portion(2 portions)
100g Chamnamul, 3g garlic, 3g leek, 1g sesame, 2g salt, 2g soy sauce, 3g sesame oil

재료 손질 材料处理方法 Clean ingredients

만드는 방법

1. 끓는 물에 소금을 넣는다.
2. 끓는 물에 데친다.
3. 찬물에 헹군다.
4. 물기 제거 후 3cm로 자른다.
5. 국간장을 2g 넣는다.
6. 다진 마늘을 3g 넣는다.
7. 소금을 2g 넣는다.
8. 다진 대파를 3g 넣는다.
9. 참기름을 3g 넣는다.
10. 통깨를 1g 넣는다.
11. 윤기 있게 완성한다.
12. 완성하여 담는다.

制作方法

1. 在烧开的水中放入盐。
2. 在烧开的水中焯一下。
3. 用凉水冲一下。
4. 控干水分后切成大小3cm。
5. 加入2g汤用酱油。
6. 放入3g蒜泥。
7. 放入2g盐。
8. 放入3g葱花。
9. 加入3g香油。
10. 放入1g芝麻。
11. 完成后盛放在盘子里。
12. 完成后盛菜。

How to cook

1. Put salt into boiling water.
2. Blanch the ingredient in boiling water.
3. Wash the ingredient in cold water.
4. Drain the water off and cut the ingredient by 3cm.
5. Pour 2g of soy sauce.
6. Put 3g of crushed garlic.
7. Add 2g of salt.
8. Put 3g of chopped spring onion.
9. Pour 3g of sesame oil.
10. Put 1g of sesame.
11. Serve the dish in a plate.
12. Finish.

TIP 建议

3. 참나물을 짧은 시간에 데쳐야 색을 살릴 수 있다.

3. 焯山芹菜的时间要短，只有这样才能保持山芹菜的颜色新鲜。

Step 3. You can keep the original color of chamnamul as you have it blanched quickly.

취나물

香蔬拌菜
Cheenamul

재료 및 분량(2인분)
불린 취나물 150g, 된장 10g, 풋고추 3g, 대파 2g, 실파 2g, 양파 2g, 마늘 3g, 통깨 1g, 참기름 3g
***육수** : 다시마 5g, 국멸치 3마리, 양파 3g

材料和份量(2人份)
泡好的香蔬 150g, 大酱 10g, 青辣椒 3g, 大葱 2g, 小葱 2g, 洋葱 2g, 大蒜 3g, 芝麻 1g, 香油 3g
***肉汤**: 海带 5g, 汤用鳀鱼 3条, 洋葱 3g

Ingredients and portion(2 portions)
150g soaked cheenamul, 10g soybean paste(doenjang), 3g green chilli, 2g leek, 2g spring onion, 2g onion, 3g garlic, 1g sesame, 3g sesame oil
***Stock** : 5g dried sea tangle, 3 anchovies, 3g onion

재료 손질 材料处理方法 Clean ingredients

만드는 방법
1. 취나물을 불린다.
2. 끓는 물에 데친다.
3. 물기를 제거한다.
4. 4cm로 자른다.
6. 멸치육수를 100g 붓는다.
7. 참기름을 3g 넣는다.
8. 통깨를 1g 넣는다.
9. 완성접시에 담는다.

制作方法
1. 将香蔬泡在水里。
2. 用烧开的水焯一下。
3. 控干水分。
4. 切成大小4cm。
6. 加入鳀鱼汤100g。
7. 加入香油3g。
8. 放入芝麻1g。
9. 完成后盛放在盘子里。

How to cook
1. Soak cheenamul in water.
2. Have it blanched in boiling water.
3. Drain the water off.
4. Cut cheenamul 4cm.
6. Pour 100g of anchovy stock.
7. Put 3g of sesame oil.
8. Add 1g of sesame.
9. Serve the dish right.

끓이기
煮粥
How to boil

5-1. 된장을 10g 넣는다.

5-1. 放入大酱10g。

5-1. Put 10g of doenjang.

5-2. 다진 마늘을 3g 넣는다.

5-2. 放入蒜泥3g。

5-2. Put 3g of chopped garlic.

5-3. 양파채를 0.2cm로 썰어 넣는다.

5-3. 放入切成大小0.2cm的洋葱。

5-3. Slice the onion by 0.2cm and put them into the pot.

5-4. 다진 대파를 2g 넣는다.

5-4. 放入葱花2g。

5-4. Put 2g of chopped spring onion.

5-5. 풋고추를 넣는다.

5-5. 放入青辣椒。

5-5. Add green chilli.

5-6. 실파를 넣는다.

5-6. 放入小葱。

5-6. Put small green onion.

TIP
建议

5-1. 된장으로 양념하면 구수한 맛을 낸다.

5-1. 用大酱调味可以使味道香酥可口。

Step 5-1. With doenjang, you can add a sweet flavor to your dish.

콩나물무침

豆芽拌菜

Bean sprouts mix

재료 및 분량(2인분)
콩나물 200g, 실파 3g, 대파 2g
***양념장** : 고춧가루 10g, 국간장 2g, 까나리액젓 3g, 다진 마늘 3g, 참기름 3g, 통깨 1g

材料和份量(2人份)
豆芽 200g, 小葱 3g, 大葱 2g
***调味酱** : 辣椒面儿 10g, 汤用酱油 2g, 玉筋鱼酱 3g, 蒜泥 3g, 香油 3g, 芝麻 1g

Ingredients and portion(2 portions)
200g bean sprouts, 3g spring onion, 2g leek
***Sauce** : 10g chili powder, 2g soy sauce, 3g sand eel sauce, 3g chopped garlic, 3g sesame oil, 1g sesame

콩나물 손질 豆芽的处理方法 How to prepare bean sprouts

만드는 방법

1. 콩나물 꼬리를 다듬는다.
2. 매실식초로 헹군다.
3. 소금을 넣는다.
4. 뚜껑을 닫고 삶는다.
6. 삶은 콩나물 찬물에 넣기
7. 물기를 제거한다.
8. 양념장과 대파 2g, 실파 3g을 넣는다.
9. 완성접시에 담는다.

制作方法

1. 去除豆芽尾。
2. 用梅子醋冲洗。
3. 放入盐。
4. 盖上盖子煮。
6. 将煮好的豆芽放入凉水中。
7. 控干水分。
8. 放入调味酱、2g大葱和3g小葱。
9. 完成后盛放在盘子里。

How to cook

1. Trim the bean sprouts on the end.
2. Wash the bean sprouts with plum vinegar.
3. Put salt.
4. Put a lid on and boil them.
6. Put boiled bean sprouts in cold water.
7. Drain well.
8. Put marinade, 2g of spring onion and 3g of small green onion.
9. Serve the dish well.

양념장 만들기 制作调味酱 How to make marinade.

5-1. 고춧가루를 10g 넣는다.

5-1. 放入辣椒面儿10g。

5-1. Put 10g of chili powder.

5-2. 국간장을 2g 넣는다.

5-2. 放入汤用酱油2g。

5-2. Put 2g of soy sauce.

5-3. 까나리액젓을 3g 넣는다.

5-3. 放入玉筋鱼酱3g。

5-3. Put 3g of sand eel sauce.

5-4. 다진 마늘을 3g 넣는다.

5-4. 放入蒜泥3g。

5-4. Put 3g of chopped garlic.

5-5. 참기름을 3g 넣는다.

5-5. 加入香油3g。

5-5. Add 3g of sesame oil.

5-6. 통깨를 1g 넣는다.

5-6. 放入芝麻1g。

5-6. Put 1g of sesame.

TIP 建议

4. 콩나물은 오래 삶으면 아삭거리지 않으므로 주의하여 짧은 시간 안에 삶는다.

4. 如果豆芽煮得太久就不会香脆。请注意煮得时间不要太长。

Step 4. Get the bean sprouts boiled quickly not to lose crispiness.

표고버섯나물

香菇拌菜

Shiitake mix

재료 및 분량(2인분)
표고버섯 3개, 풋고추 3g, 들깨가루 7g, 국간장 2g, 마늘 3g, 참기름 3g
***육수** : 다시마 5g, 국멸치 5마리

材料和份量(2人份)
香菇 3个, 青辣椒 3g, 野芝麻粉 7g, 汤用酱油 2g, 大蒜 3g, 香油 3g
***肉汤**: 海带 5g, 汤用鳀鱼 5条

Ingredients and portion(2 portions)
3 shiitake mushroom(mushroom pyogo), 3g green chilli, 7g perilla seed flour, 2g soy sauce, 3g garlic, 3g sesame oil
***Meat stock** : 5g dried sea tangle, 5 anchovies

재료 손질 材料处理方法 Clean ingredients

만드는 방법

1. 생표고는 0.2cm로 어슷썬다.
2. 믹서기에 육수를 넣는다.
3. 들깨가루를 넣는다.
4. 믹서기에 갈아 준비한다.
5. 표고버섯, 육수를 넣는다.
6. 다진 마늘을 3g 넣는다.
7. 볶는다.
8. 국간장을 2g 넣는다.
9. 참기름을 3g 넣는다.
10. 풋고추를 넣는다.
11. 윤기 있게 담는다.
12. 완성한다.

制作方法

1. 将香菇斜切成大小0.2cm。
2. 在搅拌机中加入肉汤。
3. 放入野芝麻粉。
4. 在搅拌机里搅碎后备用。
5. 放入香菇和肉汤。
6. 放入3g蒜泥。
7. 一起炒。
8. 加入2g汤用酱油。
9. 加入3g香油。
10. 放入青辣椒。
11. 完成后盛放在盘子里。
12. 完成。

How to cook

1. Cut shiitake 0.2cm diagonally.
2. Pour stock into a blender.
3. Put sesame flour.
4. Grind them in a blender.
5. Put shiitake and stock.
6. Put 3g of chopped garlic.
7. Stir-fry them.
8. Pour 2g of soy sauce.
9. Pour 3g of sesame oil.
10. Put green chilli.
11. Serve the food in a plate.
12. Finish.

TIP
建议

생표고버섯으로 나물을 하면 부드럽고, 향이 살아 있는 맛을 낸다.

用香菇制作拌菜，不仅味道柔嫩，而且香气怡人。

If you try this recipe, fresh shiitake basically smells good and feels soft.

피마자나물

蓖麻拌菜

Castor bean

재료 및 분량(2인분)
불린 피마자나물 80g, 국간장 3g, 대파 3g, 마늘 3g, 양파 5g, 풋고추 3g, 참기름 3g, 통깨 1g
***육수** : 다시마 5g, 국멸치 5마리, 양파 3g

材料和份量(2人份)
泡好的蓖麻 80g, 汤用酱油 3g, 大葱 3g, 大蒜 3g, 洋葱 5g, 青辣椒 3g, 香油 3g, 芝麻 1g
***肉汤**: 海带 5g, 汤用鳀鱼 5条, 洋葱 3g

Ingredients and portion(2 portions)
80g soaked castor bean leaves, 3g soy sauce, 3g leek, 3g garlic, 5g onion, 3g green chilli, 3g sesame oil, 1g sesame
***Meat stock** : 5g dried sea tangle, 5 anchovies, 3g onion

재료 손질 材料处理方法 Clean ingredients

만드는 방법

1. 뜨거운 물에 피마자를 불린다.
2. 끓는 물에 데친다.
3. 물기를 제거한다.
4. 4cm로 자른다.
6. 부드럽게 푹 익힌다.
7. 참기름을 3g 넣는다.
8. 통깨를 1g 넣는다.
9. 완성접시에 담아낸다.

制作方法

1. 将蓖麻泡在热水中。
2. 用烧开的水焯一下。
3. 控干水分。
4. 切成大小4cm。
6. 煮熟。
7. 加入3g香油。
8. 放入1g芝麻。
9. 完成后盛放在盘子里。

How to cook

1. Soak castor bean in hot water.
2. Blanch the bean in boiling water.
3. Remove water.
4. Cut the bean 4cm.
6. Get the beans boiled soft.
7. Add 3g of sesame oil.
8. Put 1g of sesame.
9. Serve the beans in a plate.

끓이기
煮粥
How to boil

5-1. 국간장을 3g 넣는다.

5-1. 加入汤用酱油3g。

5-1. Put 3g of soy sauce.

5-2. 다진 마늘을 3g 넣는다.

5-2. 放入蒜泥3g。

5-2. Add 3g of chopped garlic.

5-3. 양파채를 넣는다.

5-3. 放入整个洋葱。

5-3. Put slices of onions.

5-4. 풋고추를 넣는다.

5-4. 放入青辣椒。

5-4. Put green chilli.

5-5. 실파를 넣는다.

5-5. 放入小葱。

5-5. Put small green onion.

5-6. 멸치육수를 100g 붓는다.

5-6. 倒入鳀鱼汤100g。

5-6. Pour 100g of anchovy stock.

TIP
建议

말린 나물은 충분히 불려서 부드럽게 요리를 완성한다.

只有将干菜充分泡开后料理的味道才会柔嫩。

Try to get the dried castor beans fully soaked until they actually feel soft.

韩食美“味”秘方传授 100选 实践篇
Transmission of a Hundred Secrets of Korean Dishes Flavor 泡菜

한식의 맛 비법전수 100선

실기편

김치

깍두기

萝卜块泡菜

Kkakttuki

재료 및 분량(2인분)
무 500g, 실파 5g, 미나리 2g, 꽃소금 15g
***양념장** : 밀가루풀 15g, 쌀(밀)가루 죽 15g, 다진 마늘 10g, 다진 생강 2g, 꽃소금 3g, 설탕 10g, 고춧가루 20g, 새우젓 5g

材料和份量(2人份)
萝卜 500g, 小葱 5g, 水芹菜 2g, 大粒盐 15g
***调味酱** : 面粉浆水 15g, 大米(面粉)面儿粥 15g, 蒜泥 10g, 捣碎的生姜 2g, 大粒盐 3g, 白糖 10g, 辣椒面儿 20g, 虾酱 5g

Ingredients and portion(2 portions)
500g white radish, 5g spring onion, 2g water dropwort (Minari), 15g soy sauce
***Sauce** : 15g flour paste, 15g rice flour juk (or flour juk), 10g chopped garlic, 2g chopped ginger, 3g salt cooking, 10g flavory sugar, 20g chili powder, 5g salted shrimp

재료 손질 材料处理方法 Clean ingredients

만드는 방법

1. 무를 가로 2cm로 썬다.
2. 사방 2cm로 깍둑썬다.
3. 3%의 소금으로 절인다.
4. 소금을 섞어준다.
6. 실파, 미나리, 홍고추를 넣는다.
7. 깍두기를 물에 한번 씻는다.
8. 물기 제거 후 양념장을 넣는다.
9. 골고루 버무린다.

制作方法

1. 将萝卜切成宽2cm。
2. 切成边长为2cm的方块状。
3. 用3%的盐水腌制。
4. 放入盐混合。
6. 放入小葱、水芹菜和红辣椒。
7. 将萝卜块用水冲洗一下。
8. 控干水分后放入调味酱。
9. 均匀地混合搅拌。

How to cook

1. Cut the radish 2cm.
2. Cut the radish into 2x2cm of cubes.
3. Season the radish with 3% of salt.
4. Mix the radish with salt.
6. Add small green onion, water dropwort and red chili.
7. Wash kkattuki in water once.
8. Remove water and mix them with sauce.
9. Mix them well.

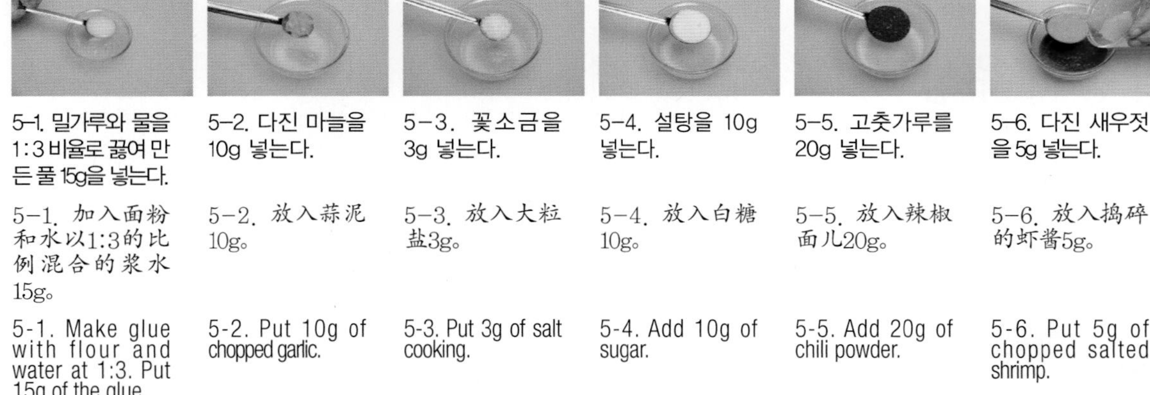

5-1. 밀가루와 물을 1:3 비율로 끓여 만든 풀 15g을 넣는다.	5-2. 다진 마늘을 10g 넣는다.	5-3. 꽃소금을 3g 넣는다.	5-4. 설탕을 10g 넣는다.	5-5. 고춧가루를 20g 넣는다.	5-6. 다진 새우젓을 5g 넣는다.
5-1. 加入面粉和水以1:3的比例混合的浆水15g。	5-2. 放入蒜泥10g。	5-3. 放入大粒盐3g。	5-4. 放入白糖10g。	5-5. 放入辣椒面儿20g。	5-6. 放入捣碎的虾酱5g。
5-1. Make glue with flour and water at 1:3. Put 15g of the glue.	5-2. Put 10g of chopped garlic.	5-3. Put 3g of salt cooking.	5-4. Add 10g of sugar.	5-5. Add 20g of chili powder.	5-6. Put 5g of chopped salted shrimp.

TIP 建议

3~4. 깍두기는 30분 정도 절인다. (너무 오래 절이면 무의 수분이 나와 맛이 떨어지거나 질겨진다.)

3~4. 将萝卜块腌制30分钟左右。(如果腌制的时间太久，萝卜会产生水分，萝卜的味道会变淡或变味儿。)

Step 3~4. Season kkattuki with the sauce for 30 minutes. (if you leave the dish marinated too long, it will be less tasty as water gets disappeared.)

무말랭이무침

萝卜干拌菜

Dried slices of radish

재료 및 분량(2인분)
무말랭이 50g, 실파 3g
***양념장** : 고춧가루 15g, 진간장 15g, 다진 마늘 5g, 실파 8g, 설탕 15g, 참기름 3g, 통깨 2g

材料和份量(2人份)
萝卜干 50g, 小葱 3g
***调味酱** : 辣椒面儿 15g, 浓酱油 15g, 蒜泥 5g, 小葱 8g, 白糖 15g, 香油 3g, 芝麻 2g

Ingredient and portion(2 portions)
50g dried radish slices, 3g spring onion
* **Sauce** : 15g chili powder, 15g thick soy sauce, 5g chopped garlic, 8g spring onion, 15g sugar, 3g sesame oil, 2g sesame

재료 손질 材料处理方法 Clean ingredients

양념장 만들기 制作调味酱 How to make sauce

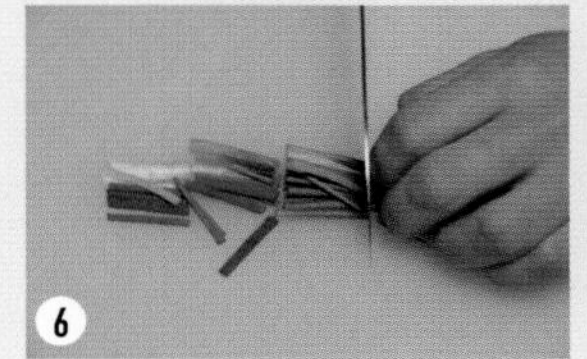

만드는 방법

1. 무말랭이를 불린다.
2. 고춧가루를 15g 넣는다.
3. 진간장을 15g 넣는다.
4. 다진 마늘을 5g 넣는다.
5. 설탕을 15g 넣는다.
6. 실파를 8g 자른다.
7. 참기름을 3g 넣는다.
8. 통깨를 2g 넣는다.
9. 불린 무말랭이의 물기를 제거한다.
10. 양념장을 넣는다.
11. 잘 버무려 담는다.
12. 완성하여 담는다.

制作方法

1. 将萝卜干泡在水里。
2. 放入15g辣椒面儿。
3. 加入15g浓酱油。
4. 放入5g蒜泥。
5. 放入15g白糖。
6. 放入8g小葱。
7. 加入3g香油。
8. 放入2g芝麻。
9. 将泡好的萝卜干的水分控干。
10. 放入调味酱。
11. 充分混合后盛放在盘子里。
12. 完成后盛菜。

How to cook

1. Get the dried slices of radish soaked in water.
2. Put 15g of chili powder.
3. Put 15g of thick soy sauce.
4. Add 5g of chopped garlic.
5. Add 15g of sugar.
6. Cut 8g of small green onion.
7. Pour 3g of sesame oil.
8. Put 2g of sesame.
9. Remove water from soaked-dried slices of radish.
10. Put the sauce.
11. Mix them well.
12. Finish.

TIP 建议

1. 무말랭이는 찬물에 서서히 불린다.

萝卜干用水慢慢的泡开。

Step 1. Get the dried slices of radish slowly soaked in cold water.

배추김치

白菜泡菜

Kimchi

재료 및 분량(2인분)
배추 반 포기 1kg, 무 50g, 미나리 3g, 실파 3g, 고춧가루 150g, 설탕 5g, 까나리액젓 5g, 통깨 3g
***양념장** : 당근 5g, 양파 5g, 생강 3g, 배 5g, 새우젓 5g, 밀가루풀 30g
***육수** : 다시마 5g, 북어머리 1개, 통마늘 2g

材料和份量(2人份)
半棵白菜 1kg, 萝卜 50g, 水芹菜 3g, 小葱 3g, 辣椒面儿 150g, 白糖 5g, 玉筋鱼酱 5g, 芝麻 3g
***调味酱** : 胡萝卜 5g, 洋葱 5g, 生姜 3g, 梨 5g, 虾酱 5g, 面粉浆水 30g
***肉汤**: 海带 5g, 明太鱼头 1个, 大蒜 2g

Ingredients and portion(2 portions)
1kg Chinese cabbage(1/2), 50g white radish, 3g water dropwort(Minari), 3g spring onion, 150g chili powder, 5g sugar, 5g sand eel sauce, 3g sesame
***Sauce** : 5g carrot, 5g onion, 3g ginger, 5g pear, 5g salted shrimp, 30g flour paste
***Stock** : 5g dried sea tangle, 1 dried pollack fillet, 2g garlic

재료 손질 材料处理方法 Clean ingredients

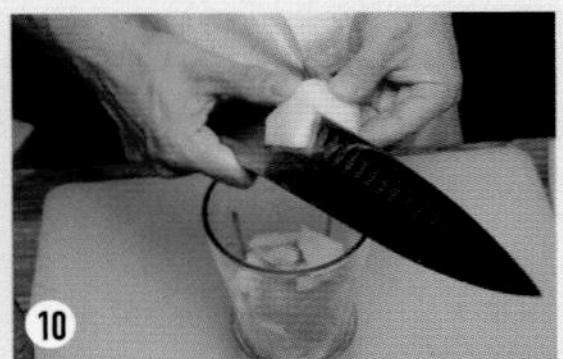

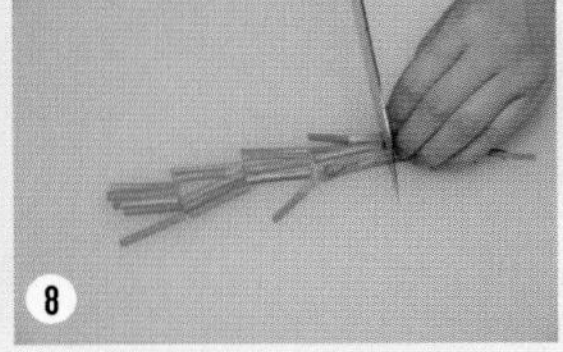

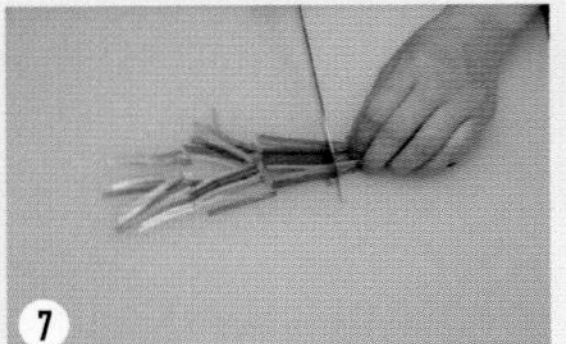

양념장 만들기 制作调味酱 How to cook sauce

만드는 방법

1. 배추는 소금 15%를 뿌려서 절인다.
2. 육수를 만든다.
3. 물과 3:1 비율로 밀가루를 넣는다.
4. 잘 섞는다.
5. 약불에 익힌다.
6. 무채를 0.2cm로 자른다.
7. 실파를 4cm로 자른다.
8. 미나리를 3cm로 자른다.
9. 생강 3g을 넣는다.
10. 배를 5g 넣는다.
11. 새우젓을 다진다.
12. 새우젓을 5g 넣는다.

制作方法

1. 在白菜上撒上15%的盐腌制。
2. 制作肉汤。
3. 将水和面粉以 3:1的比例混合后放入其中。
4. 均匀地混合。
5. 用微火煮熟。
6. 将萝卜切成大小0.2cm。
7. 将小葱切成大小4cm。
8. 将水芹菜切成大小3cm。
9. 放入3g生姜。
10. 放入5g梨。
11. 捣碎虾酱。
12. 放入5g虾酱。

How to cook

1. Season the cabbage with 15% of salt.
2. Cook stock.
3. Put water and flour at 3:1.
4. Mix them well.
5. Cook them over a low heat.
6. Cut the radish into 0.2cm slices.
7. Cut 4cm of small green onion.
8. Cut Minari 3cm.
9. Put 3g of ginger.
10. Put 5g of pear.
11. Chop up the salted shrimp.
12. Put 5g of the salted shrimp.

만드는 방법

13. 밀가루풀을 30g 넣는다.
14. 육수를 50g 붓는다.
15. 믹서기에 곱게 간다.
16. 양념장을 붓는다.
17. 고춧가루를 150g 넣는다.
18. 까나리액젓을 10g 넣는다.
19. 통깨를 넣고 잘 버무린다.
20. 설탕을 5g 넣는다.
21. 절인 배추를 한번 씻어낸다.
22. 물기를 제거한다.
23. 겉부터 양념장을 무친다.
24. 한 장씩 양념한다.

制作方法

13. 放入30g面粉浆水。
14. 倒入50g肉汤。
15. 放在搅拌机中搅碎。
16. 放入调味酱。
17. 放入150g辣椒面儿。
18. 放入10g玉筋鱼酱。
19. 放入芝麻后混合搅拌。
20. 放入5g白糖。
21. 将腌制好的白菜冲洗一下。
22. 控干水分。
23. 从外侧开始涂抹调味酱。
24. 一张一张地涂抹调味酱。

How to cook

13. Add 30g of flour glue.
14. Pour 50g of stock.
15. Grind them soft in a blender.
16. Pour the sauce.
17. Put 150g of chili powder.
18. Pour 10g of sand eel sauce.
19. Put sesame and mix the contents well.
20. Put 5g of sugar.
21. Wash the seasoned cabbage once.
22. Drain the water off.
23. Apply sauce from the outside to the inside.
24. Do it one leaf at a time.

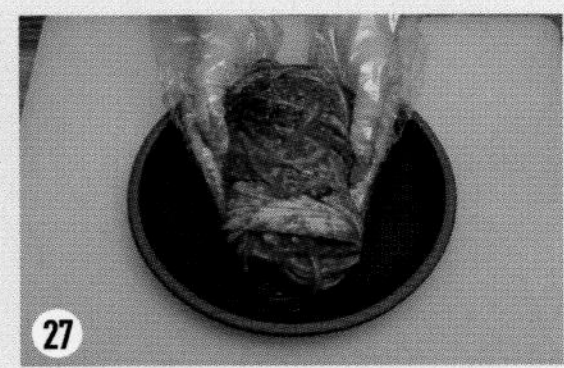

만드는 방법

25. 속까지 양념한다.
26. 겉잎으로 감싼다.
27. 모양내어 담는다.
28. 완성하여 담는다.

制作方法

25. 直至涂抹完白菜心儿。
26. 用最外层菜叶包好。
27. 美观地盛放在盘子里。
28. 完成后盛菜。

How to cook

25. Keep applying to the last leaf of the cabbage inside.
26. Cover the whole cabbage with a very outside leaf.
27. Serve Kimchi well in a plate.
28. Finish.

TIP
建议

1. 배추는 한 시간 정도 절인다.
2. 밀가루풀 만들기 → 밀가루(중력분)와 물을 1:3으로 하여 냄비에 담고 호화시킨다.

1. 白菜腌制一个小时左右。
2. 制作面粉浆水：将面粉和水以1:3的比例混合，放在锅里糊化。

Step 1. Get the cabbage seasoned with salt about one hour.
Step 2. How to make flour glue: Put flour and water at 1:3. Boil them in a pot.

부추김치

韭菜泡菜
Chives(Korean-leek) Kimchi

재료 및 분량(2인분)
부추 70g, 무 10g, 당근 3g, 양파 2g, 마늘 3g, 통깨 2g
***양념장** : 밀가루(중력분)풀 50g, 까나리액젓 5g, 다진 마늘 3g, 고춧가루 50g, 통깨 2g
***육수** : 다시마 5g, 국멸치 2마리, 통마늘 1g

材料和份量(2人份)
韭菜 70g, 萝卜 10g, 胡萝卜 3g, 洋葱 2g, 大蒜 3g, 芝麻 2g
***调味酱** : 面粉(中力粉)浆水 50g, 玉筋鱼酱 5g, 蒜泥 3g, 辣椒面儿 50g, 芝麻 2g
***肉汤** : 海带 5g, 汤用鳀鱼 2条, 大蒜 1g

Ingredients and portion(2 portions)
70g chives, 10g white radish, 3g carrot, 3g garlic, 2g sesame
***Sauce** : 50g flour paste(medium flour), 5g sand eel sauce, 3g chopped garlic, 50g chili powder, 2g sesame
***Meat stock** : 5g dried sea tangle, 2 anchovies, 1g garlic

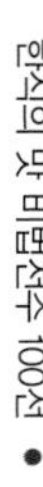

재료 손질 材料处理方法 Clean ingredients

만드는 방법

1. 육수를 끓인다.
2. 밀가루와 1:3의 비율로 물을 섞는다.
3. 약불에서 호화시켜 익힌다.
4. 무를 0.2cm로 채 썬다.
5. 양파를 0.2cm로 채 썬다.
6. 당근을 0.2cm로 채 썬다.
7. 무, 당근, 양파, 통깨, 고춧가루에 밀가루풀을 50g 넣는다.
8. 까나리액젓을 5g 넣는다.
9. 새우젓을 2g 넣는다.
10. 양념장으로 부추를 버무린다.
11. 완성접시에 담는다.

制作方法

1. 煮肉汤。
2. 将面粉和水以1:3的比例混合。
3. 用微火糊化。
4. 将萝卜切成大小0.2cm的丝状。
5. 将洋葱切成大小0.2cm的丝状。
6. 将胡萝卜切成大小0.2cm的丝状。
7. 在萝卜、胡萝卜、洋葱、芝麻和辣椒面儿中加入50g面粉浆水。
8. 放入5g玉筋鱼酱。
9. 放入2g虾酱。
10. 用调味酱拌韭菜。
11. 完成后盛放在盘子里。

How to cook

1. Cook stock.
2. Mix flour and water at 1:3.
3. Boil them over a low heat.
4. Cut the radish into 0.2cm slices.
5. Cut the onion into 0.2cm slices.
6. Cut the carrot into 0.2cm slices.
7. Put radish, carrot, sesame and chili powder into 50g of flour glue.
8. Pour 5g of sand eel sauce.
9. Put 2g of salted shrimp.
10. Mix chives with the sauce.
11. Serve the dish in a plate.

TIP 建议

부추김치는 가볍게 버무려 완성해야 풋내가 나지 않는다.

做韭菜泡菜时韭菜与调味料要轻轻的搅拌才不会产生异味。

Make sure that you mix the Chives(Korean-leek) Kimchi light so that it will not smell grass.

총각김치

嫩萝卜泡菜

Chongkak Kimchi

재료 및 분량(2인분)
총각무 900g, 양파 5g
***양념장** : 고춧가루 130g, 까나리액젓 5g, 밀가루풀 30g, 새우젓 3g, 양파즙 15g, 소금 30g, 다진 마늘 3g, 다진 생강 2g, 통깨 2g
***육수** : 다시마 5g, 국멸치 5마리, 통마늘 2g

材料和份量(2人份)
嫩萝卜 900g, 洋葱 5g
***调味酱** : 辣椒面儿 130g, 玉筋鱼酱 5g, 面粉浆水 30g, 虾酱 3g, 洋葱汁 15g, 盐 30g, 蒜泥 3g, 生姜 2g, 芝麻 2g
***肉汤** : 海带 5g, 汤鳀鱼 5条, 大蒜 2g

Ingredients and portion(2 portions)
900g chonggak radish(small radish), 5g onion
***Sauce** : 130g chili powder, 5g sand eel sauce, 30g flour paste, 3g salted shrimp, 15g onion juice, 30g salt, 3g chopped garlic, 2g chopped ginger, 2g sesame
***Meat stock** : 5g dried sea tangle, 5 anchovies, 2g garlic

재료 손질 材料处理方法 Clean ingredients

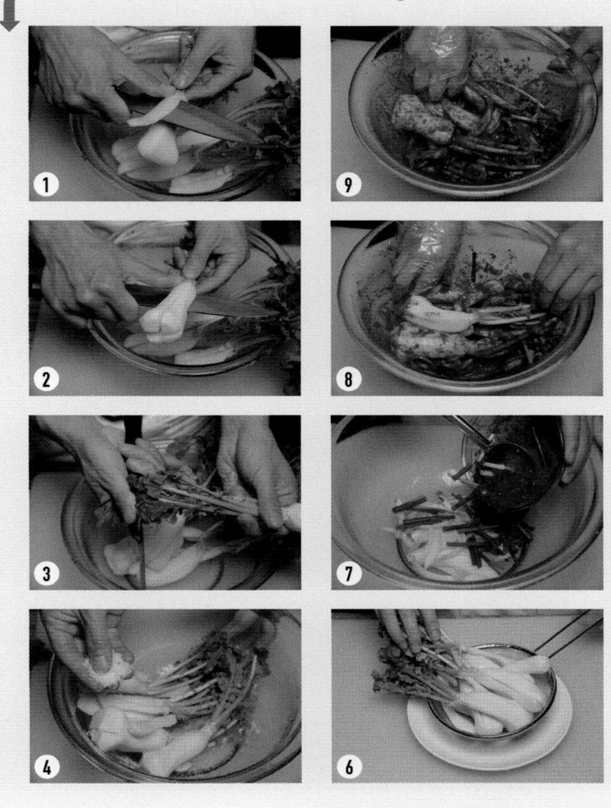

만드는 방법

1. 총각무를 잘 씻어 준비한다.
2. 무를 4등분으로 자른다.
3. 무잎을 다듬는다.
4. 소금에 절인다.
6. 총각무를 건져낸다.
7. 양념장을 넣는다.
8. 총각무와 양념장을 잘 버무린다.
9. 잘 섞어 담는다.

制作方法

1. 将嫩萝卜洗好后备用。
2. 将萝卜切成4等分。
3. 整理萝卜叶。
4. 放入盐腌制。
6. 将嫩萝卜取出来。
7. 放入调味酱。
8. 将嫩萝卜和调味酱均匀地混合。
9. 完成后盛放在容器里。

How to cook

1. Wash chonggak radishes thoroughly.
2. Quarter the radishes.
3. Trim the leaves.
4. Season the radishes with salt.
6. Take out chonggak radishes.
7. Put the sauce.
8. Mix chongkak radishes and the sauce.
9. Finish.

5-1. 고춧가루 130g에 까나리액젓을 5g 섞는다.

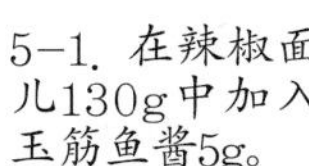

5-1. 在辣椒面儿130g中加入玉筋鱼酱5g。

5-1. Mix 130g of chili powder and 5g of sand eel sauce.

5-2. 새우젓을 3g 넣는다.

5-2. 放入虾酱3g。

5-2. Put 3g of salted shrimp.

5-3. 밀가루풀을 30g 넣는다.

5-3. 加入面粉浆水30g。

5-3. Put 30g of flour.

5-4. 다진 마늘 3g, 다진 생강을 2g 넣는다.

5-4. 放入蒜泥3g和捣碎的生姜2g。

5-4. Add 3g of chopped garlic and 2g of chopped ginger.

5-5. 통깨를 2g 넣는다.

5-5. 放入芝麻2g。

5-5. Add 2g of sesame.

5-6. 양파즙을 15g 넣는다.

5-6. 加入洋葱汁15g。

5-6. Put 15g of onion juice.

TIP
建议

4. 소금에 한 시간 정도만 절인다. (소금은 재료의 3%를 사용하여 절인다.)

4. 用盐腌制一个小时左右。(使用的盐量相当于材料的3%。)

Step 4. Get the radish seasoned with salt for an hour(with only 3% of salt comparing to the other ingredients).

오이소박이

夹心黄瓜泡菜

Cucumber sobaki

재료 및 분량(2인분)
조선오이 2개, 부추 30g, 고춧가루 60g, 까나리액젓 5g, 소금 30g, 다진 마늘 3g, 다진 생강 2g, 통깨 1g

材料和份量(2人份)
朝鲜黄瓜 2个, 韭菜 30g, 辣椒面儿 60g, 玉筋鱼酱 5g, 盐 30g, 蒜泥 3g, 捣碎的生姜 2g, 芝麻 1g

Ingredients and portion(2 portions)
2 choseon cucumber, 30g chives, 60g chili powder, 5g sand eel sauce, 30g salt, 3g chopped garlic, 2g chopped ginger, 1g sesame seed

재료 손질 材料处理方法 Clean ingredients

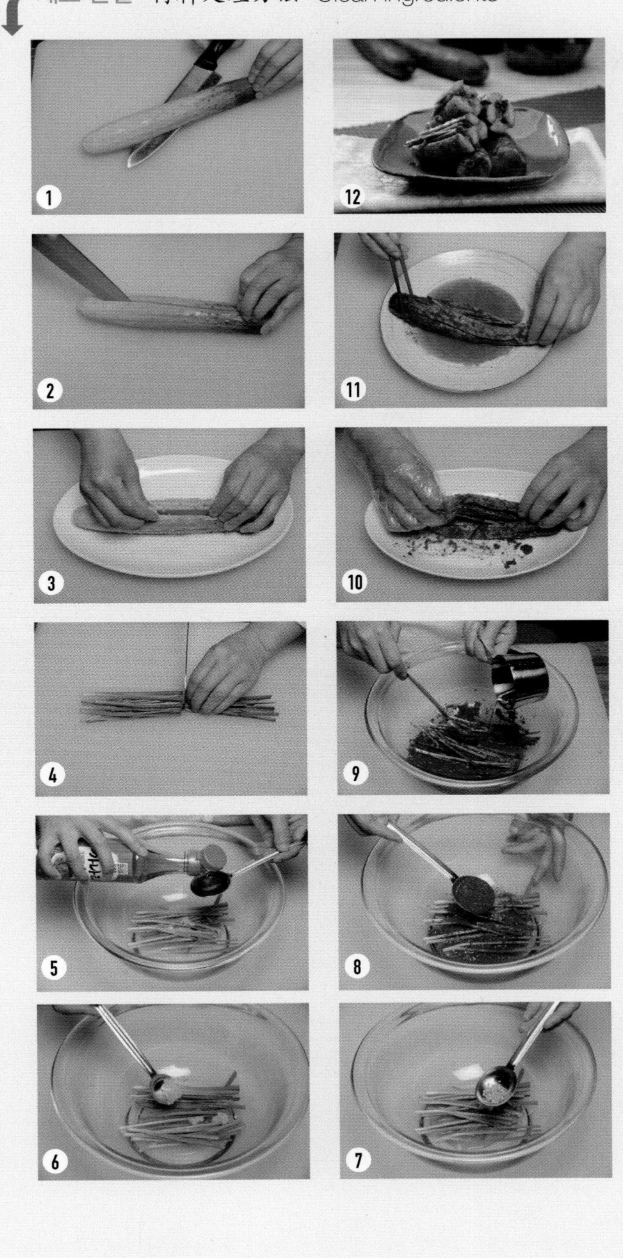

만드는 방법

1. 통오이는 안쪽으로 칼집을 낸다.
2. 길게 열십자로 칼집을 넣는다.
3. 소금에 절인다.
4. 부추를 7cm로 자른다.
5. 까나리액젓 5g으로 간을 한다.
6. 다진 마늘 3g, 다진 생강 2g을 넣는다.
7. 통깨 1g을 넣는다.
8. 고춧가루 60g을 넣는다.
9. 육수 50g을 붓는다.
10. 부추양념장을 속에 잘 넣는다.
11. 완성한 모양이다.
12. 한입 크기로 썰어서 담는다.

制作方法

1. 在整个黄瓜的内侧切口。
2. 使切口呈现十字架形。
3. 用盐腌制。
4. 将韭菜切成大小7cm。
5. 放入玉筋鱼酱5g调味。
6. 放入蒜泥3g和捣碎的生姜2g。
7. 放入芝麻1g。
8. 放入辣椒面儿60g。
9. 倒入肉汤50g。
10. 使韭菜调味酱渗入内部。
11. 完成后盛放在盘子里。
12. 完成。

How to cook

1. Make cuts in the cucumber to the inside.
2. Make a long cross-shaped cut.
3. Season the cucumber with salt.
4. Cut chives by 7cm.
5. Get the chives with 5g of sand eel sauce.
6. Put 3g of chopped garlic and 2g of chopped ginger.
7. Put 1g of sesame.
8. Add 60g of chili powder.
9. Pour 50g of stock.
10. Mix chives with sauce.
11. Done.
12. Cut them into bite-size and serve.

TIP
建议

3. 오이는 20~30분 정도 절인다.

3. 腌制黄瓜的时间最好为20~30分钟。

Step 3. Get the cucumber seasoned for about 20~30minutes.

파김치

葱泡菜

Chives(jjokpa) Kimchi

재료 및 분량(2인분)
쪽파 80g, 무 20g, 당근 3g, 양파 3g, 미나리 3g, 까나리액젓 6g, 고춧가루 60g, 설탕 5g, 다진 마늘 3g, 다진 생강 2g
***육수** : 다시마 5g, 멸치 2마리, 마늘 2g

材料和份量(2人份)
小葱 80g, 萝卜 20g, 胡萝卜 3g, 洋葱 3g, 水芹菜 3g, 玉筋鱼酱 6g, 辣椒面儿 60g, 白糖 5g, 蒜泥 3g, 捣碎的生姜 2g
***肉汤**: 海带 5g, 鳀鱼 2条, 大蒜 2g

Ingredients and portion(2 portions)
80g chives, 20g white radish, 3g carrot, 3g onion, 3g water dropwort(Minari), 6g sand eel sauce, 60g chili powder, 5g sugar, 3g chopped garlic, 2g chopped ginger
***Meat stock** : 5g dried sea tangle, 2 anchovies, 2g garlic

재료 손질 材料处理方法 Clean ingredients

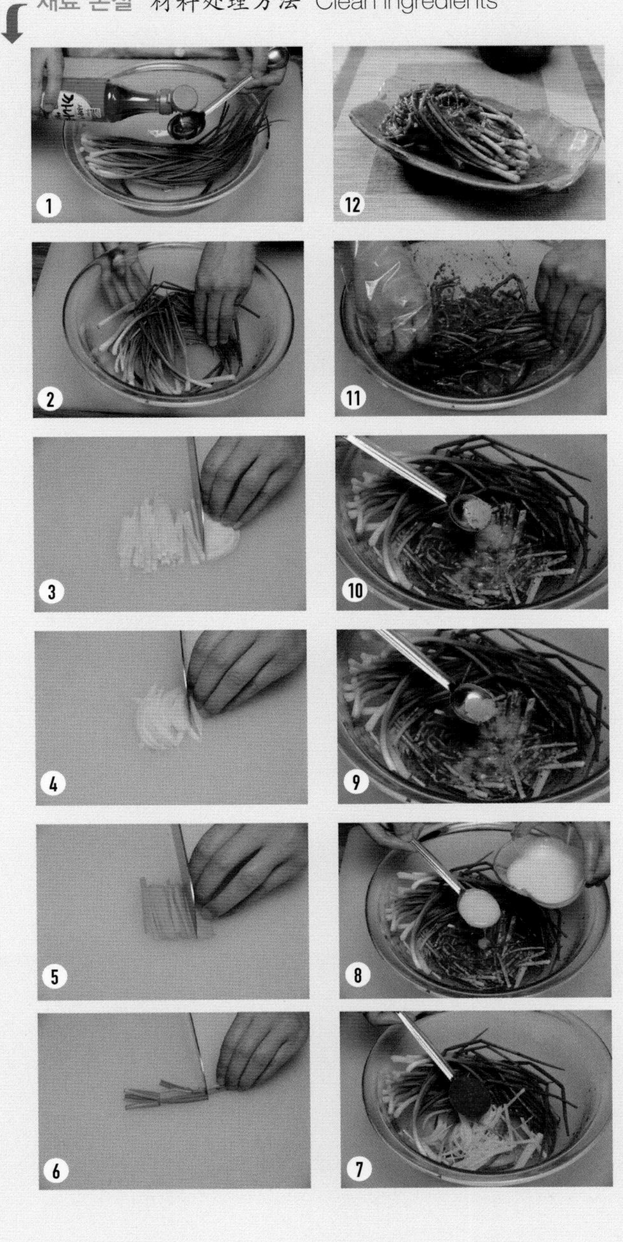

만드는 방법

1. 손질한 쪽파에 까나리액젓 6g을 넣는다.
2. 살짝 절인다.
3. 무를 0.1cm로 자른다.
4. 양파를 0.2cm로 채 썬다.
5. 당근을 0.1cm로 채 썬다.
6. 미나리를 5cm로 자른다.
7. 고춧가루를 60g 넣는다.
8. 설탕을 5g 넣는다.
9. 다진 마늘을 3g 넣는다.
10. 다진 생강을 2g 넣는다.
11. 통깨를 넣어 잘 버무린다.
12. 완성하여 담는다.

制作方法

1. 在处理好的小葱中加入6g玉筋鱼酱。
2. 稍微腌制。
3. 将萝卜切成大小0.1cm。
4. 将洋葱切成大小为0.2cm的丝状。
5. 将胡萝卜切成大小为0.1cm的丝状。
6. 将水芹菜切成大小5cm。
7. 放入60g辣椒面儿。
8. 放入5g白糖。
9. 放入3g蒜泥。
10. 放入2g捣碎的生姜。
11. 放入芝麻后混合搅拌。
12. 完成后盛菜。

How to cook

1. Put 6g of sand eel sauce to the trimmed chives.
2. Get chives slightly seasoned.
3. Cut the radish 0.1cm.
4. Cut the onion 0.2cm.
5. Cut the carrot 0.1cm.
6. Cut the water parsley by 5cm.
7. Add 60g of chili powder.
8. Put 5g of sugar.
9. Put 3g of chopped garlic.
10. Add 2g of chopped ginger.
11. Put sesame and mix them well.
12. Serve the dish.

TIP 쪽파는 까나리액젓으로 절여 완성한다.

建议 用玉筋鱼酱腌制小葱。

Get the chives seasoned with sand eel sauce.

韩食美“味”秘方传授 100选 实践篇

Transmission of a Hundred Secrets of Korean Dishes Flavor

甜点

한식의 맛 비법전수 100선

실기편

후식

식혜

韩式米酒
Sikhye

재료 및 분량(2인분)
엿기름 250g, 흰밥 150g, 생강 5g, 설탕 100g

材料和份量(2人份)
麦芽 250g, 白米饭 150g, 生姜 5g, 白糖 100g

Ingredients and portion(2 portions)
250g malt, 150g boiled non-glutinous rice, 5g ginger, 100g sugar

재료 손질 材料处理方法 Clean ingredients

만드는 방법

1. 엿기름에 물 400g을 섞는다.
2. 잘 섞는다.
3. 체에 내린다.
4. 소창에 거른다.
5. 냄비에 끓인다.
6. 잘 저어준다.
7. 밥을 넣는다.
8. 생강을 편으로 썰어 넣는다.
9. 설탕을 100g 넣는다.
10. 완성하여 차게 식힌다.

制作方法

1. 在麦芽中加入400g水。
2. 均匀地混合。
3. 准备好备用。
4. 过滤。
5. 放在锅里煮开。
6. 均匀地搅拌。
7. 放入米饭。
8. 将生姜切成片状后放入其中。
9. 放入100g白糖。
10. 完成后冷却。

How to cook

1. Mix 400g of water to malt.
2. Mix them well.
3. Strain the content through a sieve.
4. Strain the content again through a cotton cloth.
5. Boil the content in a pot.
6. Stir the content well.
7. Put boiled rice.
8. Cut the ginger into thin slices and put them in the pot.
9. Put 100g of sugar.
10. Leave sikhye to cool down.

TIP
建议

식혜는 시원하게 보관하여 음청류로 먹는 후식이다.

将米酿饮料冷藏保管，可用作饮料类餐后甜点。

Enjoy a glass of sikhye, a Korean traditional dessert, as a cool refreshing beverage.

단호박식혜

南瓜米酒

Sweet pumpkin sikhye

재료 및 분량(2인분)
단호박 50g, 엿기름 250g, 흰밥 150g, 설탕 100g, 생강 10g

材料和份量(2人份)
南瓜 50g, 麦芽 250g, 白米饭 150g, 白糖 100g, 生姜 10g

Ingredients and portion(2 portions)
50g sweet squash, 250g malt, 150g boiled non-glutinous rice, 100g sugar, 10g ginger

재료 손질 材料处理方法 Clean ingredients

만드는 방법

1. 단호박을 쪄서 익힌다.
2. 엿기름에 물 400g을 섞는다.
3. 체에 내린다.
4. 소장에 거른다.
5. 냄비에 넣고 끓인다.
6. 가볍게 저어주면서 끓인다.
7. 식은 밥을 넣는다.
8. 설탕을 100g 넣는다.
9. 거품을 걷어낸다.
10. 단호박을 체에 내린다.
11. 단호박을 넣고 완성한다.
12. 완성하여 차게 식힌다.

制作方法

1. 将南瓜蒸熟后冷却。
2. 在麦芽中加入400g水。
3. 准备好备用。
4. 过滤。
5. 放入锅里煮开。
6. 边煮边轻微地搅拌。
7. 放入冷却的米饭。
8. 放入100g白糖。
9. 去除产生的泡沫。
10. 将南瓜准备好备用。
11. 放入南瓜后完成。
12. 完成。

How to cook

1. Boil the sweet squash.
2. Mix malt to 400g of water.
3. Strain the content through a sieve.
4. Strain the content again through a cotton cloth.
5. Put the content in a pot and get it boiled.
6. Gently stir the content. Keep boiling.
7. Put cold boiled rice.
8. Add 100g of sugar.
9. Skim foam.
10. Get the sweet squash strained through a sieve.
11. Put the sweet squash and enjoy the dish.
12. Enjoy.

TIP 단호박식혜는 시원하게 보관하여 음청류로 먹는 후식이다. 단호박식혜는 보관기간이 짧다.

建议 将南瓜米酿饮料冷藏保管，可用作饮料类饭后甜点。

Keep the sweet squash sikhye cool and enjoy it as a refreshing dessert. Try to have the sweet squash sikhye right after you make it. It will go bad sooner than you expect.

韩食美"味"秘方传授 100选 附录

Transmission of a Hundred Secrets of Korean Dishes Flavor

부록

한식의 맛 비법전수 100선

간장유자드레싱

醬油柚子调味汁

Soy sauce Yuzu dressing

재료 및 분량(2인분)
진간장 110g, 유자청 90g, 설탕 10g, 매실식초 50g, 레몬 1/2개, 양파 20g, 올리고당 40g

材料和份量(2人份)
浓酱油 110g，柚子蜜 90g，白糖 10g，梅果食醋 50g，柠檬 1/2个，洋葱 20g，低聚糖 40g

Ingredients and Portion(2 portions)
110g thick soy sauce, 90g Yuzu honey, 10g sugar, 50g plum fruit vinegar, 1/2 lemon, 20g onion, 40g oligosaccharide

재료 손질 材料处理方法 Clean ingredients

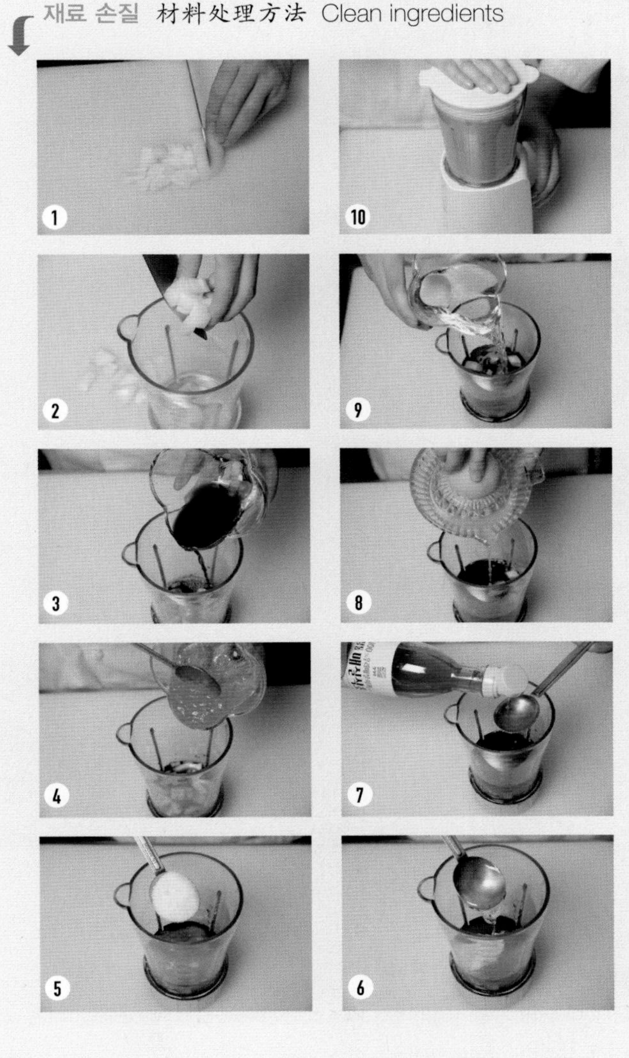

만드는 방법

1. 양파를 큼직하게 썬다.
2. 믹서기에 양파를 넣는다.
3. 진간장 110g을 넣는다.
4. 유자청 90g을 넣는다.
5. 설탕 10g으로 당도를 조절한다.
6. 올리고당 40g을 넣는다.
7. 매실식초 50g으로 새콤한 맛을 조절한다.
8. 레몬즙 15g을 넣는다.
9. 생수 150g으로 농도를 조절한다.
10. 모든 재료를 믹서기에서 곱게 갈아준다.

制作方法

1. 洋葱切成大块。
2. 搅拌机内放入洋葱。
3. 放入110g浓酱油。
4. 放入90g柚子蜜。
5. 用10g白糖调糖度。
6. 放入40g低聚糖。
7. 用50g梅果食醋调酸味。
8. 放入15g柠檬汁。
9. 用150g纯净水调浓度。
10. 用搅拌机搅细所有材料。

How to cook

1. Cut the onion into chunks.
2. Grind the onion in a blender.
3. Pour 110g of thick soy sauce.
4. Pour 90g of Yuzu honey.
5. Add 10g of sugar to adjust sweetness
6. Pour 40g of oligosaccharide.
7. Pour 50g of plum fruit vinegar to adjust sour taste.
8. Pour 15g of lemon juice.
9. Pour 150g of water to adjust consistency.
10. Grind all the ingredients finely in the blender.

TIP
建议

진간장으로 드레싱을 만들 때는 유자청으로 향을 내어 더욱 감칠맛 나게 완성한다.

用酱油做调味汁中加一点柚子蜜的话口味会更加清新。

If you want to cook dressing with thick soy sauce, you can add Yuzu honey to taste.

겨자드레싱

芥末调味汁

Mustard dressing

재료 및 분량(2인분)
겨잣가루 30g, 진간장 10g, 설탕 20g, 매실식초 30g, 오렌지주스 50g, 통잣 10g, 레몬 5g, 다진 마늘 5g, 참기름 5g

材料及分量(2人份)
芥末粉 30g, 浓酱油 10g, 白糖 20g, 梅果食醋 30g, 橙汁 50g, 松仁 10g, 柠檬 5g, 蒜泥 5g, 香油 5g

Ingredients and portion(2 portions)
30g mustard flour, 10g thick soy sauce, 20g sugar, 30g plum fruit vinegar, 50g orange juice, 10g pine nut, 5g lemon, 5g chopped garlic, 5g sesame oil

재료 손질 材料处理方法 Clean ingredients

만드는 방법

1. 겨잣가루는 40℃ 물에서 되직하게 섞는다.
2. 따뜻한 곳에서 매운 향이 날 때까지 발효한다.
3. 발효된 겨자를 넣는다.
4. 오렌지주스 50g을 넣는다.
5. 진간장 10g을 넣는다.
6. 설탕 20g으로 당도가 조절된다.
7. 매실식초 30g으로 새콤한 맛을 조절한다.
8. 통잣 10g을 넣는다.
9. 다진 마늘 5g, 레몬즙 5g을 넣는다.
10. 참기름 5g을 넣는다.
11. 모든 재료를 믹서기에서 곱게 갈아준다.
12. 완성하여 담는다.

制作方法

1. 用40℃水和芥末粉。
2. 放在温暖的场所发酵，直到散发辛辣味。
3. 放入发酵的芥末。
4. 放入橙汁50g。
5. 放入浓酱油10g。
6. 用白糖20g调糖度。
7. 用梅果食醋30g调酸味。
8. 放入松仁10g。
9. 放入蒜泥5g，柠檬汁5g。
10. 放入香油5g。
11. 用搅拌机搅细所有材料。
12. 完成后盛菜。

How to cook

1. Mix mustard flour with water at 40℃.
2. Ferment the ingredient in a warm place until it smells spicy.
3. Put the fermented mustard.
4. Pour 50g of orange juice.
5. Pour 10g of thick soy sauce.
6. Put 20g of sugar to adjust sugar content.
7. Pour 30g of plum fruit vinegar to adjust sour taste.
8. Put 10g of pine nut.
9. Put 5g of chopped garlic and 5g of lemon juice.
10. Pour 5g of sesame seed oil.
11. Grind all the ingredients finely in a blender.
12. Finish.

TIP 建议

해물류와 채소류 샐러드는 겨자드레싱으로 맛을 내어 달콤한 매운맛과 깔끔한 요리로 완성한다. (소고기편육과 채소샐러드에 다양하게 이용된다.)

海鲜和蔬菜类的莎拉用芥末调味汁调出甜辣味，并且还可以在感官上给予清新的感觉。

For sea food or vegetable salad, use mustard dressing to add a spicy-sweet taste. (You can use the mustard dressing for other salads as beef slices and vegetable salad.)

과일드레싱

水果调味汁

Fruit dressing

재료 및 분량(2인분)
사과 1개, 파인애플 80g, 키위 2개, 매실식초 20g, 황설탕 20g, 올리고당 10g, 레몬 1/2개, 사이다 20g, 소금 3g

材料及分量(2人份)
苹果 1个, 菠萝 80g, 猕猴桃 2个, 梅果食醋 20g, 黄糖 20g, 低聚糖 10g, 柠檬 1/2个, 雪碧 20g, 盐 3g

Ingredients and portion(2 portions)
1 apple, 80g pineapple, 2 kiwi fruits, 20g plum fruit vinegar, 20g brown sugar, 10g oligosaccharide, 1/2 lemon, 20g cider, 3g salt

재료 손질 材料处理方法 Clean ingredients

만드는 방법

1. 파인애플 80g을 넣는다.
2. 사과 1개를 넣는다.
3. 매실식초 20g을 넣는다.
4. 황설탕 20g을 넣는다.
5. 소금 3g을 넣는다.
6. 사이다 20g으로 농도를 조절한다.
7. 레몬즙 10g을 넣는다.
8. 위의 재료를 믹서기에서 곱게 갈아준다.
9. 키위는 따로 강판에서 갈아준다.
10. 믹서기 재료와 키위즙을 혼합한다.
11. 신선한 채소를 준비하여 샐러드를 완성한다.
12. 완성하여 담는다.

制作方法

1. 放入80g菠萝。
2. 放入1个苹果。
3. 放入20g梅果食醋。
4. 放入20g黄糖。
5. 放入3g盐。
6. 用20g雪碧调浓度。
7. 放入10g柠檬汁。
8. 搅拌机内放入所有材料后搅细。
9. 用磨擦板研磨猕猴桃。
10. 混合搅拌机材料和泥猴桃汁。
11. 准备新鲜的材料，即可完成沙拉。
12. 完成后盛菜。

How to cook

1. Put 80g of pineapple.
2. Put an apple.
3. Pour 20g of plum fruit vinegar.
4. Put 20g of brown sugar.
5. Put 3g of salt.
6. Pour 20g of cider to adjust consistency.
7. Pour 10g of lemon juice.
8. Grind all the ingredients the above finely in blender.
9. Grate the kiwi fruits on a grater.
10. Mix the ground ingredients from the blender and the kiwi juice from the grater together.
11. Finish cooking by decorating the dish with fresh vegetable.
12. Serve it.

TIP
建议

3. 식초를 넣고 함께 믹서하면 갈변이 방지된다.

3. 放入食醋后搅拌，可防止变成褐色。

Step 3. Add some vinegar for the prevention of browning.

꿀드레싱 (수삼샐러드)

蜂蜜调味汁(水参沙拉)

Honey dressing(fresh ginseng salad)

재료 및 분량(2인분)
수삼 2뿌리, 고구마 1/2개, 밤 4개, 영양부추 3g, 흰깨 · 검정깨 약간, 꿀 100g, 식초 15g, 각종 샐러드채소 20g, 소금 2g

材料及分量(2人份)
水参 2根，地瓜 1/2个，栗子 4个，营养韭菜 3g，芝麻和黑芝麻若干，蜂蜜 100g，食醋 15g，各种沙拉蔬菜 20g，盐 2g

Ingredients and portion(2 portions)
2 fresh ginseng roots, 1/2 sweet potato, 4 chestnuts, 3g chives, a pinch of white sesame seed and black sesame seed, 100g honey, 15g vinegar, 20g vegetables, 2g salt

만드는 방법

1. 수삼머리 쪽을 제거하고 씻는다.
2. 수삼을 어슷썬다.
3. 꿀 100g을 넣는다.
4. 식초 15g을 넣는다.
5. 소금 2g을 넣는다.
6. 꿀드레싱을 골고루 섞는다.
7. 밤과 고구마는 데친 후 냉각시킨다.
8. 샐러드재료를 모두 섞는다.
9. 꿀드레싱을 붓는다.
10. 검정깨, 흰깨를 약간 뿌린다.
11. 완성접시에 먹음직스럽게 담는다.
12. 완성하여 담는다.

制作方法

1. 去除水参头部后清洗干净。
2. 斜切水参。
3. 放入蜂蜜100g。
4. 放入食醋15g。
5. 放入盐2g。
6. 放入蜂蜜调味汁后搅均匀。
7. 用开水焯栗子和地瓜后放凉。
8. 放入沙拉材料后拌匀。
9. 放入蜂蜜调味汁。
10. 撒上若干黑芝麻和芝麻。
11. 装盘即可完成。
12. 完成后盛菜。

How to cook

1. Remove the top of the fresh ginseng. Clean the rest.
2. Cut the fresh ginseng diagonally.
3. Put 100g of honey.
4. Pour 15g of vinegar.
5. Put 2g of salt.
6. Mix the ingredients well.
7. Blanch the chestnuts and the sweet potato. Cool them down.
8. Hand mix all the ingredients together.
9. Pour the honey dressing.
10. Sprinkle white sesame seed and black sesame seed.
11. Serve on a plate.
12. Finish.

TIP 建议

대추채와 견과류 등으로 다양하게 재료를 선택해도 건강샐러드가 된다.

放入大枣丝和干果类，即可成为健康沙拉。

You can add sliced jujube fruit or other nuts to make healthier salad.

잣&참깨드레싱

松仁&芝麻调味汁

Pine nut & sesame dressing

재료 및 분량(2인분)
통잣 10g, 참깨 15g, 진간장 50g, 매실식초 20g, 홍초 15g, 땅콩버터 5g, 다진 마늘 3g, 양파 5g, 올리브유 30g

材料及分量(2人份)
松仁 10g, 芝麻 15g, 浓酱油 50g, 梅果食醋 20g, 红醋 15g, 花生酱 5g, 蒜泥 3g, 洋葱 5g, 橄榄油 30g

Ingredients and portion(2 portions)
10g pine nut, 15g sesame seed, 50g thick soy sauce, 20g plum fruit vinegar, 15g red vinegar, 5g peanut butter, 3g chopped garlic, 5g onion, 30g olive oil

재료 손질 材料处理方法 Clean ingredients

만드는 방법

1. 진간장 50g을 넣는다.
2. 땅콩버터 5g을 넣는다.
3. 볶은 참깨 15g을 넣는다.
4. 볶은 통잣 10g을 넣는다.
5. 다진 마늘 3g, 양파 5g을 넣는다.
6. 올리브유 30g을 넣는다.
7. 매실식초 20g을 넣는다.
8. 홍초 15g을 넣는다.
9. 생수 120g으로 농도를 조절한다.
10. 위의 재료를 믹서기에서 곱게 갈아준다.
11. 완성하여 붓는다.
12. 완성하여 담는다.

制作方法

1. 放入50g浓酱油。
2. 放入5g花生酱。
3. 放入15g炒芝麻。
4. 放入10g炒松仁。
5. 放入3g蒜泥、5g洋葱。
6. 30g橄榄油。
7. 放入20g梅果食醋。
8. 放入15g红醋。
9. 用120g纯净水调整浓度。
10. 上述材料放入搅拌机后搅细。
11. 制成后倒入即可。
12. 完成后盛菜。

How to cook

1. Pour 50g of thick soy sauce.
2. Put 5g of peanut butter.
3. Put 15g of roasted sesame.
4. Put 10g of roasted pine nut.
5. Put 3g of chopped garlic and 5g of garlic.
6. Pour 30g of olive oil.
7. Put 20g of plum fruit vinegar.
8. Pour 15g of red vinegar.
9. Add 120g of water to adjust consistency.
10. Grind the ingredients the above finely in a blender.
11. Pour the ground ingredient from the blender on a plate.
12. Finish.

TIP 建议

잣&참깨드레싱을 만들 때 과일류(사과), 채소(양파) 등을 첨가하면 맛이 새로워진다.

放入大枣丝和干果类，即可成为健康沙拉。

You can add fruits(apple) or vegetables(onion) for a new taste of the salad.

플레인요구르트 드레싱

纯酸奶调味汁

Plain yoghurt dressing

재료 및 분량(2인분)
플레인요구르트 200g, 올리고당 20g, 레몬 1/2개(과일즙: 사과 30g, 파인애플 20g, 식초 10g, 설탕 5g, 소금 2g, 키위 1개)

材料及分量(2人份)
纯酸奶 200g, 低聚糖 20g, 柠檬 1/2个(果汁: 苹果 30g, 菠萝 20g, 食醋 10g, 白糖 5g, 盐 2g, 猕猴桃 1个)

Ingredients and portion(2 portions)
200g plain yogurt, 20g oligosaccharide, 1/2 lemon(fruit juice: 30g apple, 20g pine apple, 10g vinegar, 5g sugar, 2g salt, 1 kiwi fruit)

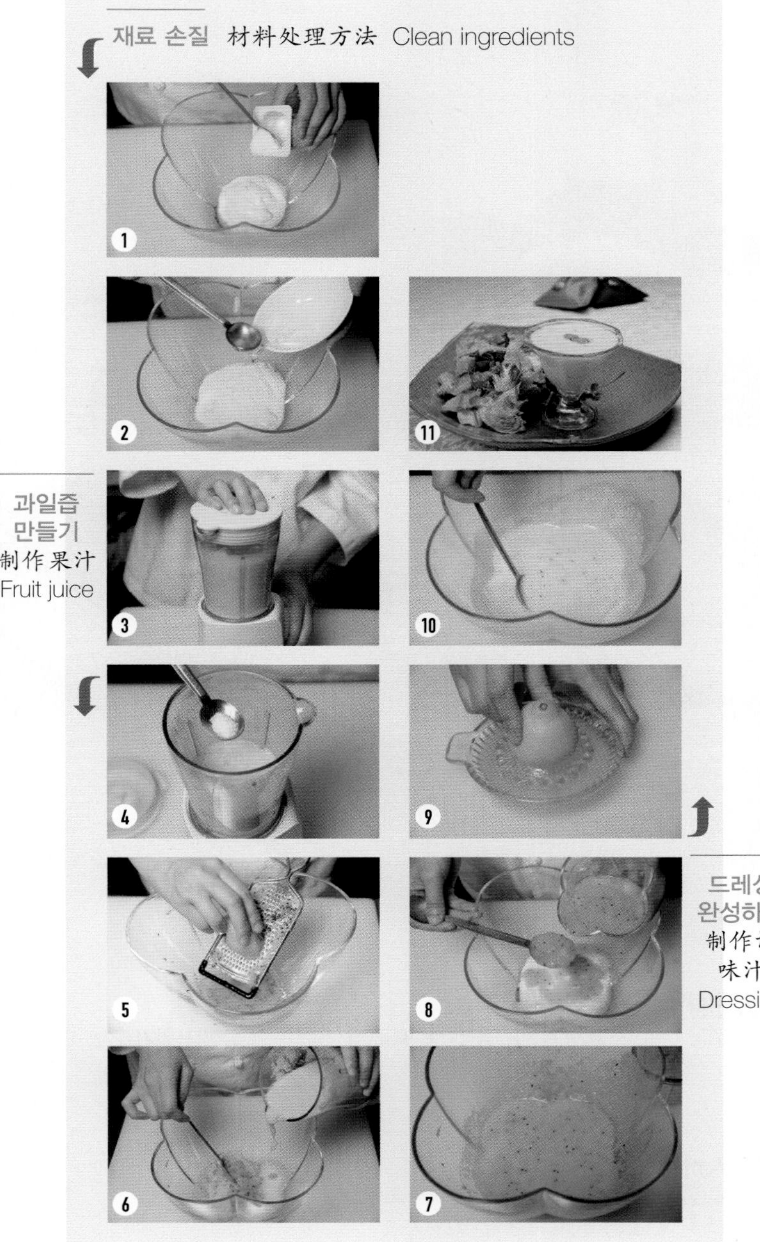

만드는 방법

1. 요구르트 200g을 넣는다.
2. 올리고당 20g을 넣는다.
3. 사과 30g, 파인애플 20g, 식초 10g, 설탕 5g을 갈아준다.
4. 소금 2g을 넣는다.
5. 키위 1개를 강판에 갈아준다.
6. 믹서기에서 간 재료와 키위즙을 섞는다.
7. 과일즙 완성
8. 플레인요구르트에 과일즙을 섞는다.
9. 레몬즙 10g을 넣는다.
10. 골고루 저어준다.
11. 완성하여 담는다.

制作方法

1. 放入200g酸奶。
2. 放入20g低聚糖。
3. 搅拌30g苹果，20g菠萝，10g食醋，5g白糖。
4. 放入2g盐。
5. 用磨擦板研磨一个猕猴桃。
6. 用搅拌机搅细的材料和猕猴桃汁拌匀。
7. 完成果汁。
8. 酸奶混合果汁。
9. 放入10g柠檬汁。
10. 均匀的搅拌。
11. 完成后盛菜。

How to cook

1. Put 200g of yogurt.
2. Pour 20g of oligosaccharide.
3. Grind 30g of apple, 20g of pineapple, 10g of vinegar and 5g of sugar.
4. Add 2g of salt.
5. Grate a kiwi fruit on a grater.
6. Mix the ground ingredients and the kiwi juice in a blender.
7. Finish the fruit juice.
8. Pour the fruit juice to the yogurt.
9. Add 10g of lemon juice.
10. Mix well.
11. Pour the dressing to fresh vegetables.

TIP 建议

플레인요구르트드레싱에 생크림을 혼합하면 신선한 맛을 느낄 수 있다.

在酸奶里倒入鲜奶油的话酸奶会更加新鲜可口。

Add whipped cream to the plain yogurt dressing for a fresher taste.

해물겨자채샐러드

海鲜芥末菜沙拉

Seafood mustard salad

재료 및 분량(2인분)
주꾸미 150g, 새우살 50g, 소고기(등심)편육 50g, 배 20g, 청오이 10g, 파프리카 10g, 밤 4개, 은행 2g, 지단 2g, 겨자드레싱 70g

材料及分量(2人份)
短蛸 150g，虾仁 50g，牛肉(里脊)肉片 50g，梨 20g，黄瓜 10g，彩椒 10g，栗子 4个，银杏 2g，煎蛋丝 2g，芥末调味汁 70g

Ingredients and portion(2 portions)
150g baby octopus, 50g shrimp, 50g boiled beef(sirloin) slices, 20g pear, 10g cucumber, 10g paprika, 4 chestnuts, 2g ginkgo nut, 2g jidan, 70g mustard dressing

재료 손질 材料处理方法 Clean ingredients

완성하기
完成
Finish

만드는 방법

1. 배는 두께 0.3cm의 골패형으로 썬다.
2. 홍파프리카를 1x4cm로 썬다.
3. 청오이를 1x4cm로 썬다.
4. 밤은 두께 0.3cm로 썬다.
5. 지단을 1x4cm로 썬다.
6. 주꾸미를 손질하여 데친다.
7. 새우살을 데친다.
8. 소고기편육은 1x4cm로 썬다.
9. 해물과 채소와 편육을 섞는다.
10. 겨자드레싱으로 간을 맞춘다.
11. 골고루 섞어서 샐러드를 완성한다.
12. 완성하여 담는다.

制作方法

1. 梨切成0.3cm厚度的骨牌形状。
2. 红彩椒切成1x4cm。
3. 黄瓜切成1x4cm。
4. 栗子切成0.3cm的厚度。
5. 煎鸡蛋切成1x4cm。
6. 用开水焯整理好的短蛸。
7. 开水焯虾仁。
8. 牛肉片切成1x4cm。
9. 海鲜、蔬菜、肉片拌匀。
10. 用芥末调味酱调味。
11. 拌匀即可完成沙拉。
12. 完成后盛菜。

How to cook

1. Cut the pear into 0.3cm wide pieces.
2. Cut the red paprika into 1cm x 4cm pieces.
3. Cut the cucumber into 1cm x 4cm pieces.
4. Cut the chestnuts into 0.3cm wide pieces.
5. Cut jidan into 1cm x 4cm pieces.
6. Clean baby octopus and blanch.
7. Blanc shrimp.
8. Cut the beef(sirloin) boiled beef slices into 1cm x 4cm pieces.
9. Mix seafood, vegetables and boiled beef slices together.
10. Season with the mustard dressing.
11. Hand mix the ingredients.
12. Serve on a plate.

TIP
建议

새우, 갑오징어, 전복 등의 재료와 신선한 채소를 다양하게 곁들여 매콤, 달콤한 겨자드레싱으로 버무려 완성한다.

虾，乌贼，鲍鱼等材料与新鲜的蔬菜和芥末调味汁搅拌后，即可成为健康沙拉。

You can use any other healthy ingredients as shrimp, cuttlefish, abalones and vegetables to add a spicy-sweet taste to your salad.

간장게장

醬油蟹醬

Crab Marinated in soy sauce

재료 및 분량(2인분)

꽃게 2마리, 진간장 200g, 양파 20g, 배 10g, 사과 5g, 레몬 3g, 홍고추 2g, 풋고추 2g, 마늘 2g, 생강 2g, 실파 2g, 대추 2개, 통계피 2g, 황기 1g, 당귀 1g, 감초 2개, 구기자 2g, 통후추 1g, 설탕 20g, 올리고당 15g, 청주 15g

材料和份量(2人份)

花蟹 2只, 浓酱油 200g, 洋葱 20g, 梨 10g, 苹果 5g, 柠檬 3g, 红辣椒 2g, 青辣椒 2g, 大蒜 2g, 生姜 2g, 大葱 2g, 大枣 2个, 整桂皮 2g, 黄芪 1g, 当归 1g, 甘草 2个, 枸杞子 2g, 胡椒 1g, 白糖 20g, 低聚糖 15g, 白酒 15g

Ingredients and portion(2 portions)

2 blue crabs, 200g thick soy sauce, 20g onion, 10g pear, 5g apple, 3g lemon, 2g red chili, 2g green chili, 2g garlic, 2g ginger, 2g green onion, 2 jujube fruits, 2g cinnamon, 1g milk vetch root, 1g Korean angelica roots(dong guai), 2 licorices, 2g chinese wolfberries(gugija), 1g whole black pepper, 20g sugar, 15g oligosaccharide, 15g rice-wine

꽃게 손질 处理花蟹 Clean blue crabs

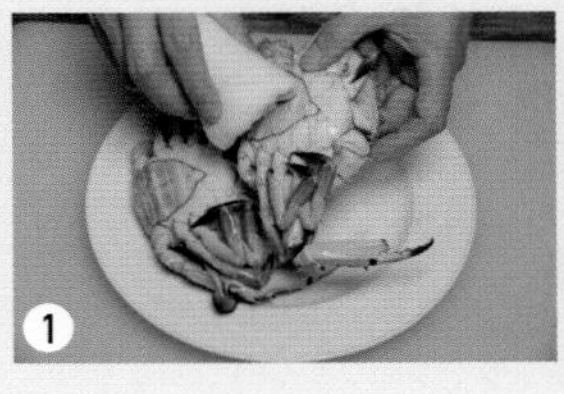

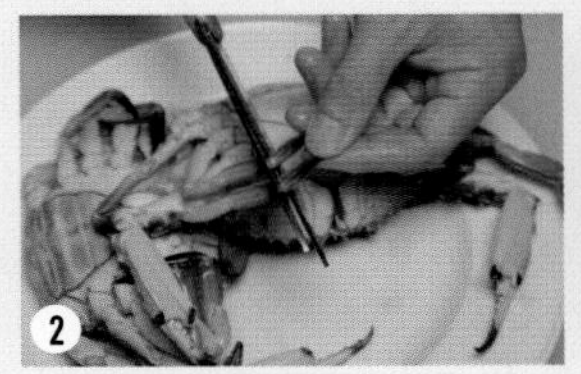

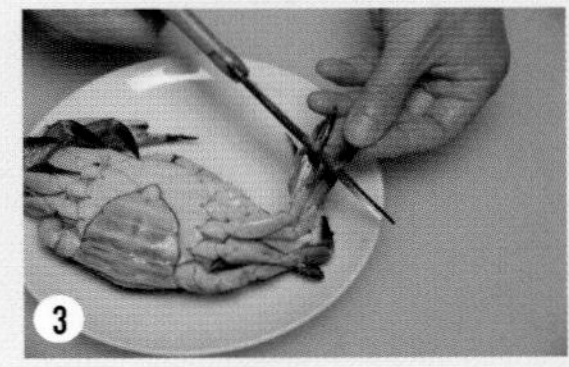

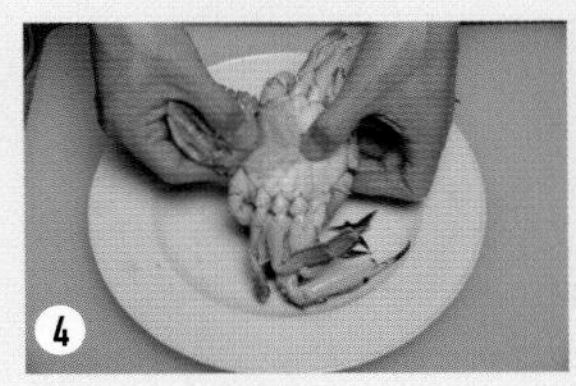

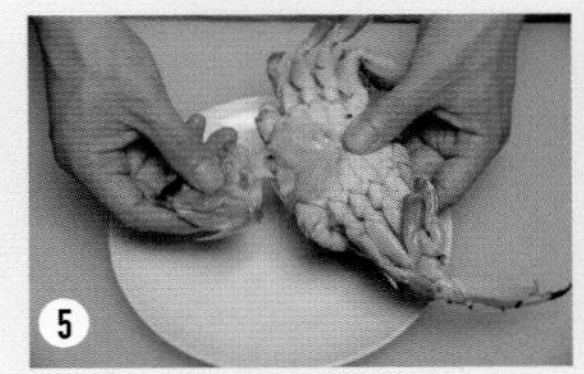

채소 손질
处理蔬菜
Clean vegetables

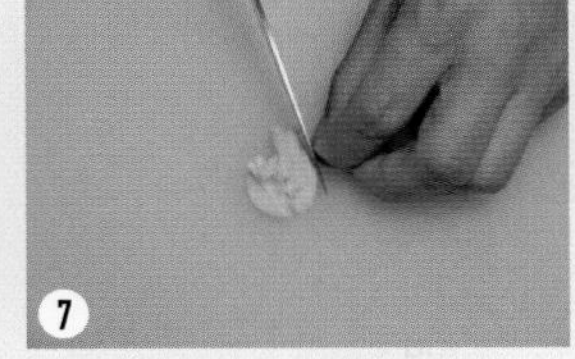

만드는 방법

1. 꽃게는 씻은 후 물기를 제거한다.
2. 꽃게의 잔다리를 제거한다.
3. 양쪽의 끝부분 다리를 모두 제거한다.
4. 배쪽을 손질한다.
5. 배쪽 모래를 제거한다.
6. 생강을 채 썬다.
7. 마늘을 채 썬다.
8. 양파를 채 썬다.
9. 실파를 5cm로 썬다.
10. 향신채소를 준비한다.
11. 별도의 향신채소와 마른 홍고추를 굽는다.
12. 향신료(통계피, 감초, 당귀, 황기, 대추, 마른 홍고추)를 준비한다.

制作方法

1. 将花蟹洗好后控除水分。
2. 切除花蟹的多余蟹腿。
3. 切除两侧末端的蟹腿。
4. 处理花蟹的肚子部分。
5. 清除肚子部分的沙子。
6. 将生姜切成细丝。
7. 将大蒜切成细丝。
8. 将洋葱切成细丝。
9. 将小葱切成大小5cm。
10. 准备好香味菜。
11. 烤另外准备的香味菜和干红辣椒。
12. 准备好香味菜(整桂皮,甘草,当归,黄芪,大枣,干红辣椒)。

How to cook

1. Wash the crabs in water. Drain water.
2. Remove all the small legs of the crabs.
3. Remove the ends of the legs on the sides.
4. Wash the stomach of the crabs.
5. Remove sands on the stomach.
6. Julienne the ginger.
7. Julienne the garlic.
8. Julienne the onion.
9. Cut the green onion into 5cm lengths.
10. Prepare condiment vegetable.
11. Grill the condiment vegetable and dried red chili.
12. Prepare spices (cinnamon, licorice, Korean angelica roots, milk vetch root, jujube fruits, dried red chili).

간장양념 만들기
制作酱油调味料
Cook soy sauce

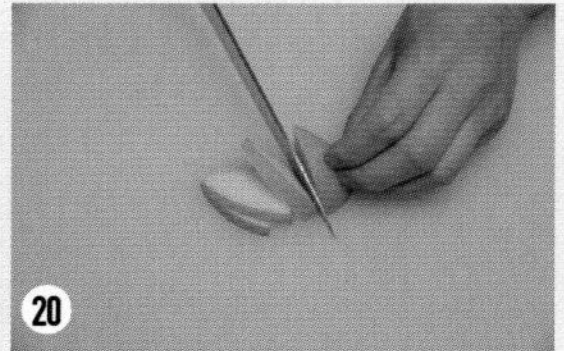

만드는 방법

13. 생수 600g을 넣는다.
14. 진간장 200g을 넣는다.
15. 생수에 붓는다.
16. 향신료를 넣는다.
17. 생강채를 넣는다.
18. 양파채를 넣는다.
19. 실파를 넣는다.
20. 배를 편 썬다.
21. 사과를 편 썬다.
22. 배와 사과를 넣는다.
23. 구운 향신채를 넣는다.
24. 청주 15g을 넣는다.

制作方法

13. 加600g水。
14. 加入200g酱油。
15. 加入生水。
16. 放入香味调料。
17. 放入生姜丝。
18. 放入洋葱丝。
19. 放入小葱。
20. 将梨切成薄片。
21. 将苹果切成薄片。
22. 放入梨和苹果。
23. 放入烤好的香味菜。
24. 加15g白酒。

How to cook

13. Pour 600g of water.
14. Pour 200g of thick soy sauce.
15. Add water.
16. Put the spices.
17. Put the julienne ginger.
18. Put the julienne onion.
19. Put the julienne green onion.
20. Slice the pear.
21. Slice the apple.
22. Put the pear and the apple.
23. Put the grilled condiment vegetable.
24. Pour 15g of clean rice-wine.

완성하기
完成
Finish

만드는 방법
25. 레몬을 편 썬다.
26. 레몬을 넣는다.
27. 설탕 15g을 넣는다.
28. 올리고당 15g을 넣고 간장소스를 30분간 끓인다.
29. 간장소스가 식은 후 손질된 꽃게에 붓는다.
30. 대나무잎을 올려 냄새를 잡는다.
31. 24시간 후에 간장소스를 다시 끓여서 식은 후에 붓는다.
32. 24시간 후에 간장소스에서 건져서 완성한다.

制作方法
25. 将柠檬切成薄片。
26. 放入柠檬。
27. 放入白糖15g。
28. 加入低聚糖15g和酱油调味料后煮30分钟。
29. 待酱油调味料冷却后将其倒入花蟹中。
30. 放上柱子叶调味。
31. 24小时后再次把烧开的酱油调味汁冷却后放入花蟹。
32. 24小时后把花蟹从酱油调味汁中取出，即可完成。

How to cook
25. Slice the lemon.
26. Put the lemon.
27. Put 15g of sugar.
28. Put 15g of oligosaccharide. Boil the soy sauce for 30 minutes.
29. Cool the soy sauce down. Pour it to the clean blue crabs.
30. Put bamboo leaves to remove smell.
31. After 24 hours, boil the soy sauce again. Cool it down and pour it.
32. After 24 hours, take the blue crabs out from the soy sauce. Finish.

TIP
建议

첫 번째 간장게장소스는 끓여서 완전히 식은 후 꽃게에 붓는다. 24시간(밀봉) 후 간장소스를 두 번째 끓여 식은 후 꽃게에 붓는다. → 24시간(밀봉) 후 바로 건져야 꽃게가 짜지 않다. 꽃게만 따로 냉장 보관하면서 시식한다.

第一次调味时酱油调味汁冷却后将花蟹放入酱油中。24小时后再次把酱油调味汁烧热，等到冷却后再次把花蟹放入酱油中。24小时后把花蟹从酱油调味汁中取出后冷藏。

Make sure that you completely cool the boiled soy sauce down, before you pour it over the blue crap. After 24 hours (keep the sauce in a container which is tightly closed), re-boil the soy sauce. Cool the sauce down and pour it to the blue crabs. -> Right after 24 hours (keep the ingredients in a container which is tightly closed), take the blue crabs out from the soy sauce. Keep the blue crabs only in a refrigerator. Enjoy the dish.

꽃게양념무침

调味拌花蟹

Seasoned blue crab

재료 및 분량(2인분)
꽃게 2마리, 레몬 3g, 홍고추 2g, 풋고추 2g, 미나리 2g, 통깨 1g
***양념장** : 고춧가루 20g, 진간장 100g, 생수 100g, 올리고당 10g, 청주 15g, 대파 2g, 마늘 2g, 생강 2g, 설탕 5g, 배즙 2g, 사과즙 2g, 양파즙 2g

材料和份量(2人份)
花蟹 2只, 柠檬 3g, 红辣椒 2g, 青辣椒 2g, 水芹菜 2g, 芝麻 1g
***调味酱** : 辣椒面儿 20g, 浓酱油 100g, 生水 100g, 低聚糖 10g, 白酒 15g, 大葱 2g, 大蒜 2g, 生姜 2g, 白糖 5g, 梨汁 2g, 苹果汁 2g, 洋葱汁 2g

Ingredients and portion(2 portions)
2 blue crabs, 3g lemon, 2g red chili, 2g green chili, 2g water dropwort(Minari), 1g sesame seed
***Sauce** : 20g chili powder, 100g thick soy sauce, 100g water, 10g oligosaccharide, 15g rice-wine, 2g leek, 2g garlic, 2g ginger, 5g sugar, 2g pear juice, 2g apple juice, 2g onion juice

꽃게 손질 处理花蟹 Clean blue crabs

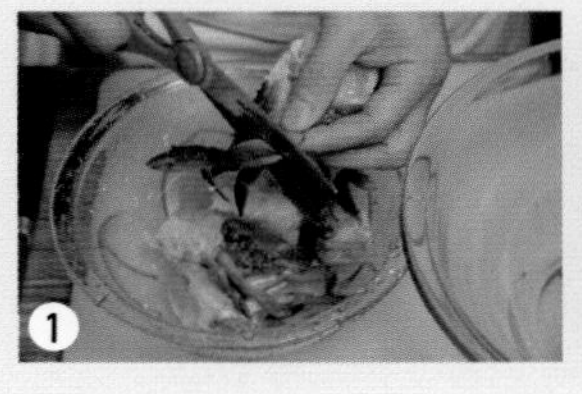

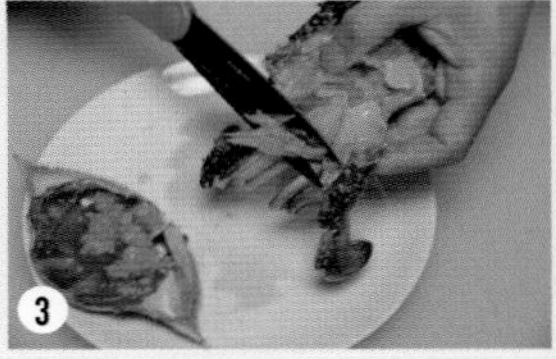

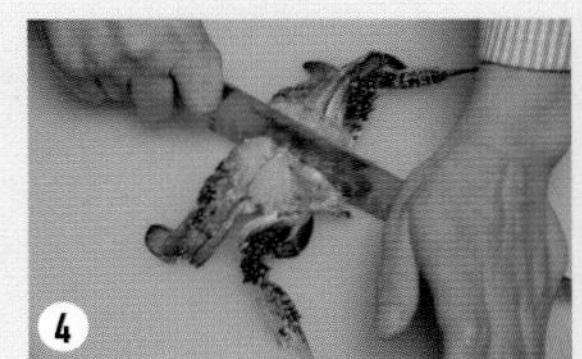

채소 손질
处理蔬菜
Clean vegetables

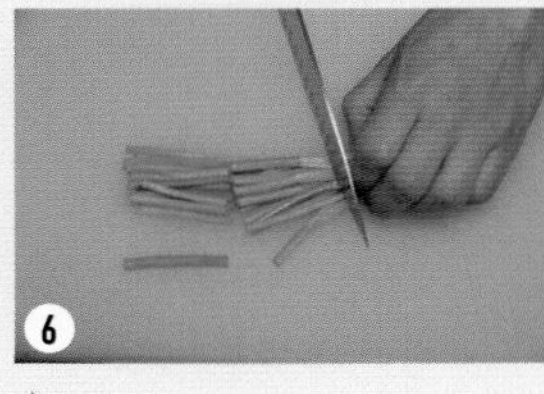

양념장 만들기
制作调味酱
Cook sauce

만드는 방법

1. 꽃게를 씻은 후 잔다리를 제거한다.
2. 등껍질과 속살을 분리한다.
3. 먹지 못하는 부위를 손질한다.
4. 먹기 좋게 자른다.
5. 4등분하여 손질한다.
6. 미나리는 길이 5cm로 자른다.
7. 대파는 길이 5cm로 잘라 반으로 가른다.
8. 마늘을 편 썬다.
9. 홍고추를 어슷썬다.
10. 풋고추를 어슷썬다.
11. 생수 100g에 진간장 100g을 붓는다.
12. 마늘, 생강 편을 넣는다.

制作方法

1. 将花蟹洗好后切除多余蟹腿。
2. 分离背壳和里面的肉。
3. 处理掉不能食用的部分。
4. 切成适合食用的大小。
5. 4等分后整理。
6. 将头长度切成长5cm。
7. 将大葱切成一半长5cm。
8. 将大蒜切成薄片。
9. 斜切红辣椒。
10. 斜切青辣椒。
11. 在生水中加入100g中加入酱油100g。
12. 放入蒜片、生姜片。

How to cook

1. Wash the crabs in water. Remove all the small legs of the crabs.
2. Separate the shell on the back and the flesh.
3. Remove inedible parts.
4. Cut the crabs into bite-size pieces.
5. Cut the crabs into quarters.
6. Cut water dropwort (Minari) into 5cm lengths.
7. Cut the leek in half. Then, cut it into 5cm lengths.
8. Slice the garlic.
9. Cut the red chili diagonally.
10. Cut the red chili diagonally.
11. Pour 100g of thick soy sauce to 100g of water.
12. Put sliced garlic and sliced ginger.

만드는 방법

13. 대파를 넣는다.
14. 설탕 5g을 넣는다.
15. 올리고당 10g을 넣는다.
16. 청주 15g을 넣는다.
17. 고춧가루 20g을 넣는다.
18. 배즙 2g, 사과즙 2g, 양파즙 2g을 넣는다.
19. 양념장을 완성한다.
20. 손질된 꽃게에 버무린다.
21. 골고루 양념이 배도록 한다.
22. 채소재료를 넣는다.

制作方法

13. 放入大葱。
14. 放入5g白糖。
15. 加入10g低聚糖。
16. 加入15g白酒。
17. 放入20g辣椒面儿。
18. 加入2g梨汁、2g苹果汁和2g洋葱汁。
19. 调味酱完成。
20. 将调味酱涂抹在处理好的花蟹上。
21. 均匀涂抹调味料。
22. 放上蔬菜材料。

How to cook

13. Put the leek.
14. Put 5g of sugar.
15. Pour 10g of oligosaccharide.
16. Add 15g of rice-wine.
17. Put 20g of chili powder.
18. Pour 2g of pear juice, apple juice and onion juice.
19. Sauce finish.
20. Hand mix the clean blue crabs with the sauce.
21. Wait for the sauce to seep in to the blue crabs.
22. Put the vegetables.

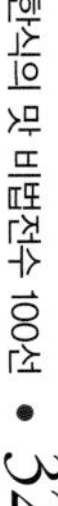

만드는 방법
23. 레몬즙, 통깨로 완성한다.
24. 완성하여 담는다.

制作方法
23. 撒上柠檬汁和芝麻后完成。
24. 完成后盛菜。

How to cook
23. Finish the dish with lemon juice and sesame seed.
24. Serve it.

TIP
建议

반드시 신선한 꽃게를 준비하여 깨끗이 손질하고 양념장은 끓여서 식은 후에 버무린다.
양념된 꽃게는 신선도 유지 때문에 오래 보관하지 않는다.

把新鲜的花蟹切好以后，倒入冷却的调料汁。
为了保持花蟹料理的新鲜度，不可将其长期冷藏食用。

Make sure that you clean the blue crabs thoroughly. Cool the boiled sauce down before you put it to the crabs.
Help yourself before the crabs go bad.

순두부찌개

嫩豆腐汤

Soft tofu stew

재료 및 분량(2인분)
순두부 100g, 소고기(양지) 20g, 바지락 10g, 애호박 5g, 대파 2g, 실파 2g, 느타리버섯 2g, 양파 2g, 홍고추 2g, 풋고추 2g, 마늘 2g, 고춧가루 2g, 국간장 2g, 새우젓 1g 또는 까나리액젓 2g, 청주 15g

***육수** : 국멸치 3마리, 다시마 2g, 대파 2g 또는 마른 홍고추 1개, 통마늘 1g, 마른 홍합살 2g 또는 마른 새우살 1g

材料和份量(2人份)
软豆腐 100g, 牛肉(牛排) 20g, 蛤仔 10g, 西葫芦 5g, 大葱 2g, 小葱 2g, 平菇 2g, 洋葱 2g, 红辣椒 2g, 青辣椒 2g, 大蒜 2g, 辣椒面儿 2g, 汤用酱油 2g, 虾酱 1g 或玉筋鱼酱 2g, 白酒 15g

***肉汤** : 汤用鳀鱼 3条, 海带 2g, 大葱 2g 或干红辣椒 1个, 大蒜 1g, 干海红肉 2g 或干虾仁 1g

Ingredients and portion(2 portions)
100g soft tofu (soft bean curd), 20g beef (brisket), 10g clam, 5g long squash, 2g leek, 2g spring onion, 2g oyster mushroom, 2g onion, 2g red chili, 2g green chili, 2g garlic, 2g chili powder, 2g soy sauce, 1g salted shrimp or 2g sand eel sauce, 15g rice-wine

* **Stock** : 3 dried anchovies, 2g dried sea tangle, 2g leek or 1 dried red chili, 1g garlic, 2g dried mussel or 1g dried shrimp.

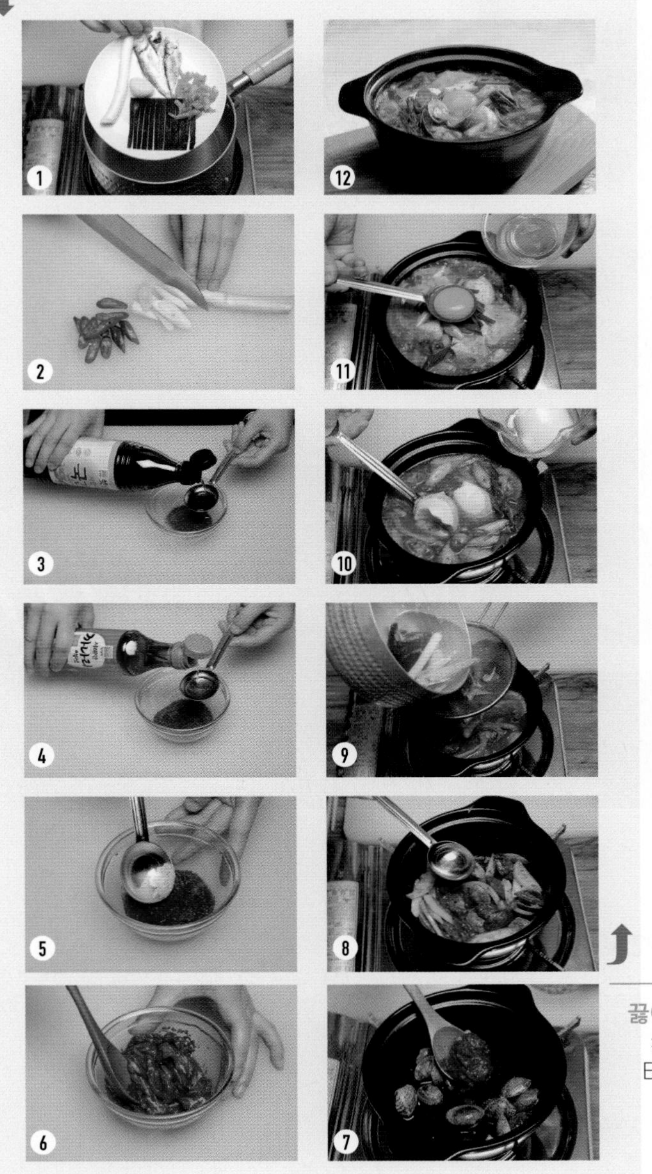

만드는 방법

1. 육수를 10분 정도 끓인다.
2. 홍고추, 풋고추, 대파는 어슷썬다.
3. 고춧가루 2g과 국간장 2g을 섞는다.
4. 까나리액젓 2g을 붓는다.
5. 다진 마늘 2g을 넣는다.
6. 소고기는 편 썰어 버무린다.
7. 뚝배기에 고추기름 → 바지락 볶다가 → 양념한 소고기를 넣는다.
8. 양파채와 청주 15g으로 냄새를 제거한다.
9. 육수는 8부 정도 붓는다.
10. 순두부는 떠서 넣고, 나머지 채소재료를 넣는다.
11. 완성된 찌개에 달걀 노른자만 올린다.
12. 완성하여 담는다.

制作方法

1. 煮肉汤10分钟左右。
2. 斜切红辣椒、青辣椒和大葱。
3. 放入2g辣椒面儿和汤用2g酱油后混合。
4. 放入2g玉筋鱼酱。
5. 放入2g蒜泥。
6. 将牛肉切成薄片后拌合。
7. 在石锅中放入辣椒油->蛤仔一起炒->放入调味好的牛肉。
8. 放入洋葱丝和15g白酒调味。.
9. 倒入肉汤80%左右。
10. 拿勺舀出软豆腐放入其中，放入其他蔬菜材料。
11. 在完成的汤上放上鸡蛋黄。
12. 完成后盛菜。

How to cook

1. Boil stock for about 10 minutes.
2. Cut red chili, green chili and leek diagonally.
3. Mix 2g of chili powder and 2g of soy sauce together.
4. Pour 2g of sand eel sauce.
5. Put 2g of chopped garlic.
6. Slice beef and mix it with the other ingredients.
7. Fry clam with chili oil in an earthen pot. Put the seasoned beef.
8. Add the sliced onion and 15g of rice-wine to remove smell.
9. Pour about 80% of the stock.
10. Put soft tofu(soft bean curd) with the other vegetables.
11. Finish the dish by putting the yolk on top.
12. Done.

TIP
建议

순두부는 부드럽기 때문에 마지막에 넣어야 모양이 부서지지 않는다.

最后放入嫩豆腐才能保持其形状。

Put soft tofu(soft bean curd) at the finish so that it would not fall apart.

청국장

清国酱汤

Cheonggukjang(fast-fermented bean paste)

재료 및 분량(2인분)
청국장 20g, 소고기 5g, 무 3g, 두부 5g, 애호박 5g, 양파 2g, 배추김치 5g, 표고버섯 1개, 대파 2g, 홍고추 2g, 풋고추 2g, 다시마 2g, 마른 홍고추 2g, 마른 새우 2g, 감초 1g, 마늘 3g, 고추씨 2g, 들기름 2g, 들깨가루 2g, 국간장 2g, 고춧가루 1g

材料和份量(2人份)
豆瓣酱 20g, 牛肉 5g, 萝卜 3g, 豆腐 5g, 西葫芦 5g, 洋葱 2g, 白菜泡菜 5g, 香菇 1个, 大葱 2g, 红辣椒 2g, 青辣椒 2g, 海带 2g, 干红辣椒 2g, 干虾 2g, 甘草 1g, 大蒜 3g, 辣椒籽 2g, 芝麻油 2g, 野芝麻粉 2g, 汤用酱油 2g, 辣椒面儿 1g

Ingredients and portion(2 portions)
20g fermented soybeans, 5g beef, 3g white radish, 5g bean curd(tofu), 5g long squash, 2g onion, 5g kimchi, 1 shiitake mushroom(mushroom pyogo), 2g leek, 2g red chili, 2g green chili, 2g dried sea tangle, 2g dried red chili, 2g dried shrimp, 1g licorice, 3g garlic, 2g pepper seed, 2g perilla seed oil, 2g perilla seed flour, 2g soy sauce, 1g chili powder

육수 끓이기 煮肉汤 Stock

재료 손질 处理材料 Clean ingredients

끓이기 煮 Boil

1 2 3 4 5 6 7 8 9 10 11 12

만드는 방법

1. 육수는 20분 정도 약불에서 끓인다.
2. 재료는 알맞은 크기로 썬다.
3. 냄비에 소고기 5g, 콩가루 2g, 들기름 2g으로 볶는다.
4. 육수와 재료를 넣고 들깨가루 2g을 넣는다.
5. 배추김치를 넣는다.
6. 다진 마늘 2g을 넣는다.
7. 청국장 20g을 국간장 2g과 섞는다.
8. 고춧가루 1g을 섞는다.
9. 양념된 청국장을 풀어준다.
10. 마지막에 홍고추, 풋고추, 대파를 넣어준다.
11. 뚝배기에 옮겨 끓여서 완성한다.
12. 완성하여 담는다.

制作方法

1. 在微火上煮肉汤20分钟左右。
2. 将材料切成适当的大小。
3. 在锅里放入5g牛肉, 2g豆油, 2g芝麻油后一起炒。
4. 放入肉汤和材料及2g野芝麻粉。
5. 放入白菜泡菜。
6. 放入2g蒜泥。
7. 将20g清国酱和2g汤用酱油混合起来。
8. 清国酱中放入1g辣椒面。
9. 冲泡开调味好的豆瓣酱。
10. 最后放入红辣椒、青辣椒和大葱。
11. 转移到石锅中煮好完成。
12. 完成。

How to cook

1. Boil the stock over a low heat for about 20 minutes.
2. Cut the ingredients into bite-size pieces.
3. Roast 5g of beef with 2g of bean flour and 2g of perilla seed oil.
4. Put the ingredients and 2g of perilla seed flour to the stock.
5. Put Kimchi.
6. Put 2g of chopped garlic.
7. Mix 20g of fermented soybeans with 2g of soy sauce.
8. Add 1g of chili powder.
9. Put the seasoned fermented soybeans.
10. Put red chili, green chili and leek.
11. Boil in an earthen pot.
12. Serve it.

TIP 建议

청국장찌개는 뚝배기에 모든 재료를 끓인 후 청국장을 넣어야 향과 영양분을 살릴 수 있으며, 이때 약불로 조절하여 넘치지 않게 주의한다.

做清国酱汤时, 先把所有材料放入石锅中煮过以后再放清国酱。放入清国酱以后用微火来煮。

Put fermented soybeans after you boil all the ingredients in an earthen pot so that you would not lose healthy scent and nutrients of the dish. Cook over a low heat.

감자탕
土豆汤
Gamja tang(Potato and pork back bone soup)

등뼈해장국
脊椎骨醒酒汤
Backbone haejangguk(hangover soup)

선지해장국
血豆腐醒酒汤
Seonji haejangguk(clotted blood hangover soup)

곰취장아찌
马蹄叶咸菜
Gomchwi jangajji(pickled Groundsel)

4년근 수삼고추장장아찌
4年生水参辣椒酱菜
4-year ginseng gochujang jangajji

4년근 수삼간장장아찌
4年生水参酱油酱菜
4-year ginseng soy sauce jangajji

무장아찌
萝卜咸菜
Daikon jangajji

깻잎장아찌
苏子叶咸菜
Sesame leaf jangajji

송이버섯장아찌
松茸咸菜
Pine mushroom jangajji

더덕장아찌
沙参咸菜
Deodeok jangajji(pickled mountain herb whose roots have restorative properties)

무, 마늘장아찌
萝卜、大蒜咸菜
Daikon and garlic jangajji

멸치소금볶음
盐炒鳀鱼
Stir-fried anchovy

식재료·양념류 표기 Glossary of ingredients and sauces 食材类·调料类

식재료

가시오가피 · Eleuthero · 刺五加
가지 · eggplant · 茄子
감자 · potato · 土豆
감초 · licorice · 甘草
검정콩 · black soy bean · 黑豆
겨자잎 · mustard leaf · 芥菜叶
계피 · cinnamon · 桂皮
고구마 · sweet potato · 地瓜
고등어 · mackerel · 鲅鱼
고사리 · dried bracken(gosari) · 蕨菜
곤약면 · Konnyaku(jelly) noodles · 魔芋粉条
골뱅이 · whelk(snail) · 海螺
구기자 · chinese wolfberries(gugija) · 枸杞子
국멸치 · dried anchovies · 汤用鳀鱼
국수(소면) · plain noodles · 面条(阳春面)
굴 · oysters · 牡蛎, 海蛎子
굵은멸치 · dried big anchovy · 鳀鱼
김 · sea laver · 紫菜
김치 · kimchi(korean spicy pickled) · 泡菜
깻잎 · sesame leaf · 苏子叶
꽁치 · pacific saury (mackerel pike) · 秋刀鱼
꽃게 · blue crabs · 花蟹
낙지 · small octopus · 八爪鱼, 章鱼
냉면(칡면) · Naengmyeon(Korean cold noodle) · 冷面(葛粉面条)
녹두 · mung bean · 绿豆
녹차 · green tea · 绿茶
느타리버섯 · oyster mushroom · 平菇
늙은 오이 · old cucumber · 老黄瓜
다시마 · dried sea tangle · 海带
단호박 · sweet squash · 南瓜
달걀 · egg · 鸡蛋
닭 · chicken · 鸡肉
닭 가슴살 · chicken breast · 鸡脯肉
닭고기 · chicken · 鸡肉
당귀 · Korean angelica roots(dong guai) · 当归
당근 · carrot · 胡萝卜
당면 · starch noodle · 粉条
대구 · codfish · 鳕鱼
대추 · jujube · 大枣
대합 · clam · 大蛤
더덕 · todok(Codonopsis lanceolata) · 沙参
도라지 · bellflower roots · 桔梗
도토리 · oak nut jelly(acorn) · 橡栗
동치미 · Dongchimi(white radish water kimchi) · 腌萝卜泡菜
동태 · frozen pollack fish · 冻明太鱼
돼지(등심) · pork(loin) · 猪肉(里脊)
돼지고기(삼겹살) · pork belly · 猪肉(五花肉)
돼지고기(앞다리) · pork (arm shoulder) · 猪肉(前腿肉)
돼지등뼈 · pork backbone · 猪里脊骨
두부 · bean curd(tofu) · 豆腐
땅콩버터 · peanut butter · 花生酱
떡국떡 · long rice cake(garae tteok) · 年糕片
떡볶이떡 · Ddukbokki(thin rice cake) · 炒年糕用年糕
레몬 · Lemon · 柠檬
마른 미역 · dried seaweed · 裙带菜干
마른 새우 · dried shrimp · 干虾
마른 표고버섯 · dried shiitake mushroom(dried mushroom pyogo) · 干香菇
말린 나물류 · dried herbs(vegetables) · 干菜
멥쌀 · non-glutinous rice · 粳米
멸치 · dried anchovies(anchovy) · 鳀鱼
목이버섯 · black mushroom · 木耳
무 · white radish(Dajkon) · 萝卜
묵은 김치 · over-fermented Kimchi · 陈年泡菜
미나리 · watercress(Minari) · 水芹菜
민물새우 · freshwater shrimp · 淡水虾
바지락 · clam · 蛤仔
밤 · chestnuts · 栗子
방울토마토 · cherry tomato · 小西红柿
배 · pear · 梨
밴댕이(디포리) · Big Eyed Herring · 干青鳞鱼
병어 · pomfret fish · 鲳鱼
부추 · chives · 韭菜
브로콜리 · broccoli · 西兰花
사과 · appel · 苹果
삶은 가지 · boiled eggplant · 蒸茄子
삶은 도가니 · boiled Knee cartilage of cow · 熟牛膝骨肉
삶은 소면 · boiled plain noodle · 煮好的阳春面
삼계닭 · samgye chicken · 鸡(参鸡汤用鸡)
새송이버섯 · king oyster mushroom · 杏鲍菇
새우 · shrimp · 虾
생오리 · fresh duck · 鸭子
성게 · sea urchin · 海胆
소고기(등심) · beef(sirloin) · 牛肉(里脊)
소고기(안심) · beef(tenderloin) · 牛肉(里脊)
소고기(양지) · beef(brisket) · 牛肉(牛排)
소고기(차돌박이) · beef(marbled beef) · 牛肉(雪花牛肉)
소고기(채끝살) · beef(sirloin) · 牛肉(牛上腰)
수삼 · fresh ginseng · 人参水参
숙주 · Mung Bean Sprout · 绿豆芽
순두부 · sof tofu(soft bean curd) · 嫩豆腐
시래기 · dried radish leaves · 菜叶干 (萝卜叶)
신김치 · over-fermented Kimchi · 酸泡菜, 老泡菜
실파 · green onion · 小葱
쑥갓 · edible chrysanthemum(ssukgot) · 茼蒿
아귀 · monkfish · 鮟鱇鱼
애기배추 · baby chinese cabbage · 小白菜
애호박 · squash long · 西葫芦
양배추 · cabbage · 大头菜
양송이버섯 · button mushroom · 洋香菇
양파 · onion · 洋葱
연잎 · lotus leaf · 莲叶

엿기름 · malt · 麦芽
오이 · cucumber · 黄瓜
우무묵 · agar agar jelly · 石菜花凉粉
우엉 · Burdock root · 牛蒡
은행 · ginkgo nut · 白果，银杏核
인삼 · Ginseng fresh · 人参
자연산 송이버섯 · natural button mushroom · 野生松茸
적양파 · red onion · 圆葱，红洋葱
전복 · abalones · 鲍鱼
조개(생합) · fresh clam · 蛤蜊(生蛤蜊)
조기 · white croakers fish · 黄花鱼
조랭이떡 · small rice cake (joraengyi tteok) · 料理用年糕
주꾸미 · baby octopus · 小章鱼
죽순 · bamboo shoots · 竹笋
중하 · medium shrimp · 中等蛤蜊
쥬키니호박(애호박) · zucchini · 西葫芦
참나물 · Michba(Chamnamul) · 山芹菜
참치 · Tuna · 金枪鱼
청경채 · bok choy · 油菜
청량고추 · red hot chili peppers · 青阳辣椒
청오이 · cucumber · 青瓜
청포묵 · jelled Mung-Bean · 绿豆凉粉
콩나물 · Soy Bean sprouts · 豆芽，黄豆芽
키위 · kiwi · 猕猴桃
토란대 · taro stem · 芋头杆
토마토케첩 · tomato ketchup · 番茄酱
파인애플 · pineapple · 菠萝
파프리카 · paprika · 大辣椒
팽이버섯 · mushroom winter · 金针蘑
포북어 · dried pollack fillet · 干明太鱼
표고버섯 · shiitake mushroom(mushroom pyogo) · 香菇
풋고추 · green chili · 青辣椒
피마자나물 · castor bean leaves · 蓖麻叶
피망 · bell pepper · 青椒
한우(소갈비) · korea beef (beef ribs) · 韩牛(牛排骨)
한우뼈(사골) · koera beef bone · 韩牛骨（四骨，牛膝盖骨）
해바라기씨 · sunflower seeds · 葵花籽
호박씨 · pumpkin seeds · 南瓜籽
홍고추 · red chili · 红辣椒
홍어 · skate (fish) · 魟鱼
홍합 · sea mussel · 海虹
홍합살 · mussel flesh · 海虹肉
황기 · milk vetch root · 黄芪

양념류

간장 · soy sauce · 浓酱油
검정깨 · black sesame seed · 黑芝麻
고추기름 · chili oil · 辣椒油
고추씨 · pepper seed · 辣椒籽
고추장 · pepper red paste(gochujang) · 辣椒酱
고춧가루 · chili powder · 辣椒面儿
국간장 · soy sauce · 汤用酱油
까나리액젓 · sand eel sauce · 玉筋鱼酱
꽃소금 · salt cooking · 大粒盐
꿀 · honey · 蜂蜜
다진 대파 · chopped leek · 大葱末儿
다진 마늘 · chopped garlic · 蒜泥
다진 실파 · chopped spring onion · 小葱末儿
된장 · soybean paste (doenjang) · 大酱
들기름 · perilla seed oil · 芝麻油
들깨가루 · perilla seed flour · 芝麻粉
마늘 · garlic · 大蒜
매실식초 · plum fruit vinegar · 梅子醋
매실즙 · plum juice · 梅子汁
매운 갈비 소스 · spicy galbi sauce · 辣味排骨调味汁
물엿 · starch syrup · 糖稀
맛술 · Cooking wine · 料酒
밀가루(박력분) · flour(weak flour) · 面粉(薄力粉)
밀가루(부침가루) · flour(wheat flour) · 面粉(油炸粉)
배즙 · pear juice · 梨汁
불고기소스 · bulgogi sauce · 烧烤调味汁
새우젓 · salted shrimp · 虾酱
생강 · ginger · 生姜
설탕 · sugar · 白糖
소갈비양념 소스 · beef rib barbecue sause · 牛排调味汁
소고기다시다 · beef base · 牛肉味精，韩国牛肉粉(韩国大喜大牛肉粉)
소금 · salt · 盐
식용유 · salad oil · 食用油
식초 · Vinegar · 食醋
실고추 · red pepper thread · 辣椒丝
아몬드슬라이스 · almond slice · 杏仁片
양파즙 · onion juice · 洋葱汁
얼음 · ice cube · 冰块
올리고당 · oligosaccharide · 低聚糖
올리브유 · olive oil · 橄榄油
유자청 · Yuzu honey · 柚子蜜
전분가루 · starch flour · 淀粉
중력분 · medium flour · 中力粉
진간장 · thick soy sauce · 浓汁酱油
참기름 · sesame oil · 香油
찹쌀가루 · glutinous rice flour · 糯米粉
청국장 · fermented soybeans · 清国酱
청주 · rice-wine · 清酒
콩가루 · bean flour · 豆粉
통계피 · cinnamon · 整桂皮
통깨 · sesame seed · 芝麻
통잣 · pine nut · 松仁
통후추 · whole black pepper · 胡椒粒
튀김기름 · frying oil · 炸油
홍초 · red vinegar · 红醋
황설탕 · brown sugar · 黄糖
후춧가루 · ground black pepper · 胡椒粉
흰쌀 · non-glutinous rice · 白米,大米
흰콩(메주콩) · white beans (soy beans) · 黄豆(酱曲豆)

참고문헌

강인희, 한국식생활사, 삼영사, 1978.

강인희, 한국식생활풍속, 삼영사, 1984.

강인희, 한국의 맛, 대한교과서주식회사, 1987.

고희철, 한국조리, 학문사, 2000.

김달래, 체질 따라 약이 되는 음식 224, 경향신문사, 1996.

김덕희 외, 한국음식의 맛, 백산출판사, 2011.

김상보, 조선시대의 음식문화, 가람기획, 2005.

김상보, 한국음식생활문화사, 광문각, 1997.

김태정, 우리가 정말 알아야 할 우리꽃 백가지, 현암사, 1990.

대우증권 사회봉사단, 한국 가정요리, Bookie, 2012.

박란숙, 한국조리실습, 광문각, 2000.

박원기, 한국식품사전, 신광출판사, 1991.

손경희, 한국음식의 조리과학, 교문사, 2001.

신미혜 외, 한국의 전통음식, 백산출판사, 2006.

신미혜, 엄마도 모르는 양념공식 요리법, 세종서적, 1996.

신승미 외, 우리고유의 상차림, 교문사, 2005.

신재용, 한국인의 건강식, 동화문화사, 1990.

왕준련, 한국요리백과, 범한출판사, 1986.

유태종, 식품보감, 문운당, 1991.

유태종, 음식궁합, 도서출판 궁지, 1992.

윤국범 외, 몸에 좋은 산야초, 석오출판사, 1989.

윤서석, 한국식생활문화, 신광출판사, 2009.

윤서석, 한국음식, 수학사, 1993.

윤숙자, 조선요리제법, 백산출판사, 2011.

이규태, 재미있는 우리 음식이야기, 기린원, 1991.

이성우, 한국식생활의 역사, 수학사, 1992.

이성우, 한국요리문화사, 교문사, 1985.

전경철 외, 한식조리기능장 실기편, 크라운출판사, 2011.
전경철 외, 한식조리산업기사 실기편, 크라운출판사, 2011.
전경철, 한식조리기능사 실기편, 크라운출판사, 2012.
정혜경, 한국음식 오디세이, 생각의나무, 2007.
조후종, 우리음식이야기, 한림출판사, 2004.
조후종, 통과의례와 우리음식, 한림출판사, 2004.
차경옥 외, 한국음식, 백산출판사, 2010.
최진호, 선식의 비밀, 삶과꿈, 1998.
한국문화재보호재단, 한국음식대관, 한림출판사, 1999.
한복려 외, 조선왕조 궁중음식, 궁중음식연구원, 2003.
한복진 외, 우리생활 100년-음식, 현암사, 2001.
한복진, 우리가 정말 알아야 할 우리 음식 백가지, 현암사, 1998.
한복진, 조선왕조 궁중음식, 화산문화, 2003.
허 준, 동의보감, 남산당, 1991.
홍문화, 보약이 되는 음식이야기, 도서출판두로, 1995.
홍진숙, 한국전통음식, 예문사, 1997.
황혜성 외, 3대가 쓴 한국의 전통음식, (주)교문사, 2011.
황혜성 외, 조선왕조 궁중음식, 사단법인 궁중음식연구원, 1995.
황혜성 외, 한국의 전통음식, 교문사, 1990.
황혜성, 한국음식대관(6권), 한림출판사, 1997.

Bibliography

Cha, Kyeong-ok, et al., Korean Food, Baeksan Publisher, 2010.

Choi, Jin-ho, Secret of Seonsik, Life&Dream, 1998.

Han, Bok-jin, A hundred sorts of our Food we have to definitely know, Hyeonamsa, 1998.

Han, Bok-jin, Court Food in the Joseon Dynasty, Hawsanmunhwa, 2003.

Han, Bok-jin, et al., Our Life 100 years-Food, Hyeonamsa, 2001.

Han, Bok-ryeo, "Court Food in the Joseon Dynasty", Court Food Research Institute, 2003.

Hong, Jin-sook, Korean Traditional Food, Yemunsa, 1997.

Hong, Mun-hwa, Story about Food becoming herbal tonics, Publisher Duro, 1995.

Huh, Jun, Donguibogam, Namsandang, 1991.

Hwang, Hey-seong, et al., Korea's Traditional Food written by 3 Generations, Kyomunsa Co. Ltd., 2011.

Hwang, Hye-seong, et al., "Court Food in the Joseon Dynasty", (corp) Court Food Research Institute, 1995.

Hwang, Hye-seong, et al., Korea's Traditional Food, Kyomunsa, 1990.

Hwang, Hye-seong, Korean Food General Survey (6 volumes), Hanrim Publisher, 1997.

Jeon, Kyung-chul, Craftsman Cook, Korean Food, Volume Practice, Crown Publisher, 2012.

Jeon, Kyung-chul, et al., Industrial Engineer Cook, Korean Food, Volume Practice, Crown Publisher, 2011.

Jeon, Kyung-chul, et al., Master Craftsman Cook, Korean Food, Volume Practice, Crown Publisher, 2011.

Jeong, Hye-kyeong, Korean Food Odyssey. Thinking Tree, 2007.

Joh, Huo-jong, Our Food Story, Hanlim Publisher, 2004.

Joh, Huo-jong, Rite of Passage & Our Food, Hangrim Publisher, 2004.

Kang, In-hee, Custom of Korean Food Life, Samyoungsa, 1984.

Kang, In-hee, History of Korean Food Life, Samyungsa, 1978.

Kang, In-hee, Korea's Flavor, Mirae N Co. Ltd., 1987.

Kim, Dal-rae, "Foods 224 becoming a medicine according to constitution", Kyunghyang Daily News, 1996.

Kim, Deok-hee, et al., Flavor of Korean Food, Baeksan Publisher, 2011.

Kim, Sang-bo, Food Culture in the Joseon Dynasty, Garam Planning, 2005.

Kim, Sang-bo, Korean Food Life Culture History, Kwangmungak, 1997.

Kim, Tae-jeong, A hundred sorts of our Flowers we have to definitely know, Hyeonamsa, 1990.

Koh, Hee-cheol, Korean Cuisine, Hakmunsa, 2000.

Korea Cultural Heritage Foundation, Korean Food General Survey, Hanrim Publisher, 1999.

Lee, Kyu-tae, Our Amusing Food Story, Publisher Girinwon, 1991.

Lee, Seong-woo, History of Korea Food Life, Publisher Suhaksa, 1992.

Lee, Seong-woo, Korea Cuisine Culture History, Kyomunsa, 1985.

Park, Ran-sook, Korean Cuisine Practice, Kwangmungak, 2000.

Park, Won-ki, Korean Foods Dictionary, Shinkwang Publisher, 1991.

Shin, Jae-yong, Koreans' Health Food, Donghwamunhwasa, 1990.

Shin, Mi-hye, et al., Korea's Traditional Food, Baeksan Publisher, 2006.

Shin, Mi-hye, Seasoning Formula Cookery even Mom doesn't know, Sejong Book Center, 1996.

Shin, Seung-mi, et al., Our own Table Setting, Kyomunsa, 2005.

Social Service Corps of Daewoo Securities Co. Ltd., Cold Country Home-Cooking, Bookie, 2012.

Sohn, Kyeong-hee, Science of Cookery of Korea Food, Kyomunsa, 2001.

Wang, Jun-ryeon, Korean Cuisine Collection, Beomhan Publisher, 1986.

Yoon, Kuk-beom, et al., Mountain Wild Grass Beneficial to Body, Seogo Publisher, 1989.

Yoon, Seo-seok, Korea Food, Suhaksa, 1993.

Yoon, Seo-seok, Korean Food Life Culture, Shinkwang Publisher, 2009.

Yoon, Sook-ja, Joseon Cuisine Recipes, Baeksan Publisher, 2011.

Yuh, Tae-jong, Food Compatibility, Publisher Gungji, 1992.

Yuh, Tae-jong, Sikpumbogam, Munwundang, 1991.

찾아보기

저자 소개

전 경 철

경기대학교 대학원 외식조리관리학전공 박사졸업(관광학박사)
경기대학교 대학원 외식산업경영학전공 석사졸업(관광학석사)
경기대학교 관광대학 외식조리학과 겸임교수
그랜드인터컨티넨탈호텔서울 Chef, 사내강사 / 미국호텔협회총지배인(CHA)
국가공인조리기능장 / 대한민국조리국가대표(국제요리대회 15회 수상)
2006 러시아국제요리대회 금상, 은상 수상
2007 중국 북경 아시아요리대회 금메달 수상
2010 한식홍보대사선발 조리경연대회 대상
2010 뉴욕, 워싱턴 세계한식홍보축제 팀 단장
2011 한식홍보대사선발 조리경연대회 심사위원장
2012 제22회 중국 허난성 요리대회 팀 단장 및 심사위원
지방기능대회 요리종목 심사위원장 / 전국기능대회 요리종목 심사위원
한국산업인력공단 출제 및 검토위원 & 감독위원
KBS, MBC, SBS, EBS, 한국경제TV 등 다수 방송 출현
저서 : 한식조리기능사 실기, 한식조리기능장 실기 외 다수

김 은 영

경기대학교 대학원 외식조리관리학과 박사과정 중
경기대학교 대학원 외식조리관리학과 석사 졸업
사)한국식문화진흥협회이사장
청원생명쌀연구교육원장
국가공인조리기능장(2009)
현장한식비법전문가
대한민국조리국가대표
향토음식심의위원 및 심사위원
농림수산식품부장관상 수상
(중국)한식현지화자문위원 및 교육위원
아시아푸드페스트벌추진위원장
조리기능사, 조리산업기사, 조리기능장출제위원, 감독위원
한국, 중국전국기능대회 요리종목지도교사(금, 은, 우수상 수상)
저서 : 조리산업기사, 조리기능장 이론, 실기편, 디저트요리

전 동 철

(주)벽제갈비 서울 서래점 조리총괄주방장(차장)
(주)벽제갈비 서울 타워팰리스 조리장(과장) 역임
(주)곰바위(서울 삼성동) 조리과장 역임
(주)대한민국(서울 삼성동) 조리실장 역임
(주)춘하추동(서울 한국회관 역삼동) 조리실장 역임
뼈없는 돼지갈비 전문점(서울 역삼동) 조리실장 역임
인도네시아 자카르타 한양(주)한식당 open 총주방장 역임
중국 조리교육 및 세미나 연수
(외식산업경영학회, 외식산업대학교수협의회)
실전창업스쿨 과정 수료(서울특별시)
2008 제2회 한, 중, 일 아시아푸드페스티벌 금상 수상
2010 세계한식홍보축제 홍보대사(금상 수상)
2010 뉴욕, 워싱턴 세계한식홍보축제 부팀장
2011 한식창작요리대회(금상 수상)
한국산업인력공단 실기시험 감독위원

▣ 简历

全京喆

京畿大学研究生院厨艺管理专业博士研究生毕业(旅游经济学博士)
京畿大学研究生院厨艺管理专业硕士研究生毕业(旅游经济学硕士)
京畿大学旅游经济学院系厨艺管理专业客座教授
首尔格兰洲际酒店厨师(Chef) / 美国酒店总经理(CHA)
国家公认烹饪技能长 / 韩国烹饪国家代表(国际烹饪大赛中获十五次奖)
2006俄罗斯国际烹饪大赛获金, 银奖
2007中国北京亚洲烹饪大赛获金牌
2010韩餐宣传大使选拔烹饪大赛荣获大奖
2010纽约, 华盛顿世界韩餐宣传庆典团长
2011韩餐宣传大使选拔烹饪大赛评审委员长
2012第二十二届中国河南省烹饪大赛团长及审查委员
地方技能大赛烹饪种类评审委员长 / 全国技能大赛烹饪种类评审委员
韩国产业人力工团国家技术资格检查考试出题及审批委员&技术考试监督委员
出演KBS、MBC、SBS、EBS、韩国经济TV 等电视台节目
著书 : 韩餐烹饪技能师实技篇, 韩餐烹饪技能长实技篇

金銀永

京畿大学研究生院厨艺管理专业在读博士研究生
京畿大学研究生院厨艺管理专业硕士研究生毕业
韩国食文化振兴协会理事长
清原生命米研究教育院长
国家公认烹饪技能长(2009)
现场韩餐秘方专家
韩国烹饪国家代表
风味小吃审议委员会审查委员
获得农林水产食品部长官奖
韩食本土化 (中国) 咨询委员及审查委员
亚洲美食嘉年华促进委员长
烹饪技能师, 烹饪产业技师, 烹饪技能长考试出题委员, 监督委员
韩中全国烹饪技能大赛料理项目指导教师(获金, 银, 优秀奖)
著作 : 烹饪产业技师, 烹饪技能长理论篇, 饭后甜点

全东哲

(株)碧斋排骨首尔西来店烹饪主厨(次长)
历任(株)碧斋排骨首尔Tower Palace厨师长(课长)
历任(株)GOM BA WUI(首尔三城洞)厨师长
历任(株)大韩民国(首尔三城洞)烹饪室长
历任(株)春夏秋冬(首尔韩国会馆驿三洞)烹饪室长
历任无骨猪排店(首尔驿三洞)烹饪室长
前任印尼雅加达汉阳(株)韩餐厅open总厨师长
中国烹饪教育及研讨会进修
(餐饮服务业经济学会, 餐饮服务大学教授协议会)
实战创业学校课程结业(首尔特别市)
荣获2008第二届韩, 中, 日亚洲美食嘉年华金奖
2010世界韩餐宣传庆典宣传大使(获得金奖)
2010纽约, 华盛顿世界韩餐宣传庆典副组长
2011韩餐创作料理大赛(获得金奖)
韩国产业人力公团实技试验监督委员

▣ Profile

Jeon Kyungchul

Major in Catering Cooking Management at the Graduate School of Kyonggi University, and Graduated with Doctor's degree(Doctor of Tourism)
Major in Food Service Management at the Graduate School of Kyonggi University, and graduated with a Master's degree(Master of Tourism)
Adjunct professor in Catering Cooking Management at Kyonggi University
Chef Grand Intercontinental Hotel Seoul Chef, inside teacher, General Manager at an American Hotel(CHA)
State-authorized Master Chef / National Member of Korea Cooking/ Awarded at the Korean Cooking Contest 15 times
Won Gold Prize and Silver Prize at the 2006 Russia International Cooking Contest
Won Gold Medal at the 2007 China Beijing Asia Cooking Contest
Grand Prize at the 2010 Korean Food Honorary Ambassador Screening Cooking Contest
Leader of the 2010 NY, Washington World Korean Food Publicity Festival Team
Head Judge of the 2011 Korean Food Honorary Ambassador Screening Cooking Contest
Team Leader of the 2012 22th China Henan Province Cooking Contest
Head Judge of Local Skill Contest Cooking Event / Judge of National Skill Contest Cooking Event
Question Designer & Reviewer of National Technical Qualification Examination administered by the Human Resources Development Service of Korea & Qualifying Examination
Apperance on KBS, MBC, SBS, EBS, WOW TV, etc.
Books : Compilation of Practice for Craftsman Cook-Korean Food
Compilation of Practice for Master Craftsman Cook-Korea Food, etc.

Kim Eunyoung

Ph.D. candidate in Catering Cooking, Kyonggi University
M.A. in Catering Cooking, Kyonggi University
Board president at Korea Food Culture Promotion Association
Director at Cheongwon Life Rice Research & Education Institute
State-registered Master Craftsman Cook(2009)
Korean food cooking expert
Korean national cooking member
Judge at Rural Food Deliberation Committee
Award by Ministry of Agriculture and Forestry and Ministry for Food, Agriculture, Forestry and Fisheries
Judge and consultant Korean Food Globalization Association(China)
Chairman at Asian Food Festival Promotion Committee
Question designer for Korean Food Craftsman Cook, Industrial Engineer Cook and Master Craftsman Cook
Instructor at Korea-China National Cooking Contest(gold prize, silver prize and excellence prize in the same contest)
Author of Industrial Engineer Cook 101, Master Craftsman Cook 101 as well as workbooks for a dessert crafts cook
Author of a dessert cooking book and theory books and workbooks for industrial engineer cook and master craftsman cook

Jeon Dongchol

Chef in general charge of Cooking at Seoul Seorae Branch of Beokje Galbi Co. Ltd.
Appointed as the Cooking Head(section chief) of Seoul Tower Palace of Beokje Galbi Co. Ltd.
Appointed as Cooking Chief of Gombawi Co. Ltd.(Samseong-dong, Seoul)
Appointed as Head of Kitchen at Daehanminkook Co. Ltd.(Samseong-dong, Seoul)
Appointed as Head of Kitchen at Chunhachudong Co. Ltd.(Seocho-dong, Seoul)
Appointed as Head of Kitchen at 'Boneless Pork Rib Specialty Story'(Yeoksam-dong, Seoul)
Appointed as General Chef at Hanyang Korean Restaurant Co. Ltd., in Jakarta, Indonesia
Cooking Education in China and Training at the Seminar(Korean Academy of Food Service Industry & Management, Professor's Association of Food Service Industrial College
Completed the courses offered by Actual Start-up School(The Seoul Metropolis)
Awarded Gold Prized at the 2008 2nd Korea, China, and Japan Asia Food Festival
Honorary Ambassador for 2010 World Korean Food Publicity Festival(Awarded Gold Prize)
Vice Team leader of the World Korean Food Publicity Fesitval in NY and Washington in 2010
Won Gold Prize at Korean Creative Food Cooking Contest in 2011
Supervisory Member of Practice Examination administered by the Human Resources Development Service of Korea

韩食美“味”秘方传授 100选
2013年1月20日第1版发行
2017年1月10日第2版发行
著者 全京喆, 金銀永, 全东哲
出版社 白山出版社
京畿道坡州市西牌洞473
电话: 02-914-1621(代)
FAX : 031-955-9911
http://www.ibaeksan.kr
edit@ibaeksan.kr

价格 35,000韩元
ISBN 978-89-6183-644-9

Transmission of a Hundred Secrets of Korean Dishes Flavor
The first edition published on 20 Jan. 2013
The Second edition published on 10 Jan. 2017
Writer Jeon Kyungchul, Kim Eunyoung, Jeon Dongchol
Publisher Baeksan
370 Hoedong-gil, Paju-si, Kyeonggi-do
Tel. 02-914-1621(代)
FAX: 031-955-9911
http://www.ibaeksan.kr
edit@ibaeksan.kr

35,000won
ISBN 978-89-6183-644-9

한식의 맛 비법전수 100선

2013년 1월 15일 초 판 1쇄 발행
2017년 1월 10일 제2판 1쇄 발행

지은이 전경철·김은영·전동철
펴낸이 진욱상
펴낸곳 백산출판사
교 정 편집부
본문디자인 강정자
표지디자인 오정은

등 록 1974년 1월 9일 제1-72호
주 소 경기도 파주시 회동길 370(백산빌딩 3층)
전 화 02-914-1621(代)
팩 스 031-955-9911
이메일 edit@ibaeksan.kr
홈페이지 www.ibaeksan.kr

ISBN 978-89-6183-644-9
값 35,000원